Natural Aphrodisiacs: Myth or Reality

Ravi K. Puri, Ph.D.
Raman Puri, MD

www.plantsforlove.com
www.aphrodisiaca.com

I

First Edition: May, 2010.
Library of Congress Control Number: 2011910974
ISBN: Hardcover   978-1-4628-9440-6
                 Softcover   978-1-4628-9439-0
                 Ebook   978-1-4628-9441-3

Warning Disclaimer

Material in this book has been collected, compiled, and published for educational purpose only. We do not intend to make any recommendations for the use of any aphrodisiac discussed in this book. We are not promoting or condemning any product or formulation mentioned in this publication. Efforts have been made to survey the latest literature available and research conducted on aphrodisiacs and to reveal the facts about their merits and demerits. The authors and the publishers are not responsible for any consequences resulted from the treatment, application, or consumption of any herb or herbal formulation described in this book. Please do not attempt self-diagnosis or self treatment. Always consult a knowledgeable pharmacognosist, pharmacist, or a physician before incorporating any idea from this book into practice.

—The authors

Front Cover
Alstonia scholaris, used as aphrodisiac.

This book was printed in the United States of America.

To order additional copies of this book, contact:
Xlibris Corporation
1-888-795-4274
www.Xlibris.com
Orders@Xlibris.com
100823

Dedication

In the memory of my father who will always remain my inspiration.
One does not know the worth of water till the well is dry.

—Ravi K. Puri

Diosgenin: Precursor of Sex Hormones
Source: Dioscorea & Solanum Species

About the Authors

Ravi K. Puri is a Ph.D. in Pharmaceutical Sciences and specialized in the chemistry of natural products. He is a Post Doctorate (1980-82) from the school of Pharmacy, University of Mississippi, Oxford, where he got the opportunity of carrying out research on the microbial transformation of antimalarial drug primaquine under the guidance of eminent Professor James D. McChesney. Dr. Puri isolated and characterized novel metabolites of primaquine. He has also isolated novel glycoalkaloids from *Solanum platanifolium* during his Ph.D., and characterized them by X-ray diffraction, 2D, and ^{13}C-NMR techniques and gave one of the novel glycoalkaoids his name as *Ravifoline.*

Dr. Puri has a broad range of experience and excelled in the fields of pharmacy, biochemistry, and environmental sciences. He has also taught and conducted research at various universities in the USA and abroad including University of Mississippi, Oregon State University, University of Florida, and University of Missouri. In 1999, Dr. Puri retired as the Group Leader, at the Environmental Trace Substances Research Center, University of Missouri. Dr. Puri has published nearly a hundred research papers and scientific reviews in journals of international repute, including chapters in reference books published by CRC and Lewis Publishers. He is on the review panel of several prestigious scientific journals in the USA and abroad. His 30 years of experience in the field of natural products will benefit the readers interested in the field of natural aphrodisiacs.

Raman Puri is a MD (1996) from the School of Medicine, University of Missouri, Columbia. After his residency in Internal Medicine from his *alma mater*, he started his own practice. Dr. Puri has been successfully practicing internal medicine and critical care in Columbia, MO, for the past ten years. He is affiliated with the reputed Boone Hospital Center and also serves on its faculty. Dr. Raman is on the advisory panel of some renowned pharmaceutical companies such as Wyeth Pharmaceuticals Inc, Bayer Pharmaceutical Corporation, and Schering Plough Pharmaceuticals. He delivers invited lectures for these firms on selected topics pertaining to their products. He has also published very captivating research papers in the field of natural products in the celebrated journals of international standing. His experience and knowledge in the practice of internal medicine will certainly facilitate the consumers of natural aphrodisiacs particularly in the aspect of herb-drug interaction.

CONTENTS

Table of Figures

Acknowledgment

I am very grateful to my family for the continuous encouragement rendered to me during the duration of writing this book, and especially to my wife Mrinal, who has put up with me all these years. She stood by me through thick and thin in every walk of my life and has been a continuous source of encouragement during my difficult times. She has, as always, my undying love and gratitude.

My thanks are due to Dr. Toni Almond for reviewing the manuscript and for writing the foreword. I am equally grateful to Dr. Siobhan H. McGaughey, Urologist, for her valuable comments.

My special thanks are also to Anita Aielo, who has inspired me and encouraged me to write this text. There were many occasions when I lost interest in my writings but she kept on rejuvenating my spirit.

I extend my appreciation to my son, Dr. Raman Puri for being the co-author. He is a physician and has been practicing internal medicine for the past ten years in Columbia, Missouri. His input in this text will be most beneficial to the readers.

I am indebted to Sandra Schroeder for her artistic ideas and help in some of the illustrations as well as coordination with the publisher. Without her assistance this book would have not been published. Credit goes to Lois Sandner and Jordan Milne for their patience and time while proofreading the entire book.

Last but not the least, I appreciate my younger son, Vivek Puri, JD for his legal advice in protecting the copy right and related issues.

Ravi K. Puri, Ph. D.
Columbia, MO

Preface

The dawn of the Internet has created a new wave in the world of business. Everything sells on the Internet through supplicate email advertisement. It is easy to buy almost everything from basic consumer products to life saving drugs. About 80% of the emails are unsolicited and are pertaining to health products, weight reduction, anti-aging and aphrodisiacs herbal products, commonly known as *sex boosters*. Sales of 'natural aphrodisiacs' are booming day by day. The term *aphrodisiac* is used and advertized so loosely that it often includes remedies for impotence or erectile dysfunction and lures people to buy against false hopes to regain their lost vitality.

The majority of aphrodisiacs formulations do not have any scientific evidence and this leads to quackery. Despite the lack of scientific evidence of safety and effectiveness, the aphrodisiacs industry is flourishing. Marketers do much of their business by mail order via the Internet and are raking in millions of dollars all over the world.

The Food and Drug Administration does not have any control on aphrodisiac products since these are sold under food supplements. Nevertheless, there is no comprehensive quality control or standardization of parameters in existence for their testing. A single preparation can carry more than five herbs and each herb further comprises more than fifty ingredients. Hence, standardization of these formulations is a cumbersome phenomenon. Before consuming these aphrodisiacs, the consumers should be aware of the products, their constituents, toxicity, and interaction with the medications they are already taking. Consumers should first consult their physicians or an educated herbalist or pharmacist before making their choice to consume these products.

The use of these natural aphrodisiacs is based upon myth, folkloric or traditional background. However, very limited data on aphrodisiacs are available regarding their chemistry, pharmacology, toxicity, side effects and interaction with modern medicines. These facts are very pertinent to separate truth from myth which inspired the undertaking of the present text. Within this text, efforts have been made to avoid putting the old wine in a new bottle. The purpose of writing this text is to enlighten the readers with the latest research carried out on herbal aphrodisiacs that are vigorously sold via the Internet or over the counter in health and retail stores. Consumers should know the merits and demerits of ingredients in a 'love potion' they are going to consume.

The book is for all the consumers who are taking natural aphrodisiac products to regain their lost libido.

—The authors

Foreword

I am honored to be asked to write the foreword to this book. With more individuals taking responsibility for their health, and herbal remedies now more available, as a practicing physician I found this an informative and extensive survey of the latest research conducted on aphrodisiac herbals. The text is a fascinating review for those consumers who wish to utilize natural aphrodisiac formulations. It emphasizes differentiation of myth from reality by elucidating the origin, history of use, myth, and romance versus the chemistry, pharmacology, risk, and possible interactions. The quotations at the beginning of each chapter were enticing and the excellent hand drawings of the individual plants clearly depicting the morphological characteristics would prove helpful in the identification and verification of the plants.

The authors have painstakingly expressed the various possible merits and potential side effects in a fair and balanced discussion of the natural aphrodisiac products available. After reading this book, I not only felt well informed myself, but felt that I could give an informed opinion to anyone making inquiries about potential aphrodisiacs. The graphics in the chapters were well designed and the tabular form regarding the volatile oils presented the information in a readily accessible format perfect for quick reference. I would expect other practicing physicians and pharmacists to find this text particularly helpful in advising patients who wish to incorporate natural products into their daily health routine.

Toni Almond, MD
Magnolia Critical Care
And Internal Medicine
Columbia, MO 65203

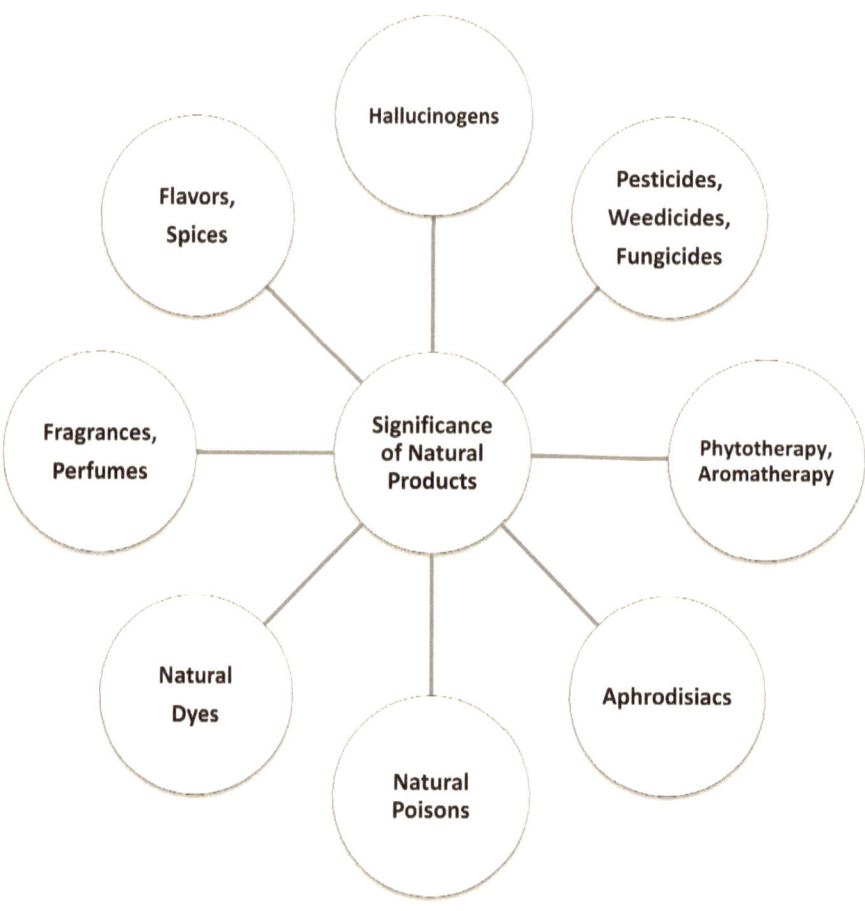

Figure 1: Significance of Natural Products

Chapter One

Introduction

We are living in the age of high technology and life is also moving at an ultra pace. Man has gone to the moon and is now exploring Mars. Distance among the planets has been overcome. Communication has become easy and rapid through cell phones and high speed Internet.

The Internet has created a new wave in the world of business, by making markets easier and simpler. Everything sells on the Internet through supplicate email advertisement. One can buy almost everything from basic consumer products to life saving prescriptions. With the merits of Internet, we have some demerits as well. False advertisements lead to exploitation of the innocent people and some of the products are nothing but scams. Nevertheless, more than 80% of the emails are unsolicited. These pertain to health products, weight reduction, anti-aging, and aphrodisiacs or commonly known as sex booster's formulations. Among these, natural aphrodisiacs are the hottest products. Sales of natural aphrodisiacs are booming and have grown to nearly $400 million dollars during the year 2006 (Pritchard 2007). Majorities of these formulations do not have any scientific evidence and lead to quackery. Despite the lack of scientific evidence of safety and effectiveness, the aphrodisiacs industry is flourishing to this day. Marketers do much of their business by mail order via the Internet and are raking in millions of dollars a year in the USA.

Quackery exists all over the world. Before, quacks had to travel from one city to another whereas now they can very easily do their job on the Internet. Plants and animal products in the form of love potions, concoctions, extracts, capsules, tablets, and creams are being sold under a 100% satisfaction guarantee, which means, the money will be returned if the products do not work. It is unsure, how many buyers get their money back. By the time they call the company, the telephone number does not exist anymore. If one tries any ordinary search engine with a query for aphrodisiacs, he can get plenty of information on these products. Many people are marketing aphrodisiacs that brag implausible claims of sexual arousal and stimulation. Love potion peddlers do not leave any stone unturned to sell their products.

Extraordinary success of Viagra has given an impetus to these people and the recent press publicization of the side effects of Viagra has given them a chance for exploitation. They have an excuse to say that here is something from a natural source that works better than Viagra without the side effects. Innocent people with low libido are buying them.

Since the time immemorial, man's quest for aphrodisiacs has led him to consume oysters, rhinoceros horns, musk, Spanish fly, tiger bones, the sex organs of animals, and many other related substances from biological kingdom. Nowdays, the quest is motivated by the sweet and false promises of quacks to make fast money at the sexual ignorance of many people. Secondly, we all want to be better lovers. A macho kind of supremacy over the female partner that he can do a better job and satisfy her sexually, leads him to consume these products. A common conception is that male performs sex to show power whereas female indulges to manipulate male. In other words aphrodisiacs tap into this basic insecurity about sexual potency and performance.

Likewise, females are also concerned about losing their sexual desire particularly during or after menopause. They also want to keep up their sensual indulgence since it influences their married life. They also feel insecure. They have low libido due to menopause, stress and many other gynaec problems. Female sexual problems had been neglected since antiquity for some basic reasons. Firstly, it is not easy to find out that what turns women on. This unsolved mystery has not only dodged husbands and lovers but also women themselves, for centuries. Secondly, women do not easily reveal their problems unless these come to the surface. However, since the time immemorial, women were also given herbal formulations and aphrodisiac potions to arouse sexual desire. The Chinese and Indian literatures reveal ample information on this subject.

The lack of desire for sex has been afflicting millions of women and enormously straining marriages around the world. Recently, a breakthrough has been noticed in this aspect and lots of herbal aphrodisiacs in the form of potions or creams to stimulate sexual desire in women are being sold on the Internet or over the counter. However, they all lack scientific evidence for efficacy. Thus, many pharmaceuticals firms in the USA such as Pfizer, Proctor & Gamble, Nastech, and Vivus are actively engaged in finding a suitable remedy to rekindle sexual desire in women with low libido.

A majority of sellers advertise aphrodisiacs formulations on their websites stating *"we have romance herbs for keeping it up and herbs for making you horny, herbs for your libido."* Survey of the literature reveals that these herbs are mentioned as folkloric medicine in the literature and are reported as sex tonic in different cultures.

Aggressive advertisements in much elaborated ways and false claims for 100% satisfaction tempt people with sexual deficiency to buy these formulations. Sex is very fascinating and an essential pleasure of human life. It is not everything but is a part of everything. People who have the money would like to spend any amount to regain their lost sexual strength.

Aphrodisiacs may act differently on the body parts based upon their active chemical constituents. Some may enhance the physical aspect of sex, other elevate moods and emotions pertaining to sexual activity and romance. In other words human sexual arousal is a very complex phenomenon and based upon the physiological, psychological, emotional, and cognitive aspects of an individual. Many of the aphrodisiacs have a physical and psychological boost, with some containing vitamins and minerals and giving stimulation in the form of tonic. There is no scientific evidence for their pharmacological action as aphrodisiacs or relating to sexual dysfunction or erectile dysfunction. Thus it is very pertinent to understand the chemistry of aphrodisiacs formulations available in the market and their uses for sexual arousal.

It is a fact that sex diminishes with the age. After puberty, sex life is inversely proportional to age. According to the statistics 50% people over the age of 50 are suffering from Erectile Deficiency in the USA and most of them are carried away by the sex stimulating advertisements and turn to these natural aphrodisiacs (Heller 1996). Moreover, these consumers believe that 'natural is better' and safe. However, 'natural' does not guarantee that the herbal product is 'innocuous.' (Eisenberg *et al.* 1998; Ernst and White 2000; Puri and Puri 2001)

The authors are not prejudice about aphrodisiac formulations or the use of traditional medicinal plants. However, before consuming these products, one needs to be aware of the toxicity of the active chemical constituents present in that particular plant product or formulation. If we are using the formulation comprising numerous herbs then it is very pertinent to find the compatibility or herb-herb interaction of the constituents present in them. These products should clear Quality Assurance and Quality Control (QA/QC) standards; otherwise these can be dangerous and deadly. Instead of prolonging ones love life in the bed, he could end up in the hospital bed. While very few deaths have been reported, the Associated Press has found records of multiple emergency room visits attributed to all-natural sex pills in Georgia, Chicago, Philadelphia, San Diego, and elsewhere.

These products are not only expensive but also lack scientific evidence for their pharmacological actions and toxic effects. There is no control by the FDA since these are sold under food supplements. There are also no comprehensive quality controls or standardization parameters in existence for their testing. A single preparation can carry more than five herbs and each herb further comprises of more than fifty ingredients. Hence, standardization of these formulations is a cumbersome phenomenon. Before consuming these products, the consumers should be aware of the products, their constituents, toxicity, and interaction with the medications they are already taking. They should first consult their physicians or an educated herbalist or pharmacist before making their choice to consume these products wisely, carefully and safely. If they are taking any prescription medication, they should consult their physician before ingesting these formulations. There may be a possibility to prescribe a different medication that won't interfere with their sex life.

Some patients who have been frustrated with the side effects of allopathic medications are consuming natural products without consulting their physicians. They feel that their primary care physician would not like this idea. Physicians should be more cooperative with their patients who are willing to include herbal extracts, dietary supplements, and other natural products into their therapeutic regimen. It is only possible if physicians also gain knowledge of the chemistry, pharmacolog, toxicology, and herb-drug interactions of these preparations. In the long run, it would be beneficial for both the physician and the patients to have a greater understanding of the role these substances may have in treating sexual dysfunction.

A survey of the literature on aphrodisiacs reveals that there are some texts* available on this topic. Most of the information is on their folkloric or traditional uses. However, very limited data on aphrodisiacs is available regarding the chemistry, toxicity, scientific background, side effects, and contraindication with modern medicines. These facts are very pertinent to separate truth from the myth. So it was thought, desirable to undertake the present text. Every effort has been made not to put the old wine in a new bottle. The purpose of writing this text is to enlighten the readers for the merits and demerits of natural aphrodisiacs that are vigorously sold via the Internet or over the counter in health and retail stores. Consumers should know the ingredients of *seduction in a bottle* or *love potion*. A comprehensive description of these aphrodisiacs, comprising their active constituents, clinical studies, dosage, availability, side effects including chemistry and the interaction with the modern medicines, food, and related herbs is given in this text. The information in the text will help the consumers to make the right decision. The book has not been written only for the consumers of natural aphrodisiacs but also for the physicians, pharmacists, scientists, and graduate students of health sciences who are interested to explore further the field of natural aphrodisiacs.

* Details are given in references

> *"No one can give you sexual ecstasy: it comes from within."*
>
> —*Margo Anand*

Chapter Two

> *"Ignorance has no beginning but an end;*
> *Enlightement has a beginning but no end."*
> —Ma Ananda Sarita

Natural Aphrodisiacs

During recent years, there has been a surge in the practice of alternative medicine. People are assiduously looking for natural herbal and related products for their ailments. Consumer sales of botanical dietary supplements have grown many folds. Despite the dramatic advances in high technology medicine such as gene therapy, laser and plastic surgery, high-resolution body scanning and scientific miracle like vital organ transplantation, it is amazing that natural healing methods are enjoying a remarkable resurrection. Among the various herbal products, aphrodisiac formulations are being sold like hot cakes via email on the Internet or over the counter at natural food stores in the USA and around the globe. Viagra revolution and its side effects spurred interest in natural aphrodisiacs; however, these products have a great concern regarding their risk-benefit profile.

Definition and Origin

Natural aphrodisiacs are believed to arouse sexual desire and enhance sexual performance. The term *Aphrodisiac* comes from the myth of the Greek goddess of love, Aphrodite, who presumably sprang naked from the sea on an oyster shell. Her name *Aphrodite* means "she who was born on foam." She was brought forth on a vulva like shell and carried onto the shore of Cyprus. She was the symbolic representation of love, fertility, and beauty. The myth crowns her as the epitome of sexual desire, seduction, and ecstasy. Artist Sandro Botticelli's painting portrays the birth of Venus. This painting is one of the most recognizable images in the history as timeless masterpiece of art. In this image, the goddess of love is brought vividly to life in a sinuous and supple form. She stands fastidiously in the center of the shell, modestly covering her nude body with her elegant hands and locks of golden hair. She looks seductive, coy, and breathtakingly beautiful in the painting (Figure 5).

The aphrodisiacs were the tools of Aphrodite, substances that stimulate or enhance sexual activity. The dove, dolphin, peaches, apricot, roses, lilies, fennel, cardamom, cinnamon, myrrh, myrtle, poppies, pomegranates, and blue water lily were considered sacred to her. That is why most of them have been used as aphrodisiacs. The term *aphrodisiac* was attributed to the goddess of love.

There are parallels of *Aphrodite* in many of the ancient civilizations. Each culture has his god or goddess who inspired sensuality and love. The Romans for example called her Venus, the famous armless statue known as the Venus de Milo and Cupid was her messenger. As Venus is the epitome of feminine beauty, the Hindus in India describe the Padmani or the lotus woman as the perfect representation of the same according to *Kama Sutra*. Similarly, there are examples from different cultures such as The Ephesians Artemis, the goddess of fertility and sexuality worshiped in the Near East and Greece. Ishtar, in Babylonian times, the goddess was considered as the Queens of heaven. Inanna was the Sumerian goddess of sexual love and fertility. Likewise, goddess Astarte was to the Phoenicians. Oshun was the African goddess of sex and fertility. Honey was her favorite offering and considered as an aphrodisiac. Similarly, Cleopatra was known as living goddess of beauty in Egypt, born in 60 BC. She was an expert in the art of love and seduction. Her beauty and charm was unique. Neroli oil of orange blossoms comprising haunting fragrance was one of her favorite fragrances. The neroli oil is still used as an aphrodisiac.

Some of the earliest love potions were based on honey, following the Babylonian practice of bride and groom drinking mead for a month after their wedding. Mead is honey wine and because the Babylonian calendar was lunar this 'honey month' was often referred to as the 'honey moon.' Mead, first brewed in Babylon more than 4,000 years ago making it one of the world's oldest alcoholic drinks, has long been believed to increase fertility and sex drive. Surprisingly, even today, Royal Society of Chemists in the UK is trying to prove this myth by preparing wine with honey fruits and spices. The society received overwhelming response from the newlyweds' honeymooners to participate in this 30 day study (Chattoe 2003).

Greek physician and philosopher Hippocrates, the father of Western medicine, also mentioned honey as a remedy for sexual vitality. The classical Indian sex manual *Kama Sutra* also recommended honey as a sexual tonic and utilized its use in some recipes.

Likewise, oysters have enjoyed a great reputation as an aphrodisiac, attributed to Casanova who is said to have consumed 50 raw oysters a day in order to uphold his legendary sexual prowess. However, oysters have in fact been reputed to be a food of love since the time of the Roman Empire, where they were a great favorite of many of the emperors themselves.

Another aphrodisiac much favored by Casanova was chocolate, although the first person associated with chocolate as an alleged aphrodisiac was the Aztec ruler Montezuma, who was said to have drunk 50 cups of hot chocolate a day in order to fully service his harem of 600 women. Such was the reputation of chocolate at that time, that the Aztecs and also the Mayans celebrated the harvest of the cocoa bean with festivals of orgies.

In brief, aphrodisiacs are a part of every culture and most of our current knowledge about aphrodisiacs is derived from myths, folklore, and traditional anecdotal evidences. Thus, it becomes very pertinent to know the history of natural aphrodisiacs.

History of Natural Aphrodisiacs
Aphrodisiacs in the East

A substantial portion of the knowledge that exists about aphrodisiacs evolved from the East. China and India have ample information on this subject. *Kama Sutra* of India known as 'Aphorisms on Love', the earliest available text on erotica was written by a Hindu sage Mallinaga Vatsyayana in AD300-AD400 during the first century (Charles 1990). Originally, it was written in Sanskrit, the literary language of ancient India. He described very interesting information on the art of love comprising eye contact, touch, embracing, kissing, coital techniques, posture in love, delaying and controlling the orgasm in man. Most of the guidance in eroticism was written for men. The text was one sided. There was lack of information or suggestions for women. It shed some light on the aphrodisiac ointments, philtres, and food to stimulate sex desire. Vatsyayana suggestions were also the use of cosmetics and perfumes to increase attraction. There were also some aphrodisiac recipes mentioned in *Kama Sutra* by Vatsyayana comprising milk, honey, ghee, sugar, and animal products.

Though *Kama Sutra* was written over 2000 years ago yet the information given on the art of love was amazing. Without a doubt, it is a celebrated classic of Indian literature on eroticism. Even to imagine such techniques and sex postures at that time was incredible. Sir Richard Burton translated it unexpurgatedly into English in 1883. Even today, Vatsyayana is still considered one of the greatest authorities on sexology. He had described sex as sacramental. Temples of Khajuraho built in Madhya Pradesh, India are the living example of sacred sexuality in Hindu culture. The sensual erotic sculpture and art work in the ancient Hindu Temples of India is the reflection of the sacramental sexuality. Sex was not only considered procreation but also recreation in ancient India. In other words these erotic sculptures acted as visual aphrodisiacs. During Muslim and British rule in India, freedom of the erotic art or visual aphrodisiacs in texts and its various displays was forbidden, though the British citizen Sir Richard Burton had the freedom and privilege to translate the erotic Sanskrit text *Kama Sutra* to English language. The publication was acclaimed as an educative sex manual throughout the world. There is no copyright on this text and number of authors throughout the world wrote books pertaining to sex by copying material from *Kama Sutra* and minted money.

Later, another Hindu poet Kalyana Malla wrote *Ananga-Ranga* during AD 1450-1525. It has been lithographed in Sanskrit, Marathi, Gujarati, and Hindi. Sir Richard Burton along with Arbuthnot as co-author also translated it into English in 1885. *Ananga-Ranga* contained large number of recipes comprising aphrodisiacs. The *Ananga-Ranga* surpassed both *Perfumed Garden* and *Kama Sutra*. The text is more focused on the details of sex problems and their remedies. Some very interesting aphrodisiac recipes were included for stimulating male and female libido.

The *Perfumed Garden* written by Arab writer Cheikh Nefzaoui is still popular today (Lovell 1999). It is a very eminent sex manual that also comprised a chapter on aphrodisiacs. Some erotic herbal recipes for improving sex vigor are given. Most of the aphrodisiac preparations described by him were composed of plant and animal products. Some of these are ridiculous and a few may be dangerous. It was originally written in Arabic between the years 1394 and 1433 and first translated to French by a staff officer of the French Army during 1850. It was first published in an edition of twenty-five autograph copies in 1876; later Sir Richard Burton translated it from French to English in 1886. This text would remain one of the world's great classic literatures on eroticism. The other two most important Arabic works on eroticism were *The Kitab al-Izah fi ilm al Nikah* (The book of exposition in the science of coition) a work attributed to the historian Jalal al Din al Siyuti and *Kitab Ruju a al-shaykh ila sabah fi l-kuwwat al-Bah*. The first was translated into English in 1900 the other in 1928. The later was translated into English in two volumes, containing chapters on the various kind of impotence in man, the philosophy of physical love, aphrodisiacs, their nature, and use (Walton 1958).

In China, Han Dynasty (AD25-220) a group of Taoist philosophy sexual expression dominated the culture. The school of Taoists wrote volumes of sex manuals including *Su Nu Ching (*Manual of Lady Purity), *Yu Fang Mi Chueh (*Secrete Code of Jade Room) *the art of the Bedchamber* and *Yu Fang Chih Yao* (Important Guidelines of the Jade Room). The Chinese represent one of the oldest civilization cultures and use of aphrodisiacs in their culture is also very old. One of the earliest known beneficiaries was Huang Ti, the Yellow Emperor, who lived around 2600 BC. He was very fond of aphrodisiacs and promoted research in this field. He was provided with a potion made from 22 herbal ingredients mixed with wine by his sex advisors. Love potions gave him an amazing sexual stamina and he was known to be one of the greatest lovers of his time. Also, the oldest sex manuals in the world are Chinese *Handbooks of Sex* written by the Yellow Emperor The Yellow Emperor is said to have had three female sex advisors, the Plain Girl (Su-nui), the Mystery Girl (Shuen-nui) and the Harvest Girl (Tsai-nui), and one male sex doctor, Pong Tsu. Their conversations and suggestions pertaining to sex were compiled into a book entitled *Su-Nui Ching*, named after Su-nui. This book was accepted as the sexual and medical manual for many generations. *Modern Day Encyclopedia of Chinese Materia Medica* was published in 1977 and listed about 6000 drugs; of which 4800 were of plant origin (Wyk and Wink 2005).

Opium has been used in China as a sex stimulant to a greater extent than anywhere else. Opium smoking induces strange sexual fantasies and a sense of voluptuous pleasure with longer erection. However, continued opium smoking leads to total impotence. In women, opium smoking intensifies the sexual instinct to a greater extent. It is a dangerous aphrodisiac when used regularly, leading to impotence and many more serious health complications.

Aphrodisiacs in the West
From the dawn of history love potions and aphrodisiacs have been employed throughout Europe. The love-philtres were of various types. Most of them were from oriental preparations used in India, Egypt, and Arabia. Nutmeg, cinnamon, and related species employed in similar way to those recommended by Vatsayayana and Nefzawi. The first book of mainly erotic contents published in Christian Europe was the *Hermaphrodite* of Antonio Beccadelli in 1426. The most famous sex manual in Rome

Ars Amatoria of the art of Love was written by Ovidius Naso about the time of Christ (43 BC-AD 17). It was a guidebook for the art of seduction. Three volumes were written; two volumes for males and the third for females. It was believed that the work celebrates extramarital sex and it was one of the reasons that Ovid was banished by King Augustus, who was trying to promote a more austere morality. The book was described as an *immoral book* representing the art of love as the adulterer's art rather than the husband's art by Irish writer H. Montgomery Hyde in *A History of Pornography*. The book won the praise of Renaissance Humanists. It begins with the words - *"Should anyone here not know the art of love read this, and learn by reading how to love? By art the boat's set gliding, with oar and sail, by art the chariot's swift: love's ruled by art."* Touch, smell, sight, scents, and sounds were considered as aphrodisiacs in *Ars Amatoria-the art of love* (Moore 1965).

Venette had published some texts on aphrodisiacs in the middle of the seventeenth century in Europe. According to Venette there are other factors besides drug and food that affect a man's appetite and coital ability such as temperament, age, climate, mood, and state of mind. He mentioned about the use of opium, cannabis, and stramonium as aphrodisiacs (Walton 1958). The variety and need for aphrodisiacs reached its peak during eighteenth century, especially in France and England. Spanish fly preparations were the hot aphrodisiacs at that time. Cantharidin, the main constituent of Cantharides (Spanish fly) was utilized not only extensively but carelessly. Many people died or suffered severe illness through the abuse of careless dose of cantharidin.

A large number of foods such as fish, oysters, crabs, caviar, mushrooms (trifles), wine, and various kinds of cheese were consumed in Europe to stimulate sex appetite. Pliny, a great Greek philosopher and herbalist mentioned about the use of asparagus at that time. Another famous personality of eighteenth century who believed in the value of food as an aphrodisiac was Gian Giacomo Casanova who used to start his breakfast with oysters.

William (1931) described the treatment of sexual impotence in men and women. He believed strongly in the value of food in stimulating sex desire. The book comprised one section on aphrodisiacs. He mentioned that the person with low libido should eat plenty of raw eggs, oyster, and raw and fried fish. It is quite apparent that awareness about the binding of biotin with cholesterol and high cholesterol level in blood was not of any concern there at that time.

All these authorities on the art of sex such as Vatsyayana, Nefzawi, and Taoist etc have compiled ample information on aphrodisiacs in one way or the other. Chinese and Indians are not the only ones that have a monopoly on aphrodisiacs. Every culture and society seems to have its own type of aphrodisiacs, some of which are specific to the plant or animal species that naturally exist in that geographical region. Greeks, Romans, Egyptians, Arabians, Ethiopians, Scandinavians, Assyrians, Incas, Spanish, Amazon, and Aztec used aphrodisiacs in their festivities and religious drinks.

The types of preparations ranged from the useless to the most toxic one which sometimes even caused death. These preparations were based on myth and hit and trial methods with results mostly varying from person to person. A majority of them do not make any sense when it comes to evidence based medicines. On account of their limited information and lack of scientific evidences these preparations have been accepted only as folkloric medicines until today. Ayurvedic, Chinese, and Homeopathic systems of medicine comprise enormous information on this subject.

Natural Aphrodisiacs in Different Systems of Medicines
Aphrodisiacs in the Ayurvedic System of Medicine
Ayurveda is derived from the Sanskrit words *Ayur* (life) and *Veda* (knowledge) and therefore means *the knowledge of life*. Following the system would help to ensure a long and happy life. The Ayurvedic System of Indian Medicine is about five thousand years old. The origins are obscured in the mist of time. Ayurveda as a total health system sprang from *Vedas*, the religious texts of Hinduism. The *Vedas* are timeless and have no human authorship. The *Vedas* comprises four parts; *the Rig Veda*, *Sam Veda*, *Yajur Veda*, *Atharva Veda*, the earliest of which dates back to 2000 BC. The principles of Ayurvedic medicines and usages of herbs are given in thousands of Shlokas (poetic hymns) in the *Rig Veda*. In India, knowledge was passed from one progeny to the next through songs and Shloka (poems) in Sanskrit language. Scholar and physicians had to learn by heart and recite. The ancient Ayurvedic texts that still exist are *Charaka Samhita* and *Susruta Samhita*. The written texts are dated from 1-200 AD. The health system is designed to understand the physiological, mental, and spiritual nature of an individual. Ayurvedic approach is holistic and achieves better results by using special plants that not only promote vitality to the reproductive organs but also act on the other organs of the body and balance the disturbed equilibrium (*Doshas* known as constitutional faults in the body). *Rauwolfia serpentina* and *Withania somnifera* are burning examples from Ayurveda that command respect even today in modern medicine. For example *Withania somnifera* (Ashwagandha) is not only an aphrodisiac but also a nerve tonic. Likewise, reserpine from *Rauwolfia serpentina* (Sarpgandha) not only controls hypertension but also used as a tranquilizer in small doses.

The Ayurvedic System is further classified and the branch of Ayurveda dealing on aphrodisiacs is called *Vajikarna*-science of Aphrodisiacs. It is a systematic approach for treating the sexual dysfunctions. It is a rejuvenating and nourishing treatment that requires strict physical and mental discipline. The body and the mind are cured during the treatment and one has to refrain from sex during the duration of the therapy. After the cure, strict diet, exercise and moderate sex is advised. Some aphrodisiacs are also recommended as tonic before sexual intercourse. So combination of the concentrated extracts of such species as *Anacylus pyrethrum* (*Akarkarabh*); *Asparagus racemosus* (*Shatawari*); *Cantella asiatica* (*Brahmi*); *Celastrus paniculatus* (*Jyotishmati*); *Chlorophytum borivilianum* (*Safed Musli*); *Glycyrrhiza glabra* (*Licorice*); *Mucuna pruriens* (*Atamagupta-Velvetbeans*); *Myristica fragrans* (*Nutmeg*); *Piper longum* (*Maghdi*); *Pueraria tuberosa* (*Indian Kudzu*); *Tinospora cordifolia*(*Guduchi*); *Tribulus terrestris* (*Gokshura*); *Vitex agnus-costus* (Nirgudhi) and *Withania somnifera* (*Ashwagandha*) along with some animal and minerals products are used as aphrodisiacs for both the sexes. Indian queens, princesses and princes had been using various Ayurvedic herbal preparations mixed with honey, milk and other nutrients to maintain their beauty, vigor, and sexual activities for centuries. At present, Ayurvedic aphrodisiacs of Indian origin in the forms of tablets and capsules are being sold like hot cakes in Germany and other parts of Europe. Some of the hot cakes are Afrodet, Afrol capsules, Vigorous and Dhathupushti for enhancing sexual vitality.

Aphrodisiacs in the Traditional Chinese System of Medicine (TCM)
The Chinese also maintain a similar view in their system as Ayurveda. The ancient Chinese Taoists used plants and animals products along with minerals to increase sexual energy. Their approach was also holistic and believed the flow of energy within the human body as the basis of physical and mental health. Channeling and accentuating the flow of sexual energy was considered the key to immortality. This concept is further expressed in *Yin and Yang* symbol representing the feminine and masculine forces as the basic principle of cosmic creation. For a healthy body yin and yang should be balanced and result in free flow of chee (qi) a vital force. The Chinese system often uses the genitals of animals to increase the strength of the reproductive organs. Tiger bones, rhino horns, bear bile, deer penises, bull testicles, and seal penises are used in aphrodisiac formulations in the Traditional Chinese Medicine even today (Tiquia 1999).

Like other races, the Chinese have their erotic potion of aphrodisiacs. It is said that one of the most powerful of Chinese aphrodisiacs is the celebrated Bird Nest Soup prepared from the nests of the sea swallow. The nests are made from an edible sea-weed, leaves being stuck together by the spawn of fish. This fish spawn is extremely rich in phosphorus has a powerful action increasing both desire and erection. Larger amount of soup can be toxic (Walton 1958). Opium in small doses had been smoked in China as an aphrodisiac too. Ginseng, dong quai, and related herbs are the burning examples of herbal tonic from China, which are still used in many aphrodisiac preparations.

Aphrodisiacs in the Homeopathic System of Medicine
There has been resurgence in the popularity of homeopathy medicine in the USA. Today the homeopathic drug market has swollen to a multimillion dollars industry in the United States, with a significant increase in both import and domestic market of homeopathic products. A most recent review article by Shang *et al.* (2005) has thrown some light on the system of homeopathy. It is concluded that the efficacy of Homeopathic System may be little more than a placebo. However, it is a long way to go and to prove that the System of Homeopathy is more than a placebo. Despite the skepticism, homeopathy is gaining popularity in the USA as an alternative treatment to many ailments (Andrew and Nichola 2000).

About 200 years ago a German physician and chemist Samuel Hahnemann (1775-1843) founded the Homeopathic System of Medicine. The word *homeopathy* is derived from a Greek word *homios,* meaning *like* and *pathos* means *disease.* Homeopathy is a system of medicine that works on the basis of the principle of *similae* or *simile*. In

Latin, *Similia similibus curantur*, meaning *like cures like*. The system is based on the principle that if any substance obtained from plant, animal and mineral source that produces symptoms of sickness in healthy people will have a curative effect when given in a very dilute and potentized form to sick people who exhibit those same symptoms. The more diluted a remedy is, the more potent it is. Hahnemann performed experiments on himself using Cinchona bark which contain quinine, a malarial remedy. According to him Cinchona produced the symptoms of malaria including fever and chills in healthy person. When Cinchona was given in a diluted or potentized form to a person suffering from malaria, he was cured. For example, the effects of peeling an onion are very similar to the symptoms of acute coryza (flu-type cold). The homeopathic remedy prepared from *Allium cepa* (the Red Onion) is used to treat flu-type colds. In another example, the symptoms and signs of acute arsenic poisoning are very similar to the symptoms seen in certain cases of gastroenteritis. According to the principle of simile, homeopathic potencies of arsenicum are used to treat gastroenteritis. That means we take the medicine that induces similar symptoms in the body. This is not a new approach to combat diseases. Vaccinations and allergy shots to imbibe immunity strongly support the concept and theory of the Homeopathic System of Medicine. More than 4,000 substances from the plant, animal, and mineral kingdom have been tested since Hahnemann first developed his theory. Homeopathy materia medica contains about 2,000 remedies (Lockie 2000).

The Homeopathic System also sheds some light on aphrodisiacs. Though, the term *aphrodisiac* is not used in the Homeopathy System of Medicine, yet there are remedies that are used to cure sexual deficiency. Treatment in this system is symptomatic. According to Homeopathy System of Medicine, sexual deficiency occurred on account of imbalance in the body that must be focused holistically. When balance is restored in the body emotionally, physically and spiritually—a rejuvenated sexual life often follows. The choice of remedies depends on individual's case with the same problem. Two persons with same sexual deficiencies may find success with different remedies on account of their individual nature. The process of finding the right medicine for an individual is trial and error. The wrong remedy will not cure the patient. Some computer programs are available to tally the symptoms with the remedy and are gaining popularity. Some of the common remedies for erectile deficiency or related sex problems in Homeopathy are *Agnus castus, Calcarea carb, Causticum, Conium maculatum, Damiana, Gelsemium, Graphite, Ignatia, Lycopodium, Medorrhinum, Natrum muriaticum, Onosmodium, Phosphoric acid, Selenium, Sepia, Staphysagria and Yohimbe.* The remedies are obtained from the plant, animal, and mineral kingdom. A preparation known as the mother tincture results from the maceration of plants and animals products in alcohol is further diluted to make potency. Strength of the potency depends on dilution of the mother tinctures comprising the active principles. The potency of the homeopathy medicine is directly proportional to the dilution of the extract. The remedies come in the form of tablets and mother tinctures with certain potency depending upon its dilution. The remedies are to be taken 15-30 minutes before meal or after meal. Aromatic substances such as camphor, menthol are not allowed while on homeopathy medication. The intake of coffee, tea, and nicotine is also not permitted as these may interact with the medication. How much useful these homeopathic preparations are as sex tonic or to cure sexual dysfunction in men and women is not scientifically known or proved so far.

Natural Aphrodisiacs on the Market and their Brands

There are various aphrodisiac preparations being sold over the counter and on the Internet as food supplements or tonic in open market. Sellers are raking in several million of dollars annually from people who are hoping against hopes to regain their lost vitality. FDA has no control on them. In accordance with FDA regulation, *"one does not make any therapeutic claims for any Dietary Supplements in accordance with the Dietary Supplement Health and Education Act."* Sellers exploit this statement.

Most of the products are formulations of herbs. Very few products are consisted of a single herb. These herbs comprise chemical constituents of different chemical nature. These constituents are complex organic molecules and are known as secondary metabolites of plants. They are pharmacologically active and can be toxic if taken more than the recommended quantity. There is no set dose of these crude drugs. Dosage varies from person to person. Even a small quantity can be toxic to some individuals. The formulations are composed of the plant, animal, and mineral products. These are complicated love potions and can be harmful. Individual herbs are powdered and mixed together and dispensed in the form of capsules and tablets. There is no indication of their incompatibility or disintegration study. Quality control and quality assurance of these preparations is very poor.

Numerous preparations of aphrodisiacs such as sex pleasure enhancers, sex boosters, and erection stimulator are sold by mail orders and unsolicited e-mail offer for both the sexes. It is distinctively mentioned in each advertisement *the FDA does not have information about this product.* It is really difficult to mention all these formulations in this text, thus some of the prominent products are given in Appendix A and Appendix B. The data have been collected from the websites. Most of these products have been reviewed by Consumer Health Digest and considered among top 25 sexual enhancers. However, the authors do not recommend the use of any of these products.

> *"For some, sex leads to sainthood:*
> *For others it is the road to hell –*
> *All depends on one's point of view."*
> *—Henry Millers (The World of Sex)*

Chapter Three

Male Sexual Dysfunction

Aphrodisiacs are considered as sex stimulants and may enhance low libido or fix some psychological inhibition pertaining to sex. However, these are not a panacea or elixir for sex life. If there are problems like impotence, frigidity, and physiological defects, clinical evidence suggests that aphrodisiacs cannot fix these problems. In other words, aphrodisiacs are not a *'quick fix'*. The term *aphrodisiac* is used and advertized so loosely that it often includes remedies for impotence. In this chapter, we will discuss the most important aspect of sexual problem: erectile dysfunction (ED). ED is a condition that has baffled most people and forced them to buy aphrodisiacs or bizarre formulations with hopes for vitality.

A widely publicized study reported that sexual dysfunction is highly prevalent in both the sexes, ranging from 10% to 52% of men and 25% to 63% of women (Laumann *et al*.1999). More than 18 million American men suffer from erectile dysfunction according to latest report by Selvin *et al.* (2007).

Impotence is a general term and should not be confused with ED (Erectile Dysfunction). ED may be defined as the inability to achieve or maintain an erection sufficiently rigid for sexual intercourse and ejaculation. Impotence comprises other problems that interfere with sexual intercourse and reproduction such as lack of sexual desire or libido, defective sperm, problems with ejaculation, and the inability to achieve orgasm. Impotence may be used to define male sexual dysfunction and covers ED, whereas ED is a specific term pertaining to erection only. Likewise, frigidity and fear of coitus are found in women. The term *impotence* is derived from Latin word, meaning *loss of power*. Impotence and frigidity have been considered as a curse since the time immemorial. The ecstasy of life in the materialistic world is through sex. Inability to perform sexual intercourse during youth is nothing but a kind of curse inflicted by Mother Nature to that individual. Ovid explained the term impotence in his book *The Art of Love*. His expression was very simple as follow:

Impotence
"If war's a gamble, love's a lottery
both have ups and down.
In both apparent heroes can collapse
so think again if you think love is a soft option
it calls for enterprise and courage
take it from me, lovers are all soldiers in Cupid's private army."
Ovid, Amores

There can come a time in a man's life when he experiences impotence. This generally has a physical cause such as disease, injury, or side effects of drugs. We will focus on impotence pertaining to erectile dysfunction. Any problem that impairs blood flow into the penis can cause lack of erection. Erectile dysfunction cases increase with age, and about 40% of men at the age of 40 carry some degree of ED and 48% of men at the age of 50 had some degree of ED and 70% by age of 70 experienced some degree of ED (Feldman et al. 1994). Ayta et al. (1999) revealed that there would be a possibility of worldwide increase in ED by the year 2025. They estimated that 152 million men worldwide have experienced ED. The projection for the year 2025 shows a prevalence of 322 million with ED, an increase of nearly 170 million men. The largest projected increases were found in the developing countries such as Africa, Asia, and South America.

A recent study by Selvin et al. (2007) in a sample of 2100 men, age 20 and older who took part in a national wide survey in the year 2001-2002 revealed that 18.4% of men over the age of 20 suffered erectile dysfunction. Older men had more problems, 70% of men of the age of 70 or over reported erectile problems compared with 5% of men at the age of 20 to 40. According to Guay et al. (2003) during aging, the absolute number of Leydig cells decreases by 40% and the decrease of pulsatile luteinizing hormone release is also reduced. This results in the decrease of free testosterone level by 1.20% per year and slowly leads to sexual dysfunction in the aged. Caretta et al. (2005) also confirmed that testosterone plays a significant role in the erectile dysfunction of aging men. Hypogonadism in them is associated with low levels of testosterone.

ED is treatable in all age groups. Awareness of this fact is increasing day by day after the 'Viagra-revolution'. Sexual dysfunction in both sexes is age related and progressive. Female sexual dysfunction affects more than 43 % of American women. In other words more than 40 million of women at some point of their lives suffer some kind of sexual dysfunction. This data is five years old. The current figures may have changed significantly. Male sexual dysfunction has become a popular subject for study during the past 25 years. Some of the deep-rooted problems in a relationship between a man and woman are contributed by impotence to some extent. An expert committee of the WHO (1975) concluded that *problems in human sexually are more pervasive and more important to the well-being and health of individual in many cultures than has previously been recognized.*

Physiology of Erection
It is very pertinent to understand the mechanism of erection. This will allow the readers to understand the medicinal or pharmacological action of aphrodisiac formulations.

500 years ago, Leonardo Da Vinci was the first scientist to conceive that during an erection the penis fills with blood. Da Vinci wrote, "The penis does not obey the order of its master, who tries to erect or shrink it at will, whereas instead the penis erects freely while its master is asleep. The penis must be said to have its own mind, by any stretch of the imagination" (Goldstein 2000). The phenomenon of erection is extremely complicated. It includes the interaction between the nervous, vascular, and endocrine

systems. There are numerous chemical transmitters involved in this, including epinephrine, norepinephrine, acetylcholine, prostaglandins, and nitric oxide. Thus erection involves the neural, hormonal, and vascular system. It occurs when blood rapidly flows into the penis vascular system. Let us discuss the vascular system of the male organ.

Vascular System of Male Organ
The Vascular system of the male organ is composed of three different regions as described below and shown in cross section Figure 2.

Structure of the Male Organ
A pair of parallel spongy columns called the *corpora cavernosum* and the central chamber *corpus spongiosum* that contains the urethra, which carries urine. All three regions are made up of erectile tissue. The *corpora cavernosa* are surrounded by a membrane called *tunica albuginea*. The *corpora cavernosa* nerves enter the penis near the prostate gland and enhance the natural erection. The cavernous nerves are branches of the pelvic plexus and inervate the penis. In the flaccid or normal penis, the muscles of the *corpora cavernosa* are in the contracted stage. The small arteries leading to the *cavernous sinuses* contract reducing the inflow of blood to penis. The smooth muscles regulating the many tiny blood vessels within the penis also contract. This is maintained by the sympathetic nervous system using norepinephrine, which binds to alpha-1-adrenergicreceptors.

Erection
The mechanism of erection is shown in the flowchart of Figure 2. Sexual arousal through touch, visual or stimuli of sexual nature stimulate the brain. The body releases chemicals that direct the flow of blood to the sex organs of both male and female. Parasympathetic nerve fibers, which go to the penis or clitoris, release acetylcholine. Acetylcholine helps cavernous smooth muscle to relax and more blood flows to the penis. The spongy chambers almost double in diameter due to the increase in blood flow. The veins surrounding the *corpora cavernosum* and *corpus spongiosum* are squeezed almost completely shut by the pressure of the erectile tissue; these are unable to drain blood out of the penis, causing it to become rigid. The *tunica albuginea* helps to trap the blood in the *corpora cavernosa* to sustain erection. Erection is reversed when muscles in the penis contract. It stops the inflow of blood and opens outflow channels. Likewise, in woman, arousal occurs when there is an increase of pelvic blood flow, resulting in vaginal lubrication. Nitric oxide plays a significant role in stimulating clitoral cavernosal smooth muscle, thereby increasing clitoral blood flow and resulting in genital engorgement.

Role of Neurotransmitters in Erection
Neural input from the brain is very essential. All stimuli through vision, touch, sound, smell, and taste that arouse the body sexually are processed in the brain through neurotransmitters. Numbers of chemicals in the brain and body such as amino acid, coenzymes, hormones, and peptides act as neurotransmitters responsible for mood, alertness, and sex arousal. Most common neurotransmitters are serotonin, norepinephrine,

dopamine, acetylcholine, and GABA (gamma-aminobutyric acid). Among these neurotransmitters, dopamine plays a major role in obtaining and retaining of the male erection. Acetylcholine helps in the release of nitric oxide, which is responsible for the relaxation of the smooth muscle and engorgement of the genital organs leading to erection. Norepinephrine from sympathetic nerves contracts the arteries and smooth muscles of *corpora cavernosum* in the penis keeping the penis soft.

It is needless to go in details about the physiology and pharmacology of neurotransmitters that is beyond the scope of this book. However, it is pertinent to throw some light on their action involved in sex response. We are discussing neurotransmitters in detail in chapter five (Neurotransmitters as Natural Aphrodisiacs)

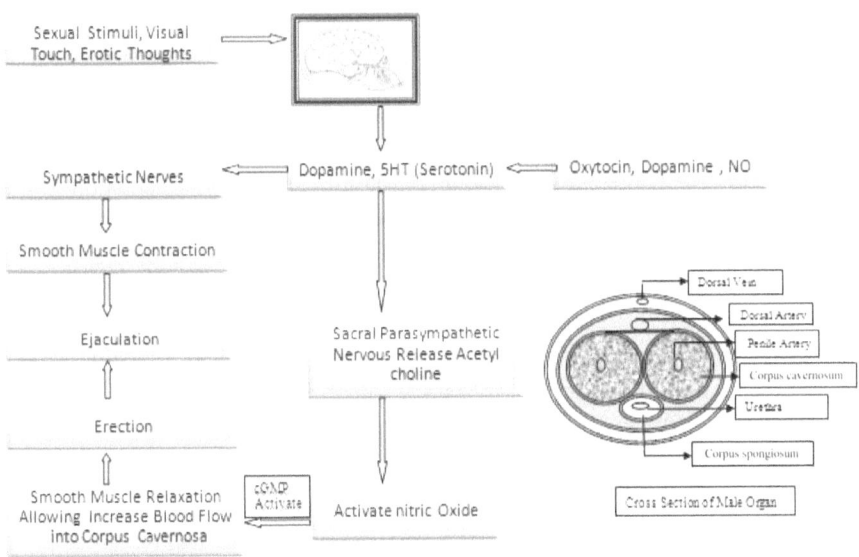

Figure 2: Physiology of Erection

Causes of Erectile Dysfunction (ED)

There are many causes of erectile dysfunction such as physical, mental, socioeconomical and side effects of many drugs as shown in Figure 3.

A. Physical

1. Oxygen Deprivation and its Causes

Occurrence of ED is most common when the male organ is deprived of oxygen-rich blood. When oxygen levels to the penis are low, an imbalance occurs in two important substances, TGF-β1 (transforming growth factor) and Prostaglandin E1. TGF β1 levels increase which trigger production of collagen, a tough layer that forms connective and scar tissue. In addition there is a reduction in prostaglandin E1, a chemical that suppresses collagen production and relaxes the smooth muscles to allow blood flow in an erection. This imbalance reduces the blood flow to the penis. Diseases such as diabetes, hypertension, peripheral vascular, hyperlipidemia, and hyper-cysteinemia can cause oxygen depreciation. Plasma transforming growth factor-β, levels in patients with erectile dysfunction has been studied by Ji-Kan *et al.* (2004).

2. Blockage of Blood Vessels

Blockage of blood vessels due to atherosclerosis also reduces the blood flow to the sex organs and cause impotence and frigidity. Atherosclerosis and vascular disease contribute about 70% of impotence. The other physical cause of impotence is smooth muscles and fibrous tissues.

3. Hypertension

ED is more common and severe with men suffering from high blood pressure. Many of the drugs for the treatment of hypertension have side effects, particularly ED. It is reversible when the drugs are stopped. Newer antihypertensive agents including angiotensin-converting enzyme (ACE) inhibitors and angiotensin-receptor blockers (ARBs) do not cause ED.

4. Parkinson's disease

It is estimated that one third of men with Parkinson's disease (PD) experience impotence. The physical cause of Parkinson's related impotence is due to an impaired nervous system. Dopamine plays a significant role in this phenomenon. Parkinson's disease is thought to be caused by lower levels of dopamine in the brain -a chemical that transmits messages from the brain's *relay center* to its nerve cells, enabling physical movements. It is possible that lack of dopamine may also cause low libido.

5. Diabetes

Diabetes is the most common cause of sexual dysfunction. It has been estimated that up to 50-60 % of diabetic men have erectile dysfunctions because the disease can damage nerve and arteries, making it difficult to achieve an erection.

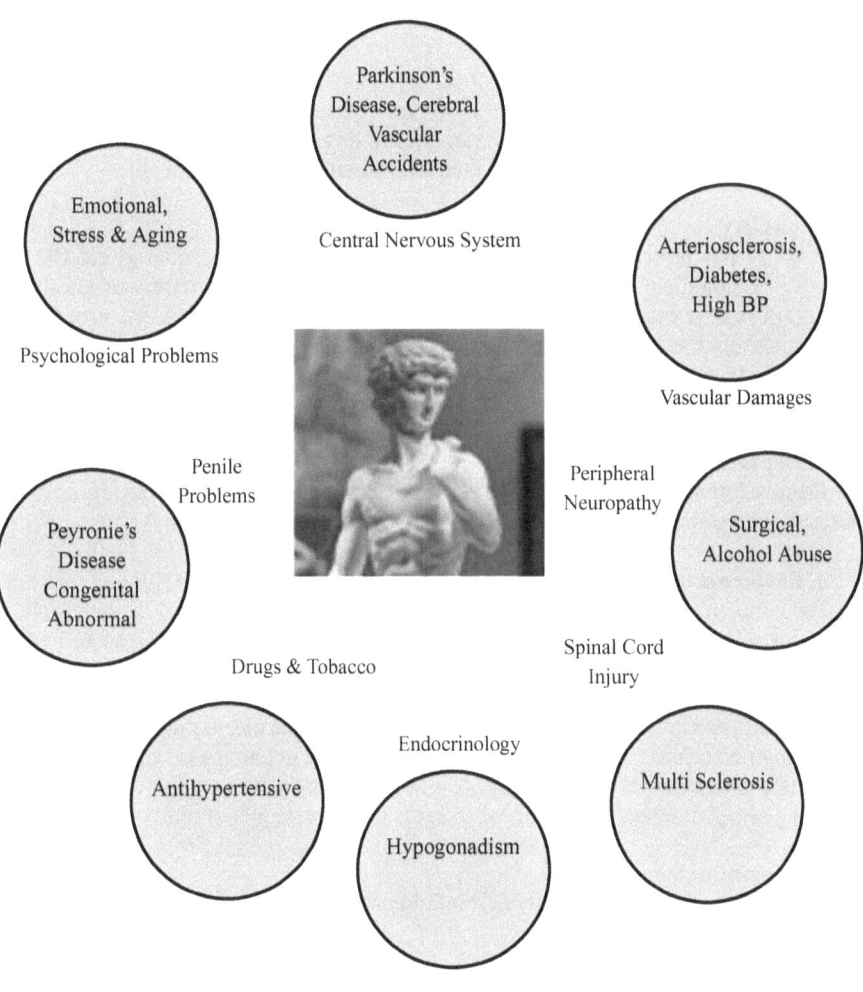

Parkinson's Disease, Cerebral Vascular Accidents

Central Nervous System

Emotional, Stress & Aging

Psychological Problems

Arteriosclerosis, Diabetes, High BP

Vascular Damages

Penile Problems

Peripheral Neuropathy

Peyronie's Disease Congenital Abnormal

Surgical, Alcohol Abuse

Drugs & Tobacco

Spinal Cord Injury

Endocrinology

Antihypertensive

Hypogonadism

Multi Sclerosis

Figure 3: Causes of Erectile Dysfunction

6. Multiple Sclerosis
People suffering from multiple sclerosis also have erectile dysfunction.

7. Obesity
Obesity is a problem all over the world. It can influence the sexual activity to a great extent.

B. Emotional disorders
Emotional disorder such as stress, anxiety, guilt, depression, low-esteem, and fear of sexual failure also contribute a major factor in causing sexual deficiency.

1. Anxiety
Anxiety has both emotional and physical consequences that can affect ED and leads to psychological impotence. Anxiety means performance anxiety or an intense fear of failure in sexual activity. The individual feels doubt on his performance. During stress or anxiety the brain releases chemicals known as neurotransmitters that constrict the smooth muscles of the penis and its arteries. This constriction reduces the blood flow into and increases the blood flow out of the penis.

2. Depression
Depression also causes impotence. The majority of the people suffering from ED also had depression. In certain cases ED causes depression.

3. Problems in a Relationship
Problems in relationship are one of the major causes of impotence. If the problems persist for a long time, these lead to ED. There are many factors involved in healthy relationship such as emotional compatibility, right mood, and intensity of love. Both the partners should be interested with the same enthusiasm for the sexual activity. They should resonate with the same frequency. If one partner is thinking something else during performance, the other loses interest. Some get angry or upset and lead to incomplete and unsatisfactory ending. Nagging spouses can never have a healthy and satisfying sex. An unhealthy relationship leads to depression and ultimately to ED.

4. Overindulgence of Sex
Virile impotence can occur by overindulgence in sex or sexual addiction.

C. Socioeconomic Issues

1. Alcohol
In small quantities alcohol acts as an aphrodisiac, but in large quantities (more than one drink) it can depress the central nervous system and impair sexual function.

2. Smoking
Smoking is also considered one of the causes that add to impotence in both the sexes, since nicotine is a vasoconstrictor and it affects blood flow in veins and arteries

D. Lack of Frequent Erection

Infrequent erections deprive the penis of oxygen-rich blood that increases the collagen production and ultimately form tough tissues that prevent the blood flow. Frequent erection experience in sleep may be a natural protection against this process.

E. Side Effects of Drugs

Many drugs such as antihistamines, antidepressants, tranquilizers, appetite suppressants, and cimetidine are also responsible for impotence as described by Roger (1996). Figure 4 shows the various classes of drugs responsible for ED.

F. Miscellaneous Causes

Surgical treatments for prostate; bladder removal of cancer; urethral stricture; and urinary surgery; carcinoma of the penis; priapism; renal transplantation; colon surgery; radiation; and penile amputation can also cause impotence by damaging nerves, smooth muscles, arteries, and fibrous tissues of the *Corpora cavernosa.*

Treatment of Erectile Dysfunction (ED)

First of all, it is very imperative to determine the cause of ED since the cause is going to dictate the treatment. Sometimes, it is due to some pre-existing illnesses or side effects of the medications that can be treated by changing the course of treatment. Most of the cases of ED particularly found in younger people, are psychological and need psychotherapies. If it is physical cause due to lack of blood circulation then of course, medication is required like sildenafil (Viagra). If the person cannot take sildenafil due to its side effects then there are other options available such as medication injected into the penis, vacuum devices, intracavernosal injection therapy, and invasive procedures such as surgery. At present, majority of people are turning to Complementary and Aletrnative medicines (CAM) or natural products.

Natural Aphrodisiacs Used in Erectile Dysfunction

Pharmacological approach of the aphrodisiac formulations available on the market is holistic and the treatment takes its own time depending upon the quality of the products. Natural herbal aphrodisiacs, used in male sexual dysfunction including ED available on the market are shown in Appendix A. A majority of these preparations are being sold on the basis of myths.

> *"All lovers swear more performance than they are able."*
> — *William Shakespere*

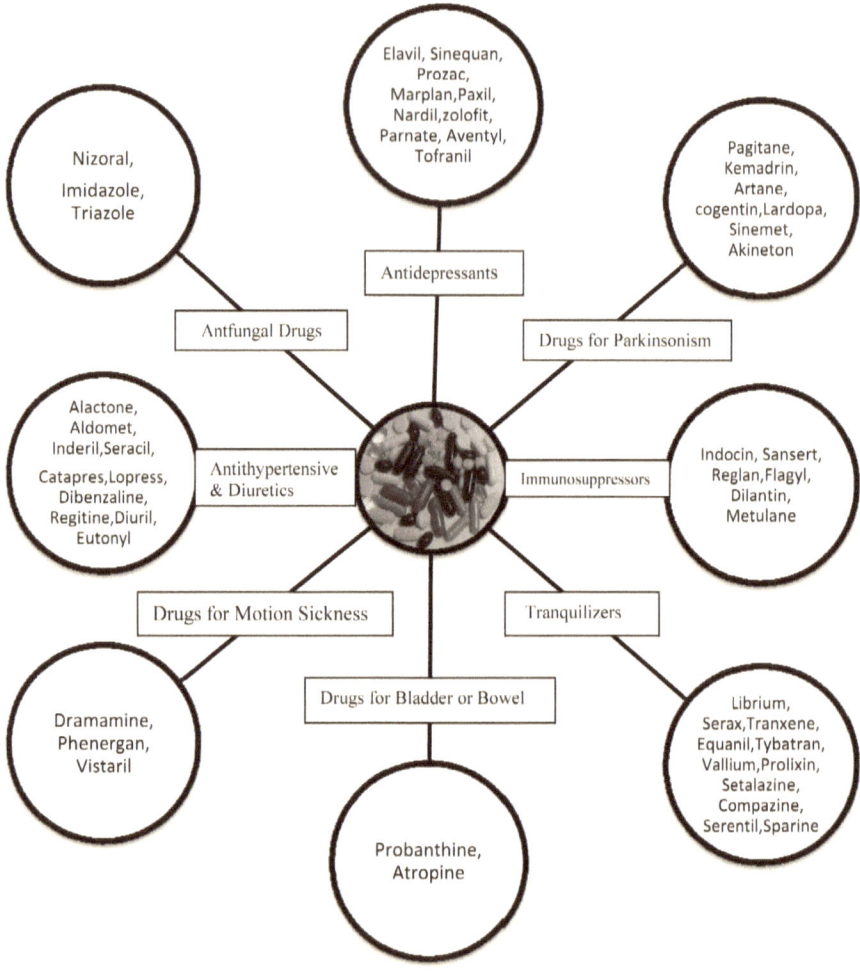

Figure 4: Drugs that Affect Erectile Dysfunction

Female Sexual Dysfunction

Frigidity may be defined as inability of a woman to perform adequately in term of sexual desire, sexual arousal and orgasm. A majority of the health professionals believe that there is no such condition as frigidity but it is a term used quite commonly in a derogatory or abusive way for women who are not sexually responsive, particularly when their partners want them to be. Impotence is an unfortunate physiological problem, while *frigidity* indicates a defective personality and a failure to live up to a wife's marital obligations. She is cold, emotionless, unresponsive, has a lack of interest in sex or rejection, unavailable, or if indulging, not taking part in the act. However, the term is not used today because of its negative connotation, being replaced by FSD (Female Sexual Dysfunction). If conditions such as low libido, aversion to sexual activity, difficulty in arousal, inability to achieve orgasm or pain with sexual activity or intercourse exist in females, it is called FSD.

FSD affects as many as 43% of all women and occurs most frequently in women aged 25 to 50 years. It is more prevalent in postmenopausal females due to age-related decline of hormones. There are several types of sexual dysfunction in women characterized by problems with arousal, desire, orgasm, and pain (dyspareunia, noncoital sexual pain and vaginismus). Women suffering from FSD are indifferent to sex or have little or no response to sexual stimulation. Other factors, including psychological, physical and relationship factors, all contribute to female sexual response and sensual enjoyment.

Female Sexual Response
There are four stages of a woman's sexual response cycle, which are interrelated and must occur to ensure an enjoyable sexual experience.

1. The first stage is **Desire or Excitement**, which is a result of physical or mental stimulation resulting in vaginal swelling and lubrication due to more blood flow to the genitals.

2. Prolonged excitement leads to stage two, **Arousal**-characterized by increased vaginal swelling, enlargement of the breasts and the nipples becoming erect. The clitoral glans moves inward under the clitoral hood.

3. The third stage is **Orgasm or Release**. Orgasm is the plateau phase of the sexual act. A feeling of well being or immense physical experience controlled by involuntary or autonomic nervous system. A female

experiences full orgasm when her uterus, vagina, anus, and pelvic muscles undergo a series of rhythmic contractions. However, the sensations are extremely pleasurable and felt throughout the body causing a transcendental mental state. Some enjoy these contractions more than the others; most women find these contractions very pleasurable.

5.   The fourth stage is **Resolution or Relaxation** follows the final stage. After the orgasm, the clitoris returns to its normal size.

Female sexuality is much more complicated than male sexuality with multiple factors concerning desire, including such diversified items as level of education, past sexual experiences, sexual expectations, cultural and religious beliefs, availability of a right partner and of course, the individual's hormonal status.

In brief, female sexual response depends on a complex interplay of physical and emotional aspects of the partners. Imbalance of any of these leads to FSD. Very little is known about the Female Sexual Dysfunction, though FSD affects more than 43% of American women (Laumann *et al.*1999). This means 40 million females at one point in their lives experience some kind of sexual dysfunction. In the underdeveloped countries of the world there is no data available on FSD. This aspect is ignored due to unknown reasons. More women than men experience sexual dysfunction due to many reasons. One of the major causes in female sexual dysfunction is considered to be stress and individual factors. However, sexual dysfunction in both sexes is also age related. After decades of neglect, pharmaceutical companies are closer than ever to finding a prescription for millions of women who lack sexual desire.

Diagnosing Female Sexual Dysfunction
Diagnosis of female sexual dysfunction is very complicated. It depends upon mulitple factors. According to recent findings (Meston and Bradford 2007) there are four classifications for diagnosing FSD.

A. Sexual Desire
An individual is not active sexually because of lack of sexual thoughts or receptivity to sexual stimulation. This is psychological.

B. Sexual Disorders
In this case there is poor vaginal lubrication, less genital sensation and poor vaginal smooth muscle relaxation. This is physiological and may result from medications, pelvic disorder, or neural and peripheral vascular disease.

C. Orgasmic Disorder
A persistent or recurrent loss of orgasmic potential after sufficient sexual stimulation and can be due to spinal cord injury or pelvic surgery.

D. Sexual Pain
There is constant genital pain associated with non-coital sexual stimulation during foreplay.

FSD can be physical or physiological. Some women think FSD is a normal part of life like menopause but that is not the case. They should consult their physician if they develop any of the above symptoms.

In healthy women arousal occurs due to an increase of pelvic blood flow that results in vaginal lubrication. Nitric oxide plays a significant role in stimulating clitoral cavernosal smooth muscle, thereby increasing clitoral blood flow and resulting in genital engorgement. One study showed that 10 to 100 mg of Viagra improved sexual response in women (Basson *et al.* 2002).

Causes of Female Sexual Dysfunction
Female sexual dysfunction is a very complicated phenomenon and depends upon many factors. Both physiological and psychological issues can be involved. Figure 5 illustrates the various causes of FSD. Figure 6 depicts the various drugs responsible for FSD.

Cure for Female Sexual Dysfunction
Limited scientific data is available on FSD. Sexual dysfunction is one of the few health issues in which the male has dominated, and female sex problems were always kept on the back burner. The reason is not known, perhaps, drug companies had other priorities.

" . . . . . . *the sexist bias in this research is clearly apparent, as pharmacological effects on female sexuality have been almost totally neglected by most investigators.*"
Rosen and Ashton, Pro sexual Drugs.

However, the Viagra revolution ignited the interest in FSD. Suddenly, there have been medical breakthroughs in treating female sexual arousal dysfunction (FSAD). Some well-known pharmaceutical companies are in later-stage of patient trials in search of 'Pink Viagra' for women. However, their efforts are focused on the synthetic products rather than searching an alternative to viagra from natural sources.

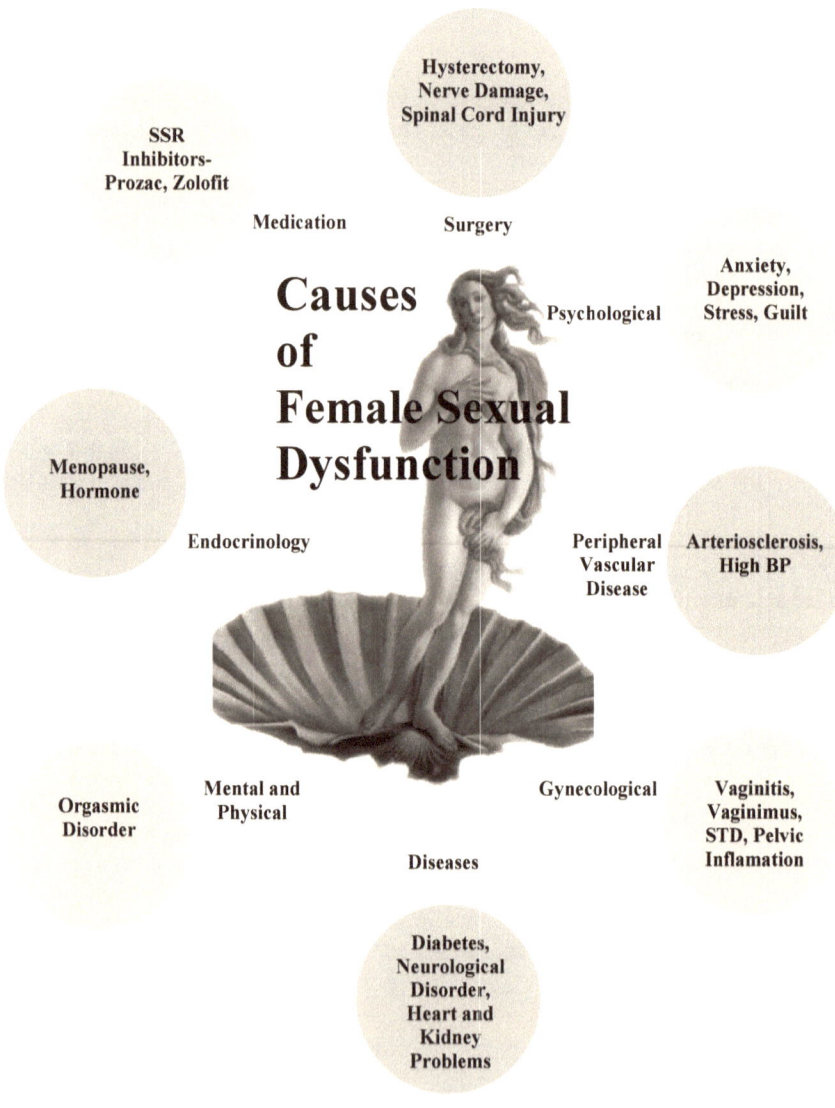

Figure 5: Causes of Female Sexual Dysfunction

Vivus, Inc a pharmaceutical company is engaged in the research and development and commercialization of products to restore sexual function in men and women. It is signing up women for final-phase trials of their product Alista, used for the treatment of female sexual arousal disorder.

Intrinsa, the Proctor & Gamble testosterone patch was the first treatment for female desire treatment. The product was developed jointly with Watson Pharmaceuticals. P&G's patch is applied to the abdomen twice a week and releases 300 micrograms a day of testosterone into the bloodstream, matching the level produced by many women in their peak reproductive years. The testosterone is believed to stimulate cells in both the genitals and the brain, prompting a physical response as well as desire and sexual fantasies. Clinical studies on the patch have been successful in normal and menopausal woman.

Pfizer studied the drug Viagra in women. The clitoris is analogous to the penis and the vaginal wall microscopically and functionally is similar to the spongy tissue of the penis. The medications that have been developed to affect the spongy tissue of the penis will also affect the spongy tissue of the clitoris and vaginal wall. The study utterly flopped. Pfizer found that although Viagra did indeed increase blood flow to the pelvic plexus, it was not quite sufficient to produce more desire. Viagra would tend to address the body but not the mind. After eight years of trials, Pfizer gave up on the idea of Viagra for women. Likewise, Eli Lilly & Co. and ICOS tested their new Viagra rival, Cialis, on women but abandoned the research in 2001.

Vivus and NexMed Incorporated are looking at creams that directly affect the genital tissue. The main ingredient in each, alprostadil, is a synthetic version of the blood vessel dilator prostaglandin E-1, which also is found in semen. Vivus, aiming at surgically menopausal women, found a 64% increase in *satisfying* sexual activity in a small trial of premenopausal women and is now launching a larger late-stage test in postmenopausal women. NexMed's cream, Femprox, is in late-stage trials in China for both pre-and post-menopausal women. The study is double blind, placebo-controlled and randomized and designated to test the efficacy and safety of Femprox cream in women diagnosed with female sexual arousal disorder (FSAD). The study on 400 patients was completed in the year 2005 and the results were encouraging.

Nastech Pharmaceutical in Bothell, Washington is studying a nasal spray form of apomorphine, which may work both above the neck and below the waist. It is believed to stimulate dopamine receptors in the hypothalamus, initiating a neural cascade that starts the cycle needed to boost blood flow to the vagina and to increase, in the genitals, the level of nitric oxide.

Palatin Technologies, in collaboration with King Pharmaceuticals, is conducting trials to find another blockbuster pharmaceutical nasal-sprayed for female sexual dysfunction. However, limited efforts have been made by these pharmaceutical

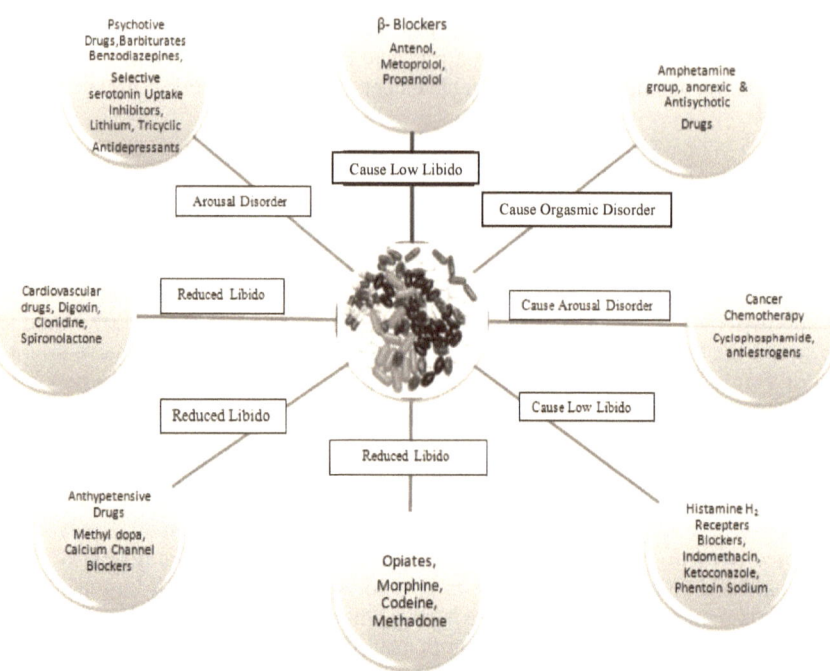

Figure 6: Drugs that Affect Female Sexual Dysfunction

companies to find a *Blue or Pink Viagra* from the natural sources. On the other hand, the herbal industry is busy in bottling *love potions* comprising herbs, on the basis of myths.

Meanwhile, women with female sexual arousal dysfunction search for their own remedy. Some women are using *natural* or *bio-identical* estrogen and testosterone. There are contradictory reports on these herbal formulations. Some feel quite comfortable with these herbal products where the others are not.

Natural Aphrodisiacs used in Female Sexual Dysfunction
Natural products for FSD have become an increasingly popular option particularly with the side effects of the synthetic drugs. This triggers the use of natural products in improving the quality of a woman's sex life. During recent years, tremendous amount of research has been carried out to understand the various factors responsible for the sexual excitement and orgasm in females. To address female sexual dysfunction and restore sexual vitality, it is very pertinent to focus certain physiological factors responsible for libido such as energy level, hormonal balance, blood flow, vascular, and nervous system functions. The herbal industry claims that there are a number of effective standardized herbal extracts, amino acids, vitamins and other substances that can contribute in different ways in assisting women regain their sexual vitality and mitigate some of the female sexual dysfunctions.

There are numerous female libido enhancement products presently being offered to female consumers. In order to make an informed decision the consumer health digest has compiled a list with the most straightforward, up-to-date information on the very best female sexual enhanced products on the market today. They have analyzed hundreds of products based on the criteria of ingredient quality, ability to enhance libido, the ability to increase sensitivity and energy, feedback from a wide range of sources including online surveys, forums, discussion groups, and independent customer reviews. They compiled, analyzed, and presented the data ranking the products accordingly. Some of these products are shown in Appendix B.

Natural Aphrodisiacs used in FSD Produced by Antidepressants
About 24% of women in the United States suffer from depression (Kessler *et al*. 1994). Use of antidepressants has affected sexual dysfunction such as low libido, delayed orgasm, and anorgasmia in up to 71% of patients using these drugs (Grimes *et al*. 1996; Montejo *et al*. 1996; Couper-smart and Rodham 1973). In 2003, estimated sales of antidepressants worldwide were nearly $20 billion out of which Zoloft accounted for $3.4 billion.

Antidepressant-induced FSD is related to the therapeutic activity of selective serotonin reuptake inhibitors (SSRIs). One of the functions of the central nervous system (CNS) in human sexual response is to supress erections through the sympathetic nervous system and a cluster of neurons known as the para gigantocellular nucleus (PGN). The PGN neurons transmit signals down their axons to the erection-generating center in the spine. There the PGN neurons liberate serotonin that behaves as a chemical messenger within the erection-generating center that represses erections by inhibiting the effects of pro-erectile neurotransmitters. As a result, nitic oxide synthase is inhibited and nitic oxide released into penile smooth muscle is reduced. Millions of people take SSRI drugs that work in part by increasing CNS levels of serotonin. It has been proposed that, by increasing the level of serotonin in the CNS, pro-erectile physiologic mechanisms are inhibited.

Natural aphrodisiacs such as Ginkgo, Yohimbe, Ginseng, Damiana, St. John's Wort, Valerian, Sage leaf, and Catuaba bark have shown excellent results in reversing the symptoms that resulted from antidepressant drugs.

Ginkgo *(Ginkgo biloba)*
During recent years, Ginkgo has gained great acceptance for treatment of antidepressant induced sexual dysfunction for female (Kleijen and Knipschild 1992). Elison and Deluca (1998) reported that Fluoxetine-induced genital anesthesia was relieved by *Ginkgo biloba*. Earlier, Cohen (1996) confirmed beneficial effects of Ginkgo on antidepressant induced sexual dysfunction. In other clinical studies, Cohen *et al*. (1998) showed that *Ginkgo biloba* extract with 63 patients was effective in 84% of patients with ASD. However, it needs more data to confirm its use.

Yohimbine *(Pausinystalia yohimbe)*
Yohimbine is obtained from the bark of *Pausinystalia yohimbe*. It increases blood circulation to erectile tissue. It is no longer FDA approved for treating male erectile dysfunction. Although several studies have demonstrated some efficacy in men, there are relatively few clinical trials including women. Gitlin reported (1997) that Yohimbine has been found to lessen ASD (Antisepressant-induced Sexual Dysfunction) in men and women. Jacobson (1992) and Price and Grunhaus (1990) reported that Yohimbine can be used to treat ASD. Side effects may include hypertension, headache, increased heart rate, dizziness, urinary urgency, and sweating. On account of its toxicity, the FDA discouraged the use of yohimbine.

Ginseng *(Panax ginseng)*
Ginseng root is derived from *Panax quinquefolius* (American ginseng) or *Panax ginseng* (Asian ginseng) comprises a large group of ginsenosides glycosides as the main constituents (Gilis 1997). Ginseng can improve sexual dysfunction such as absence of orgasm and low libido in-patients taking antidepressants. Ginseng is also known to possess phytoestrogen activity. The herb is believed to be as an adaptogen. Side effects may include hypertension, insomnia, vomiting, headache, and epistaxis. Siegel (1979) suggested that ginseng should not be recommended for use in-patients with bipolar disorder.

Damiana *(Turnera diffusa)*
Damiana is an aromatic herb that calms anxiety and induces a relaxed state of mind. It is considered a potent and very useful aphrodisiac for women. The plant contains alkaloids that supposed to stimulate the sex organs. Native Mexican women drank an infusion of the herb a couple of hours before retiring to the bed.

St. John's Wort (*Hypericum perforatum***)**
St. John's Wort (*Hypericum perforatum*) is used as an alternative to SSRIs if ASD occurs. The efficacy and safety of St. John's Wort for mild to moderate depression is better among the other available herbs. The mechanism of action is not known, but St John's Wort may have SSRIs (Selective Serotonin Reuptake Inhibitors) or MAOIs (Monoamine Oxidase Inhibitors) properties (Suzuki *et al.* 1984; Raffa 1998).

Valeriana *(Valeriana officinalis)*
Valerian root has a calming effect on the nervous system. It is good for the entire circulatory system and soothes muscles, nerves, and blood vessels. Additionally, valerian root helps promote normal sleep patterns.

Sage Leaf (*Salvia officinalis***)**
Sage leaf has a long history for its use as a medicinal herb. Sage leaf helps relieve mental exhaustion and enhances the ability to concentrate. Refreshed and alert women are better able to enjoy the physical and emotional pleasures of being with their partners. Sage also contains estrogenic substances and is a natural antiperspirant. It has been traditionally used during menopause to help the body deal with hormonal changes, hot flushes, and depression.

Catuaba bark (*Erythroxylum catuaba***)**
Catuaba Bark is derived from *Erythroxylum catuaba*, a famous Brazilian aphrodisiac plant. It is a nerve tonic with aphrodisiac properties. It contains alkaloids such as catuabine, A, B, C and D along with phytosterols and cyclolignans.

Natural Aphrodisiacs used in Female Sexual Dysfunction Caused by Menopause
Levels of estrogen and testosterone decline during menopause, resulting hot flashes, night sweats, insomnia, and sexual dysfunction. Absence of estrogen has been linked to vaginal dryness, vaginal tissue atrophy and irritation, and decreased libido. Vermeulen (1976) reported that testosterone levels in women decline by about 30 percent during menopause and lead to low libido. A number of women who want to avoid hormone replacement treatment because of its side effects as breast cancer look for an alternative treatment such as herbal preparations (Laucella 1997). These products are sold with elaborate advertisements full of hopes and people are buying them though they lack scientific data for their efficacy. The following herbs have been used on the basis of incomplete or preliminary scientific data for sexual dysfunction occurred by menopause.

Black Cohosh Root (*Actaea racemosa, synonym Cimicifuga racemosa*)
Black Cohosh is derived from the plant *Cimicifuga racemosa*. American Indians have originally used it during menopause as a folkloric medicine. Black Cohosh root is believed to contain a natural estrogen that is similar to human estrogen and helps with joint function, blood pressure, stress, menstruation and menstrual cramps, and estrogen levels in women. A low level of estrogen is linked to loss of sexual desire in women. Black cohosh root is the most popular herbal supplement in Europe for women experiencing premenopause symptoms. It provides isoflavones and other constituents that help support a woman's health during this phase of her life.

Stolz (1982) carried out the cimicifuga extract trial on human and found that 80% of menopausal women experienced improvements in physical and psychological menopausal symptoms within six to eight weeks. The plant is known to have estrogenic properties inhibiting LH release. However, it needs more clinical data to support its merits. Newton *et al* (2006) in their study of 351 menopausal and postmenopausal women has found that black cohosh did not appear to lessen their hot flashes. The women were randomly assigned to receive (a) black cohosh, (b) herbal containing black cohosh and 9 other ingredients, (c) a multibotanical plus dietary counseling about using soy products, (d) an estrogen product, or (e) a placebo. At 3, 6, and 12 months, the patients who received the herbal interventions did no better than those who received the placebo. The patients who received estrogen had significantly fewer symptoms. The authors concluded that black cohosh used alone or as part of a multi-herb regimen, *shows little potential as an important therapy* for relief of the hot flashes associated with menopause.

Chasteberry (Vitex agnus-castus)
Chaste tree berries are derived from *Vitex agnus-castus* and have been used to treat premenstrual syndrome. Chasteberry has been used to treat many hormone-related gynecologic conditions since the time immemorial. It has been shown to lower prolactin levels in women and found to be effective in treatment of menopausal symptoms. Chasteberry helps in hormonal imbalances and may help restore a more normal estrogen to progesterone balance. It can alleviate adverse effects related to PMS, including irritability, anxiety, nervousness, and mood swings. It may reduce hot flash symptoms in menopausal women. It is used on the basis of folkloric information.

Licorice Root (*Glycyrrhiza glabra*)
Licorice root is known to act as a sexual stimulant, bringing oxygen to the genital area and increasing endurance and vitality. The herb contains plant estrogens, producing mild estrogenic effects that have been helpful for women with menopause and for women dealing with menstruation.

Potency Wood (*Muira puama*)
It is a Brazilian herb employed around the world in herbal medicine. It is used as a neuromuscular tonic, to treat PMS, menstrual cramps, improve libido and as an aphrodisiac (da Silva 2002).

Wild Yam (*Dioscorea species*)
Dioscorea species contain steroidal sapogenins. The major constituent is diosgenin. Preparation available on the market contains extract of wild yam obtained from the tubers of dioscorea species on a cream base. The formulations are known as natural progesterone. Diosgenin can be converted to progesterone in the laboratory involving several steps. The absorption of diosgenin through the skin and then conversion into progesterone in the system is very doubtful.

Maca (*Lepidium meyenii*)
Used traditionally as an aphrodisiac to increase sexual desire, arousal and to promote vitality. It may increase fertility and correct hormonal imbalances.

Oat (*Avena sativa*)
Oat is used as a natural aphrodisiac. It is considered to increase the activities of neurotransmitters and enhance low libido.

Soybean Isoflavones (*Glycine max*)
Soybeans contain phytoestrogens isoflavones and phytosterols. If estrogen levels are low, these isoflavones are reported to act as "weak estrogens," and prevent the vaginal dryness and atrophy that diminishes the sexual pleasure during menopause. The North American Menopause Society (NAMS) recommends including soy foods in the diet to help reduce menopause symptoms. However, the research is in progree to investigate whether soy increases or decreases the risk of developing breast or uterine cancer in menopause women.

Red Raspberry Leaf (*Rubus idaeus*)
Red raspberry has been known as the "Vine of Female Vitality." The leaf of the vine is the part that is most often used. The chemical constituents of Red Raspberry include flavonoids, alkaloid fragrine, citric acid, malic acid, vitamin C, iron citrate, calcium, magnesium, and manganese. Traditionally, midwives and herbalists claim that it also helps tone the ligaments of the uterus to ease childbirth and regulate menstrual flow. Red Raspberry extract aids the menstruating woman as it helps to balance the menstrual cycle and reduces excessive bleeding. It has been used in the past in the recovery of sexual trauma and abuse. Raspberry Leaves are considered an excellent tonic for pregnant women to consume during their nine months to tone the uterus, nourish themselves and the growing baby, prevent a miscarriage, and facilitate birth and placental delivery. When used after birthing, leaves can decrease uterine swelling and minimize postpartum hemorrhaging. However, all of these are folkloric evidences.

Southern Bayberry Bark (*Morella cerifera*)
Bayberry bark, is a strong stimulant, brings a sense of heat and vitality to the body. Bayberry has many useful properties for the body, assisting with circulation and toning the muscles. It is especially beneficial for uterine and vaginal health.

Cayenne Pepper (*Capsicum annuum***)**
Cayenne Pepper has been used for centuries to enhance blood circulation. Cayenne contains capsaicin, a chemical that is supposed to stimulates various neurotransmitters. A small amount of capsaicin excites the nervous system into producing endorphins, which promote a pleasant sense of well being and stimulate low libido.

Kudzu Root (*Pueraria lobata)* **and Red Clover (***Trifolium pratense***)**
Root extract of Kudzu and flower extract of red clover are used in menopause (Fugh-Berman and Kronberg 2001; Tice *et al.* 2003). Isoflavones such as daidzein, puerarin, formononetin and biochnin, found in them provides estrogen like compounds acting to normalize estrogen levels. It is believed to be effective in the production of sex hormone-binding globulin. However, no scientific data is available to support this assumption.

Horny Goat Weed (*Epimedium sagittatum***)**
Horny Goat Weed is obtained from the species *Epimedium grandiflorum and E. sagittatum* commonly found in China and Japan. Their principle use is as an aphrodisiac. Their leaves contain a variety of flavonoids, polysaccharides, and sterols, as well as an alkaloid called magnaflorine. It has shown success for many years in boosting sexual desire, and fighting fatigue. Epimedium has testosterone-like effects. It stimulates sexual activity in both men and women, stimulates the sensory nerves, improves libido, and is known to restore sexual vitality. It is currently marketed as a sexual stimulant for both men and women and also as a treatment for menopausal symptoms. However, there is no reliable clinical data available to support these evidences.

All the herbs described above for the ASD, menopause and sexual arousal dysfunction in females do not have substantial scientific data to prove their efficacy; these are being used on the basis of either folkloric information or limited scientific data. However, these herbs offer excellent research projects for the scientists who want to explore the field of estrogenic activity in the plants.

Ayurvedic System of Medicine in Female Sexual Dysfunction
Vajikarana-Science of Aphrodisiacs, a branch of the Ayurvedic System of Medicine also recommends some treatment for female sexual dysfunction. According to Vajikarana, female sexual arousal problems are very complicated and depend upon many symptoms. The ayurvedic approach is holistic, first to clean the body and than creat a sound mind. Plenty of recipes involving the use of herbs, dietry nutrients, and minerals used in *vajikarana* or aphrodisiacs therapy are mentioned in the Ayurvedic ancient texts *Charka Samhita* and *Sushrita Samhita*. They have a significant impact on the semen, the ovum and genital organs of both male and females. Some of the Ayurvedic formulations for female sexual dysfunction described by Dash and Ramaswamy (2001) are as follow:

Artava or Rajas*-Increases the production of ova
Artava Janaka-Induces menstruation/ovulation
Artava Sodhka-Purifies the ovum
Vrsya-Promotes virility in females
Artava Rodhaka-Inhibits excessive flow of menstruation
Yoni Sodhaka-Cleans the vagina
Yoni Daurgandhyanasaka-Destroys foul smell from the vagina
Yoni Sravakaraka-Lubricates the vagina
Yoni Srava stambhaka-Prevents excessive secretion from the vagina
Yoni Dardhyakaraka-Strengthen the loose vaginal muscles

*Formulations names are in Sanskrit, the primitive language of the Vedas.

Following are the plants of great potential in FSD on the basis of the information available in Ayurvedic System of Medicine and need further investigation.

Albizia lebbeck; Asparagus racemosus; Butea monosuperba; Hibiscus rosasinensis; Ipomea digitata; Mimosa pudica; Mucuna pruriens; Myristica fragrans; Santalum album; Shilajit(Mineral pitch); Sida cordifolia; Solanum indicum; Solanum xanthocarpum; Tinospora cordifolia; Tribulus terrestris; Trigonella Foenum-graecum (Fenugreek or Methi) and Withania somnifera.

> *"To resist the frigidity of old age, one must combine the body, the mind, and the heart. And to keep these in parallel vigor one must exercise, study and love."*
>
> *—Bonstettin*

Chapter Five

Neurotransmitters as Natural Aphrodisiacs

Scientists believe that chemistry exists between lovers. Chemicals and hormones are released in the body by the activity of the brain and that leads to love. The role of hormones in sex life will be discussed in chapter seven. In addition to hormones there are cascades of brain chemicals that play a significant role in sex chemistry. Without them, sex is not possible. By their help, the brain connects different aspects of sex such as physical, psychological and emotional, which can stimulate or inhibit sexual performance. These chemicals are known as *neurotransmitters*. Erotic thoughts or fantasies give arousal via neurotransmitters.

Neurotransmitters transmit information across the junction from synapse to the other nerve cell, and are stored in the axon of a nerve cell. When an electrical impulse traveling along the nerve reaches the axon, the neurotransmitter is released and travels across the synapse, either prompting or inhibiting continued electrical impulses along the nerve.

The number of accepted neurotransmitters has significantly increased during recent years with now more than 300 known neurotransmitters. Some of the prominent neurotransmitters are dopamine, serotonin, acetylcholine, norepinephrine, adenosine triphosphate, and nitric oxide. Neurotransmitters transmit information within the brain and from the brain to all the parts of the body. These chemicals play a significant role in human physiology. Parkinson's disease, Alzheimer's disease, amyotrophic lateral sclerosis, and clinical depression, are found to be connected with abnormalities, malfunctioning and imbalance of certain neurotransmitters.

Types of Neurotransmitters

a.   Acetylcholine
b.   Aminoacids: Glutamate, γ-aminobutyric acid, aspartate and glycine
c.   Monoamines: Norepinephrine, phenylethylamine, dopamine, serotonin, histamine and melatonin
d.   Purines: Adenosine, ATP and GTP

Neurotransmitters also act as natural aphrodisiacs and do play a vital role in our sex life. Among the neurotransmitters, dopamine, serotonin, endorphins, norepinephrine, and phenylethylamine (PEA) cause sexual attraction. PEA is considered the infatuation-inducing stimulant. Acetylcholine and nitric oxide are responsible for erection. Serotonin and acetylcholine regulate the orgasm. They all work in unison.

Temporary elevation of dopamine levels often lead to an improvement in mood, alertness, and libido. The initial phase of the sexual act is libido that means sexual desire. Dopamine promotes libido. Estrogens also promote libido where as prolactin reduces libido. Acetylcholine and nitric oxide control arousal of genital tissues, lubrications of the genitals, an erection in men and swelling of clitoris in women. Serotonin and norepinephrine regulate the orgasms in both the sexes.

Norepinephrine and epinephrine are responsible for the flaccid stage of the male organ. Frequent release of norepinephrine contracts the arteries in the penis and contracts the smooth muscles of the *corpora cavernosum* and keeps the penis soft (Stahl 2000, 1998b; Meston and Frohlich 2000).

Endorphins as Neurohormones

Endorphins are polypeptide compounds produced by the pituitary gland and hypothalamus during strenuous exercise. Technically, endorphins are not neurotransmitters but their action is similar to neurotransmitters. They are related in function to both neurotransmitters and hormones. They can be termed as *neurohormones*. Endorphins are produced in the brain in response to a variety of stimuli. John Hughes and Hans Kosterlitz discovered them in 1975 in the brain of a pig and named them endorphins *enkephalins* (from the Greek *egkephalos, in the head*). Several other types of endorphins were discovered later. At least 20 types of endorphins have been demonstrated in humans. The word endorphin itself is abbreviated from *endogenous morphine*, which means morphine produced naturally in the body.

Types of Endorphins

There are four main types of Endorphins: alpha, beta, gamma, and sigma (α, β, γ and Σ). These contain 16, 31, 17 and 27 amino acids respectively. The most effective and potent endorphin is the β-endorphin that gives the most euphoric effect to the brain. It is composed of 31 amino acids.

Endorphins have a close functional resemblance to morphine, and are "enodegnous morphin like substance in man." These are polypeptide compunds that have similar pharmacological activity as corticosteroids and can serve as anti-stress hormones that relieve pain naturally. It may be nature's gift to cure for high levels of stress. Chemically, these are a small-chain of peptides, which activate opiate receptors, producing a feeling of euphoria resulting in insatiable desire. Endorphins also affects mood, perception of pain, memory retention, and learning. The actual receptors in the brain that the morphine connects to are endorphins. Endorphin neurotransmitters function in the transmission of signals within the nervous system. Pituitary and hypothalamus produce endorphins.

Mechanism of Action of Endorphins

The mechanism of action of these compounds is not fully understood. However, it is considered that endorphins bind to the opioid receptors in the brain and spinal cord that result in disinhibition of the dopamine pathways, causing more dopamine to be released into the synapses.

Endorphins regulate feelings of pain and hunger and are connected to the production of sex hormones. Stress and pain are the two most common factors leading to the release of endorphins. Endorphins interact with the opiate receptors in the brain to reduce our perception of pain, having a similar action to drugs such as morphine and codeine. Unlike drugs, however, activation of the opiate receptors by the body's endorphins does not lead to addiction or dependence addition to decreased feelings of pain. Secretion of endorphins leads to feelings of euphoria, modulation of appetite, release of sex hormones, and enhancement of the immune response.

According to some reports, laughter also releases endorphins in the brain. So besides widening the blood vessels, suppressing the production of stress hormones and raising antibody levels in the blood, laughing would thus also have an analgesic effect.

One theory of why some people find BDSM (different patterns of sexual behavior such as bondage and disciplines, domination and submission, sadism and masochism) activities pleasurable is that these activities stimulate endorphins in a controlled way.

Some scientists presume that people who suffer from severe headaches have lower levels of endorphins. High endorphin levels decrease pain and stress. Endorphins have been indicated as modulators of the so-called "runner's high" that athletes achieve with prolonged exercise. While the role of endorphins and other compounds as potential triggers of this response have been extensively controversial in the literature, it is considered that the body does produce endorphins in response to prolonged, continuous exercise. However, some researchers question the mechanisms at work believing the high comes from completing a challenge rather than just through the exertion.

Release of Endorphins
Release of endorphins varies among individuals. Two people who exercise at the same level or suffer the same degree of pain will not necessarily produce similar levels of endorphins. Certain foods, such as chocolate or chili peppers, can also lead to enhanced secretion of endorphins. In the case of chili peppers, the spicier the pepper, the more endorphins are secreted. The release of endorphins upon ingestion of chocolate explains the comforting feelings that associate with this food and the craving for chocolate in times of stress.

Even if you don't participate in strenuous athletics, you can also try activities that increase your body's endorphin levels. Studies of acupuncture and massage therapy have shown that both these techniques can stimulate endorphins secretion. Sex is also a potent trigger for endorphin release. The practice of meditation can increase the amount of endorphins released in your body. Alpha sound frequency music also tends to stimulate endorphins and gives pleasure. Ultraviolet light may also stimulate the release of endorphins. Some people enjoy sun tan and feel more relaxed due to the release of endorphins. This also explains Sun Worship during sun rise early in the morning in India.

Functions of Endorphins

Endorphins play significant role in a human body. Some of the functions are described as follows:

1. Enhance immune system
2. Block the lesion of blood vessel
3. Have anti-aging effects
4. Are anti-stress hormones
5. Have a pain-relieving effect
6. Help improve memory
7. Relieve pain
8. Enhance low libido

Nitric Oxide as Neurotransmitter

Nitric Oxide (NO) plays a significant role in sexual arousal and acts as a natural aphrodisiac. It is responsible for erection and engorgement of the sexual organs. During recent years, this molecule has awakened great interest in the scientific community (Koshland 1992). With increased work in this field of scientific research, nitric oxide has gained great recognition; it has become known as the *magic gas,* the *wonderful molecule* and the *secret messenger.* Viagra acts via the release of the nitric oxide (Stahl 1998a, b, c)

Robert Furchgott, Louis Ignarro and Ferid Murad, who demonstrated the role of nitric oxide in the process of cellular communication, received the 1998 Nobel Prize in Medicine.

Before the findings of nitric oxide, scientists were baffled about the complexity of dilating and constriction of blood vessels. Research has revealed the existence of a chemical messenger-the nitric oxide molecule. It is this molecule that gives instructions for the blood vessels to dilate. The endothelium cells are responsible for the production of the nitric oxide messenger. Every nitric oxide molecule lasts about 10 seconds. It is designed to communicate its message within this short time to the relevant recipients and do this perfectly without fail.

Role of Nitric Oxide in Erection

The messenger nitric oxide molecules secreted by the endothelium cells are dispersed with great speed in every direction. Those that are directed towards the smooth muscle cells enter the membrane of these cells. The smooth muscle cell membrane acts as a selector giving entry to the nitric oxide it recognizes. Without wasting any time, the nitric oxide molecule that enters the smooth muscle cells finds a special enzyme called GC (Guanylate cyclase) enzyme and communicates its vitally important message. As a result, a series of complex chemical reactions occur within the cell. When GC enzyme in the smooth muscle cells receives the message brought by nitric oxide, it begins its activity by converting the GTP (Guanosine triphosphate), the energy-carrying molecules, into cGMP (3, 5 cyclic Guanosine monophosphate). Many reactions occur between these stages that are still unknown (Stahl 1997a, b).

At the end of the activity of the enzymes, the concentration of calcium in the muscle cell diminishes, causing a separation in the fibers and the relaxing of the muscle cells. As a result of this, the vessels dilate and more blood flows to the penis. The spongy chambers almost double in diameter due to the increase in blood flow and cause erection.

Conclusion

Neurotransmitters play a significant role in controlling sex drive since these are involved in physical, mental, psychological, and emotional aspects of life. Though neurotransmitters are produced in a very small quantity but carry very strong effects. Dopamine, serotonin, norepinephrine, and endorphins have great influence on sex. If their levels are inadequate, they can cause lots of physical and mental problems in the body. It is very pertinent to keep them in balance. Neurotransmitter levels can be determined by a simple urine test. Monitoring ones neurotransmitter levels can help to prevent some diseases from occurring in the future. Some natural preparations are being sold as nutrient supplement on the Internet as **BrainBalance** formulations for balancing neurotransmitters. These preparations are composed of aminoacids and vitamins. The ingredients are advertised as precursors of neurotransmitters such as norepinephrine, epinephrine, dopamine, and GABA etc. Consumers should consult their physicians before incorporating these preparations in their daily regime otherwise consumption of these formulations can leads to some serious complications.

"I used to think that the brain was the most wonderful organ in my body. Then I realized who was telling me this."

—Emo Phillips

Chapter Six

Herbs as Natural Aphrodisiacs

This chapter covers a variety of famous herbs that have been used as aphrodisiacs over the years and sold either as a single preparation or in a formulation comprising five to ten herbs along with, animal or mineral products (Figure 7). Efforts have been made to review the latest scientific literature available on these herbs to enable the readers to make the right decision about their usage. Classification of these natural aphrodisiacs cannot be done on the basis of pharmacological actions owing to inadequate clinical data available, so it was thought desirable to arrange them in alphabetical order. Tentative pharmacological classification is given in Figure 8. Natural aphrodisiacs described in this chapter are based on the following criteria.

Description of the Commonly Used Natural Aphrodisiac

Product(s): Formulation available on the market containg the specific herb[1].
Common Name(s): Standard name first, followed by other common name of the herb.
Latin Name: Botanical name.
Synonym(s): A Latin name that also applies to same taxon. A listing of synonyms often contains names that for some reason did not make it as formal or official name
Family: Plant family.
Geographical Distribution: Plant's native range and availability.
Habitat: Describes the ecological conditions for growth.
Part(s) Used: Part of the plant used as an aphrodisiac.
Description of the Plant: The morphological characteristics of the plant are described. Pencil drawing of the plant depicts various morphological characteristics of the plant.
Chemical Constituents: Mostly the secondary metabolites of the herb are described.
Aphrodisiac Myth: Traditional /folkloric uses of the plant as aphrodisiacs.
Scientific Evidence: Clinical studies on animals or humans for aphrodisiac activity.
Additional Benefits: Significant uses of the plant besides aphrodisiac properties.
Dosage: Daily intake of the specific herb's part or its extract, based on published data.
Toxicity: The degree to which an aphrodisiac is toxic or able to damage any organ.
Contraindication: A condition which makes a particular treatment inadvisable such as use of aphrodisiacs is contraindicated during pregnancy, breast feeding, and surgery.
Side Effects: Symptoms occur during the ingestion of an aphrodisiac product.
Drug Interaction: Interaction of the herbal aphrodisiacs with modern medicines.
Herbal Interaction: Interaction of the herbal aphrodisiac being ingested with other herbs and herbal products taken simultaneously.
Food Interaction: Interaction of the herbal aphrodisiac with other food products.
Future Prospects: Discussion on the future of natural aphrodisiac plant used on the basis of myth or folkloric information or preliminary clinical data.

Herb[1]-The term is used for both woody and herbaceous plants in this book.

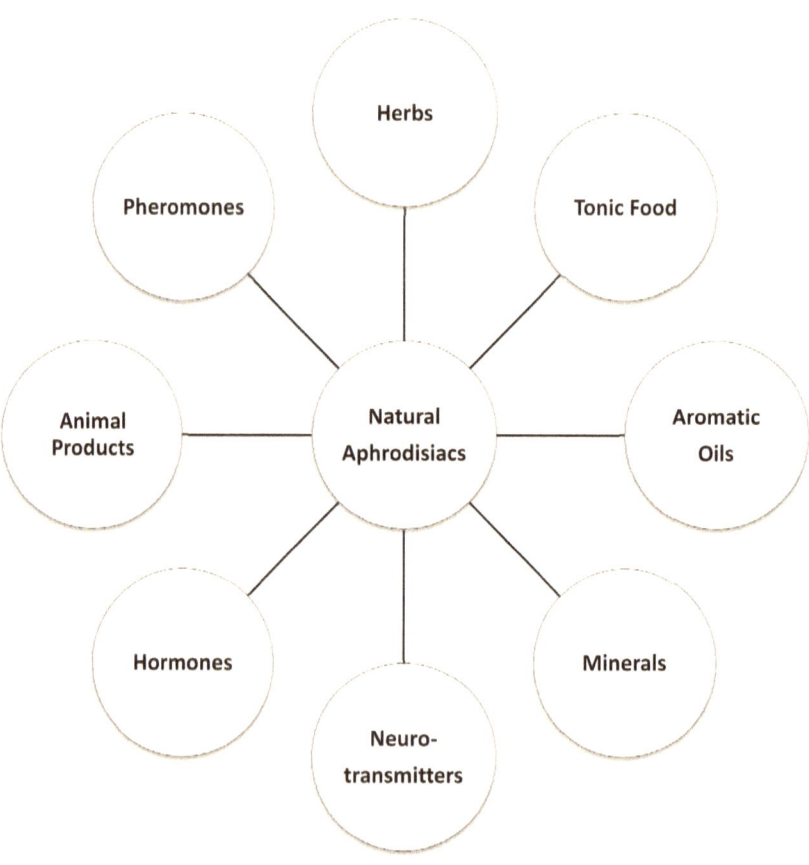

Figure 7: Natural Aphrodisiacs

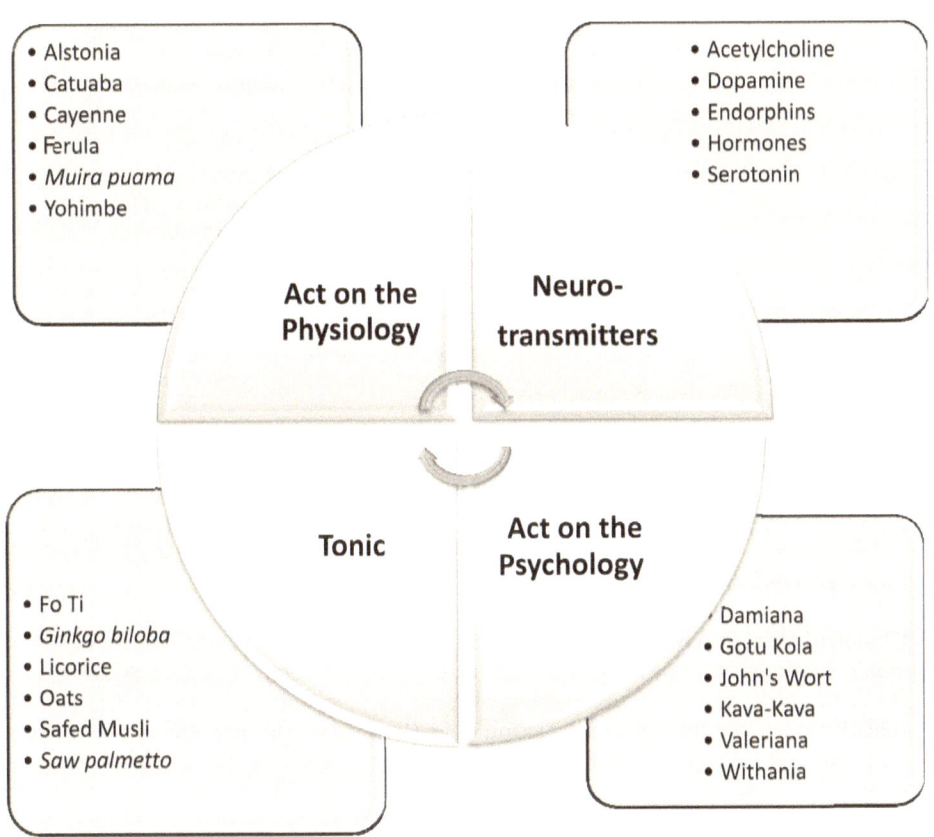

Figure 8: Pharmacological Classification of Natural Aphrodisiacs

Alstonia scholaris (Saptparn)

Product(s): It is sold as standard extract and infusion or powder in capsules.
Common Name(s): Dita Bark, Bitter Bark, Indian Devil Tree, Pali-Mara, Saptparn.White Cheesewood.
Latin Name: *Alstonia scholaris.*
Synonym(s): *Alstonia venenata, Alstonia spectabilis, Echites scholaris.*
Family: Apocynaceae.
Geographical Distribution: New South Wales and Australia, Philippines, India and Pakistan.
Habitat: Plant prefers sandy loamy soil and tropical rainforest.
Part(s) Used: Ripe fruit, seeds, bark, leaves, latex and roots. Bark is mostly used.

Description of the Plant: Tree grows up to 50-60 feet high. It has a furrowed trunk, with oblong leaves dispersed in seven whorls round the stem. That is why it is called Saptparn in *Sanskri* language, meaning seven leaves. Upper surface is glossy. Inflorescence consists of terminal or axillary cymes or compound umbels of small fragrant funnel shaped flower.

Chemical Constituents: The genus Alstonia comprises about 40-60 species rich in indole alkaloids that are present in various parts of the plant. *A. scholaris* contains O.3% alkaloids. Alstonine, alschomine, alstonamine, astonidine, ditamine, detaine or echitamine, echitenine, nareline, picrinine, and reserpine, rhazimanine, strictamine, scholarine, vincamajine, vilastonine and yohimbine are among the alkaloids (Gosh *et al.* 1988; Kam *et al.* 1997; Yamauchi 1990a and 1990b). Rahman *et al.* (1985, 1987) isolated scholaricine and vallesamine form *Alstonia scholaris.* Salim *et al.* (2004) isolated akuammiginone, echitaminic acid, echitamidine N-oxide alkaloids from the bark of this plant and elucidated their structures by spectroscopic methods. First seco-ulein alkaloids manilamine (indole) from the leaves of Philippine *Alstonia scholaris* has been isolated and characterized by MS and NMR studies by Macabeo *et al.* (2005). Cai *et al.* (2007) isolated monoterpenoids indole alkaloids from *Alstonia scholaris.* Cai *et al.* (2008) also reported an unprecedented cage-like alkaoid scholarisine A from the leaves of *A. scholaris.* Very recently, six monoterpenoid indole alkaloids scholarisines B-G together with 15 known analogues have been isolated from the bark of *A. Scholaris.* Their structures were determined by 1D and 2 D NMR and MS analyses (Feng *et al.* 2009). Alstonic acid A and B, unusual 2, 3 secofernane triterpenoids from the leaves of *A. scholaris* were isolated and their structures were established by NMR and MS techniques by Wang *et al.* (2009).

Aphrodisiac Myth: Traditionally, it is used as aphrodisiac. *Alstonia* causes minor irritation of the genitals, in the male it prolongs erection and delays ejaculation. Seeds of the plant were used as aphrodisiacs by the Tantric in India along with exercise that prolonged erection and delayed orgasm by controlling specific genital pubococcygeus muscles (Miller 1993).

Scientific Evidence: Survey of literature does not show any clinical data to confirm its role as an aphrodisiac.

Additional Benefits: *Alstonia scholaris* has shown a wide range of biological activities. A comprehensive pharmacological review of the alkaloid alstonine has been published by Elisabetsky and Costa-Campos (2006). Recently pharmacological activities of *Alstonia scholaris* have been reviewed by Arulmozhi *et al.* (2007). However, the ethanolic extract of the leaves also showed some bronco vasodilation in animals (Channa *et al.* 2005). Anticancer activity of the alcoholic extract of the plant has been shown by Baliga *et al.* (2004). In India, bark is used for malaria, diarrhea, and dysentery and in leprosy. The bark also produces a yellow dye used on cotton fabrics (Chopra 1992). The yellow color is due to delta 3-alstovenine.

Dosage: It is sold as standard extract or 10-15 grain powder in capsules.

Toxicity: Toxicity in mice treated with *Alstonia scholaris* has been found by Arulmozhi *et al.* (2007). The hydroalcoholic extract of the plant produced teratogenic effects in mice at doses greater than 240mg/Kg as suggested by Jagetia and Baliga (2003).

Contraindication: Avoid the use in pregnancy and urinary tract infection.

Side Effects: Not known.

Drug Interaction: It imparts synergistic effect when taken with berberine.

Herb Interaction: It has synergistic effect when taken with herbs such as *Withania somnifera, Rauwolfia serpentina,* and *Piper methysticum.*

Food Interaction: Not known.

Future Prospects: Owing to lack of scientific evidence in support of its aphrodisiac properties, it will stay as myth. The plant possesses a great future due to the presence of indole alkaloids. More clinical studies are suggested to confirm its myth.

> *"Life without love is like a tree without blossoms or fruit."*
>
> — *Kahlil Gibran*

Angelica sinensis (Dong Quai)

Product(s): Boom.
Common Name(s): Empress of the Herbs, Sovereign Herb for women, The Female Ginseng, Dong Quai, Dang Gui and Tang-Kuei.
Latin Name: *Angelica sinensis.*
Synonym(s): *American angelica, Angelica acutiloba (Japanese).* The related species, *Angelica acutiloba*, appears to have similar properties to *Angelica sinensis.*
Family: Apiaceae

Geographical Distribution: It is a native to China, Korea, and Japan and has been used in their culture since a long time. Angelica is also found in UK, Lapland, and Iceland. Several species are found in North America.

Habitat: This plant is commonly found in well-watered mountain ravines, riverbanks, and damp meadows. Angelica prefers bottomlands and swamps.

Part (s) Used: Whole herb, mostly root.

Description of the Plant: Plant is perennial. The one year old root is harvested for medicinal purposes. Its stem is round, grooved, hollow, branched near the top, and tinged with blue. The plant grows up to 3 to 7 feet. Compound leaves grow from dilated sheaths that surround the stem. The leaves are pinnately compound, with toothed leaflets. The plant produces greenish-white flowers in May to August. Elliptic-oblong fruit is composed of two yellow winged seeds. Seeds ripen in October and November. The name *angelica* is derived from Medieval Latin *herba angelica* meaning "angelic herb" that has special powers against contageous diseases like plague.

Chemical Constituents: Dong Quai contains various coumarins and flavonoids. Coumarin derivatives include angelol, angelicone, bergapten, oxypeucedanin, osthole, psoralen, and 7-desmethylsuberosin. Detailed phytochemical investigations showed the presence of safrole, scopoletin, β-sitosterol, isosafrole, umbelliferone, sesquiterpenes, and unsaturated and saturated fatty acids. Compounds such as carvacrol, α-pinene, p-cymene, carvacrol, linalool and borneol, vanilic-acid, cadinene, falcarindiol, falcarinol, falcarinone, ferulic acid, folacin, folinic-acid were found in volatile oil (Sun *et al.* 2006; Wu *et al.* 2005; Zschocke *et al.* 1998). Volatile constituents were also identified by supercritical fluid CO_2 extraction coupled with gas chromatography and mass spectroscopy by Kim *et al.* (2006). Five ligustilide compounds from the root of *Angelica sinensis* have been isolated and characterized by Lu *et al.* (2003 and Hon *et al.* 1990). 13 organic compounds have been identified by Lao *et al.* (2004) by GC/Ms coupled with pressurized liquid extraction. A new dimeric phthalide derivative from *Angelica sinensis* has been reported by Su *et al.* (2005). More recently 21 compounds including a new ester angeliferulate have been identified by Deng *et al.* (2006). Chen *et al.* (2007) reported new neodiligustilide from *Angelica sinensis.*

Aphrodisiac Myth: Myth says that *Angelica* was revealed in a dream by an angel to cure the deadly bubonic plague. The Chinese consider *Angelica* (dong-qui) second to ginseng. It is considered as a "female ginseng" remedies and has a powerful action on female reproductive system. A decoction of American or European *Angelica* was used by colonial women in painful menstruation. Midwives have used decoctions and syrup made from *Angelica* roots after childbirth to expel the retained placenta. It was also used for abortion. In Chinese medicine, Dong Quai is most often used in combination with other herbs. Dong Quai is considered to return the body to proper order by nourishing the blood and harmonizing vital energy. The name Dong Quai translates as *return to order* based on its alleged restorative properties (Murray 1996; Pizzorno and Murray 1996; Newell *et al.,* 1996).

Scientific Evidence: Recently, interest in Dong Quai has revived due to its proposed estrogen-like properties by Amato *et al.* (2002); Huntley and Ernest (2003); and Cercosta *et al.* (2006). However, it is unclear if Dong Quai has the same effects on the body as estrogens or blocks the activity of estrogens. Hirata *et al.* (1997) studied the estrogenic effect in Dong Quai. A well-designed 24-week human trial compared the effects of Dong Quai to a placebo in 71 women with menopausal symptoms. This study found no differences in hot flashes or in the Kupperman Index (a commonly used measure of menopausal symptoms) between Dong Quai and placebo groups. No changes occurred in blood estrogen levels, thickness of the uterus lining, or vaginal dryness. This study suggests that Dong Quai may not have short-term estrogen-like effects on the body.

Most of the studies of Dong Quai are done on animals. There is little human evidence supporting the effects of Dong Quai. Few clinical studies have been done, most of which have been either poorly designed or reported insignificant results. The herb is used in combination with other herbs so it is difficult to pinpoint its effect.

Additional Benefits: The Chinese discovered Dong Quai. It is considered a panacea for women and is known as female ginseng in China. It is used for the treatment of infertility in women. Women with infertility due to tubal occlusion received uterine irrigation with an extract of Dong Quai for 9 months. The results showed 79% recovery and significant number of patients got pregnant (Fu 1988). It is also used in ovarian cancer (Powell *et al.* 2002). Recently lau *et al.* (2005) studied the use of water extract of Dong Quai to treat pre or post menopausal symptoms in women with breast cancer. The herb is also used in amenorrhea and dysmenorrhea (Kotani *et al.,* 1997; Zhiping *et al.* 2002). Angelica oil is used in perfumery and winery.

Dosage: Standard oral dose for dried root or rhizome is 1-2 grams. The powdered root can be used in capsules or tablets.

Toxicity: Safrole, present in volatile oil of Dong Quai, can be carcinogenic. Components of Dong Quai may increase the risk of bleeding due to anticoagulant and anti-platelet effects, although there are no reliable reports of clinically significant bleeding in humans.

Contraindication: Dong Quai is not recommended during pregnancy due to possible hormonal and anticoagulant/anti-platelet properties. Cancer patients should also avoid the use of Dong quai.

Side Effects: Prolonged exposure to sunlight or ultraviolet light should be avoided while taking Dong Quai due to the presence of photosensitive chemicals such as furocoumarins, psoralen, and bergapten (Hann *et al.* 1991).

Drug Interaction: Dong Quai has been reported to interact with bleomycin an anticancer drug used to treat tumors of the cervix, uterus, testicles penis and certain types of lymphoma (Scott and Elmer 2002). Dong Quai has synergistic effect when taken with antiplatelet or anticoagulant drugs (Page and Lawrence 1999). In May 2002 the FDA added Dong Quai to the list of herbal products not to be used with sodium warfarin. Non-prescription drugs such as aspirin, paracetamol may be avoided.

Herbal Interaction: If Dong Quai is used with other herbs that affect blood clotting, bleeding may occur. Some of the most common herbal products that might inhibit blood clotting include Devil's Claw, Eleuthero, Garlic, Ginkgo, Horse Chestnut, *Panax ginseng*, Papain, Red Clover, and *Saw palmetto* (Appendix D).

Food Interaction: None reported.

Future Prospects: The results are encouraging regarding the estrogenic activity of the plant. However, lack of clinical data, contradictory studies, and side effects of Dong Quai require further investigations. Additional research is necessary in this area to draw a firm conclusion.

*Dosage: Doses in this chapter are based upon the traditional uses of the herbs and published research.

> *"The poetry of the earth is never dead."*
> —*John Keats*

Avena sativa (Oat)

Product(s): Vipro, Enzyte.

Common Name(s): Oat Straw, Oats, Rolled Oats, Wild Oats, Groats, Steel Cut Oats, Common Oats. Avoine, Hafer and Avena in French, German and Italian respectively.

Latin Name: *Avena sativa.*

Synonym(s): *Avena fatu, Avena sativa var orientalis.*

Family: Poaceae.

Geographical Distribution: It is a native of Mediterranean region, Europe, North Africa and Central Asia but cultivated all over the world.

Habitat: It grows in cool and damp climates and prefers heavy rainfall. The plant prefers sandy, loamy and clay soils. It requires well-drained and acid, neutral and alkaline soils.

Part(s) Used: Unripe green oats as aphrodisiacs.

Description of the Plant: The genus comprises about 25 species. Plant is an annual grass growing either in tufts or in single stalks. The stems grows to a height of 1.50 meters and do not branch out. The leaves are lanceolate with blade like margin, green and the glabrous sheaths rounded on the back; ligules blunt, membranous. The inflorescence is a diffuse panicle with 2-3 florets, all bisexual. The plant blooms from July to August.

Chemical Constituents: The seeds contain alkaloids, such as gramine, trigonelline, avenine, and saponins such as avenacosides A and B. β-sitosterol, stigmastadienol, cholesterol, brassicasterol, campesterol, and stigmasterol are also present. The seeds are rich in iron, calcium, manganese and zinc along with lysine and methionine. New dimeric compounds of avenanthramide phytoalexin in oats have been found by Okazaki *et al.* (2007). Phenolic compounds in oat grains of *Avena sativa* were shown by Dimberg *et al.* (2005). Wenzig *et al.* (2005) isolated three novel flavonolignans from *Avena sativa*.

Aphrodisiac Myth: Throughout the ages *Avena sativa* has been used as a natural aphrodisiac by men and women. Over 200 years ago the German Pharmacopoeia mentioned *Avena sativa* as an aphrodisiac. It is believed that it released testosterone from its bound form in the body. Testosterone is believed to be more effective as free form to stimulate libido. The green oats is believed to increase sexual desire. In women, Avena is helpful in infertility and amenorrhea, as well as in nervousness, insomnia and anxiety associated with premenstrual syndrome (Kurtz and Wall 2007). Avena is also an excellent post-partum restorative, and can help to increase breast milk production. In men, the saying of "sowing your oats", provides a clue as to the effect upon male reproductive function. Thus, Avena has been considered to be an excellent restorative in male impotence (Gibson *et al.* 2002).

Scientific Evidence: Survey of literature does not reveal any significant evidence to prove its myth to stimulate low libido.

Additional Benefits: Used in osteoporosis, cardiovascular disorder, diabetes, reduces cholesterol (Karmally *et al.* 2005) and help in coeliac disease (Butt *et al.* 2008) and smoking cessation. Bay *et al.* (1974) published contradictory reports on cigarette smoking benefits. The incorporation of oat grains and bran in the food products improves not only nutrition but is also a therapy against various conditions. Extract from the leaves, of *Avena sativa* had an LH-releasing activity in rats (Fukushima *et al.* 1976). Oat bran flour high in beta-glucan had a low glycemic response and acted as an active ingredient decreasing postprandial glycemic response of an oral glucose load in subjects with type 2 diabetes (Topala *et al.* 2005).

Dosage: 500 mg tablets twice a day to increase low libido.

Toxicity: None known.

Contraindication: None.

Side Effects: None.

Drug Interaction: None.

Herbal Interaction: Antidepressant herbs (Appendix I) and hypoglycemic herbs (Appendix E) will have synergistic effect with oat.

Food Interaction: None.

Future Prospects: Significant scientific evidence is required to accept its use as aphrodisiac. Further clinical studies are required to prove how green oats stimulate the sex drive since Avena has been mentioned in many aphrodisiac formulations available on the market. It is very pertinent to find out the aphrodisiac activity in the green unripe and mature seeds.

> *"Look upon a woman as a goddess whose special energy she is, and honor her in that state."*
> — *Uttar Tantara*

Capsicum annuum (Cayenne)

Product(s): Male Boost, Magna RX.
Common Name(s): Red Pepper, Chili Pepper, Garden Pepper, African Chilies, Tabasco Pepper.
Latin Name: *Capsicum annuum.*
Family: Solanaceae.
Synonym(s): *Capsicum minimum, Capsicum frutescens.*
Geographical Distribution: It is found throughout the globe. It is a native of South America and Zanzibar.
Habitat: It prefers tropical and subtropical climates. It requires sunny location and well-composted soil.
Part(s) used: Fruits and seeds, the fruit is eaten raw or cooked.
Description of the Plant: It is a tropical perennial shrub with long, pod like berries. These berries are green when unripe. On maturity it turns to purple red orange or yellow.

Chemical Constituents: The active constituent is capsaicin alkaloid. Other forms of capsaicin are dihydrocapsaicin, nordihydrocapsaicin, and homodihydrocapsaicin. It also contains capsanthin, capsorubin, capsaicinoids, salicylates, β-carotene, lutein, flavonoids, resins, volatile oil, saturated, unsaturated fatty acids, and vitamin C (Barnes *et al.* 2002). Ten new sesquiterpenes from *Capsicum annuum* have been reported by Kawaguchi *et al.* (2004).

Aphrodisiac Myth: It is used in combination with other herbs for its aphrodisiac effect. Aphrodisiac effect is attributed to dilation of blood vessels, and thus causing erection. Capsicum triggers the release of endorphins by the brain. However, there is no scientific evidence for these statements.

Scientific Evidence: No scientific evidence is found in the literature regarding its aphrodisiac property or myth.

Additional Benefits: Cayenne causes the brain to secret more endorphins, a natural analgesic. Capsaicin obtained from cayenne has proved so effective at relieving pain that its active ingredient in the over the counter cream. Zostrix is being recommended by physicians for arthritis, psoriasis, post surgical pain, diabetic foot pain, pain of the shingles, and cluster headaches. It reduces cholesterol level and the risk of internal blood clots that trigger heart attack. However, modest reduction in appetite has been found in healthy Japanese women and white men when they have consumed 10 grams of cayenne along with meals in a double blind trial. A similar trial found that cayenne could increase metabolism of dietary fats in Japanese women. These trials suggest cayenne may help in the treatment of obesity (Yoshioka *et al.* 1998, 99; Chu *et al,* 2002). Lee *et al.* (2007) found that capsaicin could have an anti-ulcer protective effect on stomachs infected with *H. pylori.*

Cayenne can also cure prostate cancer. UK prostate experts say capsaicin could be the basis of a future drug (BBC 2006).

Dosage: It is sold as capsules, tincture, and extracts. As an aphrodisiac, it is taken thirty minutes prior to sex. A dose of 500 mg is recommended as capsules.

Toxicity: Not known.

Contraindication: Never to be used by pregnant or lactating women.

Side Effects: When consumed as food in large quantity for many years, cayenne may increase the risk of stomach cancer (Lopez-Carrillo *et al.* 1994) A different study found that the people who ate the most cayennes actually have lower rates of stomach cancer (Buiatti *et al.*1989). Overall the current scientific data is contradictory. Biological variations in tolerance of cayennes are possible. People with ulcers, heartburn or gastritis should not use cayenne product that may aggravate the symptoms.

Drug Interaction: Capsaicin has supportive interaction with aspirin. Capsaicin reduces the harmful effects of aspirin in the stomach (Abdel *et al.* 1995). Interaction with anticoagulant warfarin can cause excessive bleeding. Capsicum has the ability to normalize both low and high blood pressure and is often used to equalize blood pressure; hence capsaicin may interfere with the blood pressure regulating function of ACE inhibitors (Appendix P). The production of Substance P by cayenne has the known effect of dilating the arteries, thereby lowering the blood pressure. Capsaicin may aggravate cough due to ACE-inhibitors drugs. Monoamine Oxidase Inhibitors (Appendix S) may increase sedation and catecholamine secretion.

Herbal Interaction: Herbal products that effect the blood pressure should not be used (Appendix G, H). Avoid the use of anticoagulant herbs (Appendix D).

Food Interaction: Capsaicin has been found to increase the absorption and bioavailability of theophylline (Fugh-Berman 2000). Those who are on theophylline prescription for asthma should consult their physician before consuming cayenne. Tea and coffee consumption should be less with cayenne intake since both contain theophylline.

Future Prospects: It can be used as synergistic with other herbs in aphrodisiac formulations. As a single preparation its future as an aphrodisiac is doubtful. Some kind of clinical evidence is required to prove its use in aphrodisiac formulations.

Chlorophytum borivilianum (Safed Musli)

Product(s): Vita-Ex (Gold) Musli Pak, Kama Raja and Kama Rani.
Common Name(s): Haqaqule in Arabic, Swetha Musli (Sanskrit) and Safed Musli in Hindi.
Latin Name: *Chlorophytum borivilianum.*
Synonym(s): *Chlorophytum tuberosum and C. malabaricum, C. arundinaceum.*
Family: Liliaceae (Agavaceae)
Geographical Distribution: Native to Indian subcontinent.

Habitat: It prefers wide range of temperature and rainfall. Sandy loamy soil with proper drainage system stimulates its growth.

Part(s) Used: Tubers of the plant.

Description of the Plant: Leaves are linear and slightly yellowish. White flowers with 6 petals are arranged on the flowering stalk which emerges from the center of the plant. About 20-25 flowers inflorescence appear in July. The seed is very small and black in color. Number of tubers varies from plant to plant. 5-30 tubers per plant are observed. Tubers are white, and hence it is called Safed Muslim. Safed means white in Hindi language. There are about over 250 varieties of *Chlorophytum* found in the world and 17 among them are found in India. Among these, *Chlorophytum borivilianum* has good market throughout the world. The Medicinal Plants Board of India has recognized Safed Musli as the sixth important herb to be protected and promoted.

Chemical Constituents: Safed Musli contains 27 alkaloids, steroidal saponin 2-17%, spirostanol glycosides, furastanol glycosides-polysaccharides (40-45%) carbohydrates, proteins (7-10%), minerals and vitamins. The presence of glycosidal alkaloids makes it an ideal aphrodisiac (Kothari 2004). Acharya *et al.* (2009) isolated four new spirostanes-type saponins named borivilianosides E-H from an ethanolic extract of the roots of *C. borivilianum* and their structures were determined by 2D NMR and mass spectroscopy.

Aphrodisiac Myth: Safed Musli is named in *Atharva Veda* (Indian sacred book) as a Divine herb (Divya Aushad) with unique aphrodisiac properties. It is the main ingredient in the preparation of many Ayurvedic formulations. Besides its extensive use in the Ayurvedic System of Medicine, Safed Musli is also gaining importance as a vitalizer and tonic. The Indian system of medicine 'Ayurveda', conceptualizes a category of drug activity known as 'Rasayana'. The word Rasayana is derived from words 'Rasa' meaning elixir and 'Ayana' meaning house. The word defines the property of the plant that helps to rejuvenate the system, i.e. adaptogenic activity

(Handa 1996). 'Rasayan' therapy prevents diseases and delays the aging process (Puri 2003). Safed musli is enlisted as Rasayan.

Scientific Evidence: Safed Musli is used in oligospermia. It has significant effect in increasing semen volume and total sperm count, however, no concrete scientific evidence is known. Nandan Biomatrix Ltd (NBL) recently completed clinical trials on humans and the study on safety parameters of 'Safed Musli' (*Chlorophytum borivilianum*). The company, in its clinical trials conducted at the Indian Institute of Chemical Technology (IICT), has reported positive results on the aphrodisiac parameters of Safed Musli such as sexual desire, erectile function, sperm mobility, sperm count, and frequency of coitus. The efficacy study was conducted in Wistar rats, which was studied in eight groups and results were compared with Viagra, nitric oxide and testosterone (The hindubusinessline.com retrieves June 16, 2006). The results were encouraging and the company is in the process of formulating some products for the global market.

Effects of *Chlorophytum borivilianum* on sexual behavior and sperm count in male rats were studies by Kenjale *et al.* (2008). This study was designed to evaluate the aphrodisiac and spermatogenic potential of the aqueous extract of dried roots of *Chlorophytum borivilianum* (CB) in rats. Male Wistar albino rats were divided into four groups. Rats were orally treated with (1) Control group: distilled water; (2) CB 125 mg/kg/day; (3) CB 250 mg/kg/day; and (4) Viagra ((R)) group: 4 mg/kg/day sildenafil citrate and their sexual behavior was monitored 3 h later using a receptive female. Their sexual behavior was evaluated on days 1, 7, 14, 21, and 28 of treatment by pairing with a pro-oestrous female rat. For sperm count the treatment was continued further in all groups except the Viagra ((R)) group for 60 days. At 125 mg/kg, CB had a marked aphrodisiac action, increased libido, sexual vigor, and sexual arousal. Similarly, at the higher dose (250 mg/kg) all the parameters of sexual behavior were enhanced, but showed a saturation effect after day 14. On day 60 the sperm count increased significantly in both the CB groups, 125 mg/kg and 250 mg/kg, in a dose dependent manner. Thus, roots of *Chlorophytum borivilianum* can be useful in the treatment of certain forms of sexual inadequacies, such as premature ejaculation and oligospermia. Thakur *et al.* (2009) studied the aphrodisiac acivity of *Chlorophytum borivilianum* in male albino rats and reported improvement in the sexual behavior of the rats' incuding penile erection. The obsereved effects appear to be attributable to the testosterone like effects of the extract. Their results supported the folkloric claim.

Additional Benefits: It is also used as adaptogen to restore the immune system (Thakur *et al.* 2006).

Dosage: 3-5 grams dry powered tubers with warm milk. 15 grams tubers boiled in milk is recommended twice a day for premature ejaculation and impotence.

Toxicity: None known.

Contraindication: None known.

Side Effects: None known.

Drug Interaction: None known.

Herbal Interaction: Avoid antidiabetic herbal products (Appendix E).

Food Interaction: None known.

Future Prospects: Safed Musli is being considered as natural and safe aphrodisiac. It is called a natural Viagra. Presence of glycosidal alkaloids in the plant can impart interesting results for aphrodisiac activity. There is a great demand of this herb in Ayurvedic System of Medicine and it is commercially cultivated in India and exported all over the world. The demand is over 45000MT annually where only 15000 to 20000MT is produced in India. Its supply could not meet the demand.

It has a great future. More clinical research is required to prove its potential as an alternative to Viagra.

"Flowers are the sweetest things that God ever made and forgot to put a soul into."
— *Henry Ward Beecher*

Crocus sativus (Saffron)

Product(s): Dried stigmas of flowers.
Common Name(s): Saffron flowers.
Latin Name: *Crocus sativus*.
Synonym(s): *Crocus vernus*.
Family: Iridaceae.
Geographical Distribution: *Crocus sativus* is the most versatile and popular member of the genus crocus. It is originally from Western to Central Asia. Kashmiri saffron from India has a good reputation. Saffron is also found in Eastern Asia. Saffron was known to the Sumerians almost 5000 years ago. Today, saffron is cultivated from the Western Mediterranean (Spain) to India (Kashmir). Spain and Iran are the largest producers, accounting together for more than 80% of the world's production. Saffron is sterile, and its flowers cannot produce any seeds and propagation is done only by corms.
Habitat: Plant prefers cold climate and hilly area.
Part(s) Used: The stigmata of the flowers.

Description of the Plant: Crocus is an extraordinary autumn-blooming beauty. It has a pleasing fragrance and deep lavender, purple-veined flowers. The three scarlet color stigmas of the flowers are plucked and dried. The Genus crocus comprises about 80 species. Crocus means saffron (Krokos being the Greek word for saffron).

Chemical Constituents: Saffron contains about 150 volatile and aromatic compounds. The major two specific compounds: picrocrocin and crocin are of great importance. Methods for the analysis of the saffron secondary metabolites crocin, crocetins, picrocrocin and safranal for the determination of the quality of saffron using thin-layer chromatography, high-performance liquid chromatography and gas chromatography has been investigated by Sujata *et al.* (1992). Tarantilis *et al.* (1994a) carried out the separation of picrocrocin, cis-trans crocins and safranal from saffron using high-performance liquid chromatography with photodiode-array detection. Presence of pigments imparts the intensive color of saffron. Saffron contains α-and β-carotene, lycopene and zeaxanthin, however, its color property is due to crocetine esters. Crocetine, a diester of crocin with gentiobiose, is the single most important saffron pigment. Eessential oil contains several terpene aldehydes and ketones. Monoterpene aldehydes and isophorone-related compounds of saffron were identified by Zarghami and Heinz (1971b). The major constituent is safranal which imparts the aroma. Terpene derivatives have been identified as pinene, cineole and related compounds by Zarghami and Heinz (1971a). The bitter taste of saffron is attributed to picrocrocin. On de-glucosylation, picrocrocin yields safranal. A novel glycoside from saffron has been isolated and identified by Straubinger *et al.* (1997).

Aphrodisiac Myth: Since antiquity, the saffron is used as an aphrodisiac. It has been cited in the *Kama Sutra* for its aphrodisiac potential. The name *saffron* comes from Arabic, where it is known as za'fran that name is often explained to derive from a Semitic root, signifying "be yellow" or "become yellow". This is an ancient name, as demonstrated by its Akkadian incarnation Azup ru; yet it has no Hebrew cognate. The Sanskrit names of saffron point to the ancient Indian area of saffron production: Kashmirajanman "Product of Kashmir." Indian names of saffron are Sanskrit *kesaravara*, Hindi *kesar*. Among flowers saffron crocus is the only one who has such an ancient history. It is mentioned in the Egyptian *Papyrus Ebers*, a pharmaceutical record. It is cited in the *Old Testament*, too, in the richly poetic Song of Solomon. Egyptians, Greeks, and the Minoans of Crete (2100-1600BC) all grew the plant crocus.

Saffron is reputedly an aphrodisiac, described in the 16th century Gerard's herb as stirring 'fleshly lust' (Willan 2003). It is commonly used in many food preparations in India and Italy. It is quite prominent in India for its aphrodisiac qualities. In Egypt, Cleopatra used saffron in her baths. Persian' also used saffron as an aphrodisiac (Willard 2001). Saffron imparts stamina and stigma and maintains a healthy body. A balanced consumption of saffron could create a state of physical and psychic comfort. Saffron induces sleep, relaxation and has an aphrodisiac effect had been cited by Plinius the Old. Saffron taken in hot milk is good for both partners, may be even better than champagne or cognac; it creates an atmosphere of serenity in intimate relationships. Herbalist Christopher Catton expressed "Saffron has power to quicken the spirits, and the virtue thereof pierces by and by to the heart provoking laughter and merriment.*"*

Scientific Evidence: Survey of the literature did not show any significant scientific evidence to support its aphrodisiac use. However, very recently Hosseinzadeh *et al.* (2008) studied the effect of saffron, *Crocus sativus* stigma extract and its constituents, safranal and crocin on sexual behavior in normal male rats. Though safranal did not show aphrodisiac effects yet crocus aqueous extract and its constituent crocin revealed aphrodisiac activities. Heidary *et al.* (2008) studied the effect of saffron on semen parameters of infertile men and reported that it increased motility of sperms in these people.

Additional Benefits: The clinical findings suggest that saffron is a safe and effective antidepressant. Preliminary double-blind study compared with imipramine has shown that saffron extract can treat depression (Akhondzadeh *et al.* 2004). In a randomized, double-blind study, 30 mg of saffron extract given for 6 weeks resulted in significant alleviation of depression compared to those on placebo without evident side effects (Akhondzadeh *et al.* 2005). Saffron when compared to the drug fluoxetine (Prozac); shown encouraging results in treatment of depression and epilepsy (Noorbala *et al.* 2005). Anticonvulsant activity in the extract of saffron has been found by Zhang *et al.* (1994); Abe and Saito (2000); Hosseinzadeh and Khosravan (2002). Anticancer activities have been shown by Tarantilis *et al.* (1994b); Escribano *et al.* (1996); Garcia-Olmo (1999) and Abdullaev-Jafarova (2004).

Dosage: 5-10 stigmas decoction in milk or water is recommended as an aphrodisiac.

Toxicity: In high dosage, saffron exhibits toxic qualities. It has even been used as an abortifacient in doses of five to ten grams; such amounts, are, however already severely toxic. Due to its high price and the much smaller amounts for usage, accidental saffron poisoning seems to be very rare.

Contraindication: Do not used in pregnancy and breastfeeding. Crocus stimulates menstrual flow. Large amounts (more than 5 g, which is greater than amounts used in food) have uterine stimulant and abortive effects.

Side Effects: Saffron has been shown to cause anaphylaxis and seasonable allergic symptoms. Higher doses can cause vomiting, uterine bleeding, intestinal cramping and even paralysis.

Drug Interaction: Avoid the use of barbiturates and sedatives.

Herbal Interaction: It can cause synergistic effect with *Piper methysticum (Kava Kava*) and *Hypericum perforatum* (St. John Wort) preparations.

Food Interaction: Avoid the use of alcohol.

Precautions: Adulteration in Saffron is as old as the saffron trade itself. The crocus flower's golden stamens or male flower parts, which have no culinary value, are used to adulterate a crop. Stamens are half of the size of stigma. Sometimes stigmas of closely related species are added to increase profit. The price is so high because harvesting is done by hand and over 4,000 crocus stigmas are needed to yield one ounce of saffron. So adulteration of saffron is very common. It should be stored in air tight container away from heat and light. Do not buy powdered saffron, the quality can be inferior.

Future Prospects: No significant scientific data or evidence was found in the literature for its aphrodisiac properties except one study by Hosseinzadeh *et al* (2008), so its use as such will remain as myth until proved clinically. The other study by Heidary *et al.* (2008) on the motility of semen is also very encouraging. Active constituents such as crocin and its derivatives are very interesting and its pharmacological studies can be fruitful. The plant has a potential for aphrodisiac activity.

Epimedium sagittatum (Horny Goat Weed)

> "The goats went wild sexually when they ate the leaves
> from an *Epimedium sagittatum* plant."
> *Gary S. Ross, MD.*

Product(s): Cobra, Chinese Virility pills, Stamanex, Avela, Enzyte and Vipra.
Common Name(s): Horny Goat Weed, Yin Yang Huo, Inyokaku, Fairy Wings.
Latin Name: *Epimedium sagittatum*
Synonym(s): *E. grandiflorum, E. brevicorum, Epimedium sinense.*
Family: Berberidaceae.
Geographical Distribution: It is a native of China Also found in western and eastern Asia.
Habitat: It is a woody, pungent ornamental herb found in western and eastern Asia and the Mediterranean. Various species are grown as groundcover, particularly in shady areas. The plant prefers sandy loamy, acid, neutral, alkaline, and moist soils. It can grow in semi-shade.

Part(s) Used: Leaves.

Description of the Plant: It is a prostrate plant often used as ground cover. Margin of the leaf is thorny serrulated. Mostly animal avoid them except goats. It is grown as ornamental herb in Asia and in Mediterranean region and flowers from May to June. The flowers are hermaphrodite. The Genus *Epimedium* comprises about 60 species.

Chemical Constituents: Leaves of the plant contains an alkaloid magnoflorine along with sesquiterpenes, lignans, sterols and phenethyl glycosides and a flavonoid icariin (Jin 1981; Wu *et al.* 2003). Icariin is the major constituent. Six prenylated flavonol glycosides, ikarisoside A, icarisid II, epimedoside A, icariin, epimedin B, and epimedokoreanoside-I as the active ingredients have been reported by Kuroda *et al.* (2000). Five new prenylflavones, yinyanghuo A , yinyanghuo B , yinyanghuo C, yinyanghuo D, and yinyanghuo E , along with six known flavonoids, chrysoeriol, quercetin, apigenin, apigenin 7,4'-dimethyl ether, kaempferol, and luteolin, were isolated from the leaves of *Epimedium sagittatum* and their structures were determined by spectroscopic studies by Chen *et al.* (1996). Sagittatins A and B flavonoid glycosides from *E. sagittatum* have been isolated by Oshima *et al.* (1989) and new flavolignins by Wang *et al.* (2007). Very recently, Zhang *et al.* (2008) extracted epimedin A, B, C and Icariin from *Epimedium* species by ultrasonic technique. Nine flavonoids metabolites and active compounds of *Epimedium koreanum* were identified in rats after oral administration of herbal extract of the plant by Wu *et al.* (2008).

Aphrodisiac Myth: The Chinese refer to the herb as Yin Yang Kuo, which means *licentious goat plant.* It improves sexual energy in yin (female) and yang (male). Epimedium has been used in China as a medicinal herb for thousands of years and is one of the most valued ingredients in their tonics. Epimedium first recorded use dates back to the ancient text, *Shen Nong Ben Cao Jing* (ca. 200 B.C.-100 A.D.). It was used as an aphrodisiac herb that improved male fertility (Bensky *et al.* 1992). The herb was given its popular name, *horny goat weed*, when goats grazing on the herb were observed to have significantly increased sexual drive. It is believed that horny goat weed has a testosterone-like effect which stimulates sexual activity in men by increasing sperm production, stimulating and increasing sexual desire. It is said to increase sexual desire in women (Chen 2003).

Scientific Evidence: Recently, Epimedium has been extensively used in China. Studies on animals suggest that icariin, found in Epimedium, may enhance the production of testosterone. In one study, Epimedium was used as an immuno-enhancing supplement in patients with chronic renal insufficiency. These patients reported a significant increase in their sexual response, as well as, an increase in their immune response (Liang *et al.*1997).

Brown (1995) reported that the use of horny goat weed reduced fatigue. Wang *et al.* (2004) believed that it may help to prevent adrenal exhaustion in high stress situations. It increased the production of androgens, thus increasing testosterone level resulting increase in libido or sexual drive. It stimulates the sensory nerves and helps in erectile dysfunction. Horny goat weed inhibits the enzyme acetylcholinesterase (AChE). By inhibiting AChE, it supports cholinergic neurotransmitters associated with sexual arousal. Animal studies have shown that it may influence levels of neurotransmitters such as norepinephrine, serotonin and dopamine, and reduce cortisol levels. Like Viagra, icariin, the active compound in Epimedium increases levels of nitric oxide and inhibits the activity of PDE-5 in rabbit penile tissues, has been reported by Chiu *et al.* (2006) and Ning *et al.* (2006). Estrogenic activity of a polyphenolic extract of the leaves of *Epimedium brevicornum* was found by De Naeyer *et al.* (2005). Molecular and pharmacodynamic properties of estrogenic extracts from *Epimedium brevicornum* were studied by Yap *et al.* (2007). Estrogenic characteristic of *Epimedium* species was standarized by Shen *et al.* (2007) and Yong *et al.* (2007). It was established that oral administration of lipid-based suspension of dry extract of *Epimedium koreanum* in wheat germ oil improved erectile function of aged rats (Makarova *et al.* 2007). Effect of *Epimedium brevicornum* Maxim extract on elicitation of penile erection was studied by Chen and Chiu (2006). They found the intracavernous administration of Epimedium extract may induce penile erection in the rat. Nitric oxide may be involved in this penile erection-inducing effect. Very recently Dell et al. (2008) studied the modified icarin strctures. Inhibitory concentrations for PDE-5 close to that of sildenfil were attained from the derivatives of Icarin. It increased blood flow to the penis by inhibiting the enzymes PDE-5A1 and was more selective towords that enzyme than Viagra along with few side effects.

Additional Benefits: Horny goat weed alleviates fatigue and dilation of the blood vessels. It is also used in the treatment of respiratory diseases including asthma, bronchitis, and sinusitis. It inhibits growth of breast cancer (Yap *et al.*2005) and also effective in leukemia (Lin *et al.* 2004).

Dosage: Suggested dose is 250-1000 mg/day extract in 2-3 divided doses.

Toxicity: High doses of Epimedium can cause kidney and liver damage.

Contraindication: Avoid its use during pregnancy and breast-feeding.

Side Effects: Excessive doses of the herb have been associated with mild dizziness, nausea, vomiting and nose-bleeding. Tachyarrhythmia and hypomania with horny goat weed have been reported by Partin and Pushkin (2004).

Drug Interaction: Nonsteroidal antiinflammatory drugs (NSAIDs) as shown in Appendix O interact with Epimedium. Anticoagulant and antiplatelet medicines such as aspirin, warfarin (Coumadin), clopidogrel (Plavix), ticlopidine (Ticlid), and heparin hydrochlorothiazide (Esidrix), tirofiban (Aggrastat), and eptifibatide should be avoided during the ingestion of *Epimedium*. Alpha blockers (Appendix R) such as prazosin (Minipress) and doxazosin (Cardura) also interact with this herb. ACE inhibitors (Appendix P) and Calcium channel blockers also interact with Epimedium (Appendix X). β blockers such as atenolol, bisoprolol, metoprolol, and propranolol should be avoided (Appendix Q).

Herbal Interaction: Herbs (with blood-thinning effects) such as angelica, anise, arnica, capsicum should be avoided (Appendix D).

Food Interaction: No record is available.

Future Prospects: Viagra revolution has triggered the research on many herbs including horny goat weed. Recently some studies were conducted as discussed above with various Epimedium species on the sexual behavior in rats and the results were encouraging. Comprehensive and systematic clinical studies on humans are required to prove its efficacy as an aphrodisiac. It could be the future natural Viagra for female.

Erythroxylum catuaba (Catuaba)

Product(s): Fematril.
Common Name(s): Catuaba, Cataguá, Chuchuhuasha, Tatuaba, Pau De Reposta, Caramuru, Piratancara, Angelim-rosa, Catiguá.
Latin Name: *Erythroxylum catuaba.*
Synonym(s): *E. vacciniifolium.*
Family: Erythroxylaceae.
Geographical Distribution: North of Brazil, the Amazon, Para, Pernambuco Bahia, Maranhao, and Alagoas.
Habitat: Tree grows in rain forests.
Part(s) Used: Bark.

Description of the Plant: Catuaba is a small tree that produces yellow and orange flowers and small, dark yellow, oval-shaped, inedible fruit. Three unapproved botanical names for catuaba are used incorrectly in herbal world today such as *Juniperus brasiliensis, Eriotheca candolleana, and Anemopaegma mirandum. Juniperus brasiliensis* is called as small catuaba. Eriotheca is a different species. *Anemopaegma mirandicum* is a huge tree in the Bignoniaceae family, growing to 40 m tall and called catuaba-verdadeira in Brazil. This species is now harvested and exported out of Brazil by inexperienced or unethical harvesters resulting in the adulteration of herbal products sold in the U.S. today as just 'Catuaba.' *Erythroxylum catuaba* and *Trichilia catigua* are the preferred Brazilian herbal medicine species, with the history of use as 'big and little catuaba.' Both types are used interchangeably in Brazilian herbal medicine systems for the same condition. Classification of commercial Catuaba samples by NMR, HPLC and chemometrics are now possible (Dolio *et al.* 2008).

Chemical Constituents: The chemical constituents found in Catuaba include alkaloids, tannins, aromatic oils, resins, phytosterols, cyclolignans, sesquiterpenes, and flavonoids. A mixture of flavalignans, cinchonanine has been reported. Three alkaloids catuabine A, catuabine B and catuabine C are found in Catuaba (Grafe 1978; Kilham 2006). These are tropane N-oxide alkaloids. Recently methyl pyrrole tropane alkaloids from the bark of *E. vacciniifolium* have been reported by Zanolari *et al.* (2003, 2005). Kettler *et al.* (2004) isolated and characterized catuabine D and its hydroxymethyl derivative from Catuaba.

Aphrodisiac Myth: The Tupi Indians first discovered the qualities of the Catuaba plant. They used it as a male aphrodisiac (Bernardes1984). Catuaba is also one of the most famous Brazilian aphrodisiacs. It is especially helpful in male impotence and erectile dysfunction. Catuaba is considered to increase sexual interest as well as sexual fantasy for both men and women. It is sold at health food stores in the West without knowing the source of the plant species. In European herbal medicine Catuaba is considered an aphrodisiac and nerve tonic. Tea prepared from the bark of Catuaba is used for sexual weakness, impotence, nervous debility, and exhaustion (Bartram 1995). In the United States Catuaba is used to stimulate low libido and impotence and general exhaustion. Catuaba has been used in Europe as an aphrodisiac for men (Van Straten 1994).

Scientific Evidence: Very few scientific evidences are available in support of its aphrodisiac activities. Catuabine A, B, C and D alkaloids are said to be responsible for its aphrodisiac action (Khilham 2006). Antunes *et al.* (2001) studied the relaxation of isolated rabbit *corpus cavernosum* by Catuaba and its constituents and the results were encouraging.

Additional Benefits: In Japan, studies have shown that the herb can be used to prevent infections of *E. coli* and *Staphylococcus aureus* in people suffering from HIV (Manabe *et al.* 1992). A standardized Catuaba extract could be of great interest for the treatment of depression disorder (Campos *et al.* 2005).

Dosage: Average dose is 0.5 to 1.5 grams of Catuaba bark per day.

Toxicity: None reported.

Contraindication: None known.

Side Effects: None reported.

Drug Interaction: Avoid antidepressant drugs (Appendix S).

Herbal Interaction: Avoid herbal preparations containing Valerian root, St. John's Wort, Kava-Kava, Mandrake, Damiana, and Withania.

Food Interaction: Avoid the use of alcohol.

Future Prospects: It needs evidence or further investigations to support its folkloric use as an aphrodisiac. The plant contains alkaloids that have great pharmacological potential. There is a possibility to find the active constituents from Catuaba for its aphrodisiac use. Clinical studies are also suggested. It is worthwhile to study this plant.

> *"A coward is incapable of exhibiting love;*
> *it is the prerogative of the brave."*
> — *Mahatma Gandhi*

Eurycoma Longifolia (Tongkat Ali)

Product(s): Avela, Stamanex.
Common Name(s): Tongkat Ali, Tung Saw, Pasak Bumi, Malaysian Ginseng.
Latin Name: *Eurycoma longifolia.*
Synonym(s): *Eurycoma harmandiana, Eurycoma latifolia, Eurycoma apiculata*
Family: Simaroubaceae.
Geographical Distribution: Plant is found in forests throughout Malaysia and Southeast Asia. In Indonesia, this species only occurs naturally in Sumatra and Kalimantan.
Habitat: Plant grows in areas with an average temperature of 25^0 C and 86% humidity. The plant can be grown in poor nutrients soil. Seedlings require shade, during which time they develop an extensive root system. Plant grows better in sun light. Flowering occurs from June to July while fruiting in September.

Part(s) Used: Roots used as an aphrodisiac.
Description of the Plant: The Genus *Eurycoma* comprises about 200 species. *Eurycoma longifolia* is a small evergreen tree that can grow to 12-15 m high. The flowers are dioecious (male or female flowers on different tree). Leaves compound, long, ovate, lanceolate and crowded at the tips of the branches. Flowers are borne in axillary panicles, mostly large and puberulus with short hairs. Flowers are unisexual; the male with a sterile pistil and female with sterile stamens. Fruits are ellipsoid or ovoid, green to blackish-red when ripe.

Chemical Constituents: The plant contains several compounds including β-carboline alkaloids and quassinoid-type glycosides such as quassin, neo-quassin, glaucarubin, sedrin, and eurycomanol. The presence of scopoletin, indole alkaloid, and stigmasterol has been reported recently, along with eurycomanone, eurycomanol, eurycomalactone and canthine-6-one alkaloids (Le-Von-Thoi 1970; Bedir *et al.* 2003). Three quassinoids, eurycolactone D, eurycolactone E, and eurycolactone F were isolated from the roots of *Eurycoma longifolia Jack* by Ang *et al.* (2002) and their structures were elucidated by spectroscopic methods. Eurycolactone F was further confirmed by X-ray crystallography. The known quassinoids, laurycolactone B, and eurycomalactone were also identified by them. New β-carboline alkaloids were also isolated by Kuo *et al.* (2003). Their structures were determined by comprehensive analyses of their 1D and 2D NMR and mass spectral data by them. Sixty-five compounds were isolated from the roots of *Eurycoma longifolia* and characterized by 1D and 2D NMR, and mass spectral data by Kuo *et al.* (2004). Among these isolates, four quassinoid diterpenoids were reported from natural sources for the first time, namely eurycomalide A, eurycomalide

B, 13-β, 21-dihydroxyeurycomanol, and 5-α, 14 β, 15-β-trihydroxyklaineanone. Chemistry of biological active quassinoids has been recently reviewed by Guo *et al.* (2005).

Aphrodisiac Myth: It is used as an aphrodisiac in Malaysia and Indonesia (Cyranoski 2005). Eurycoma is currently a popular aphrodisiac in Southeast Asia where all parts of the plant have been used medicinally for hundreds of years. It is supposed to enhance testosterone level in the body.

Scientific Evidence: There have been quite a few studies on the effect of Eurycoma on the sexual behavior of male rats, which support its folkloric use as an aphrodisiac. The effects of Eurycoma were studied on sexually experienced male rats, castrated rats, sexually inexperienced rats, and middle-aged rats. All studies indicated an increase in the sexual activity of rats by Ang *et al.* (1997-2004). More recently Ang et al. (2004) evaluated the aphrodisiac activity of E*urycoma longifoliia* in sexually sluggish old male rats and found interesting results. The study further supported to the use of the plant by indigenous populations as a traditional medicine for its aphrodisiac property.

Additional Benefits: Studies conducted by Jiwajinda *et al.* (2002) and Tjahjadi (2003) have shown that quassinoids, the chemical constituents of Eurycoma were found to exhibit anti-tumor and anti-parasitic activities. Cytotoxic and antimalarial beta carboline alkaloids from the roots of *Eurycoma longifolia* has been reported by Kuo *et al.* (2003). Farouk and Benafri (2007) reported antibacterial acitvity in the leaves and stem extracts of *Eurycoma longifolia.*

Dosage: 200 to 400 mg of plant extract.

Toxicity: It can be toxic with doses more than recommended.

Contraindication: Do not use in pregnancy and breast feeding.

Side Effects: None known.

Drug Interaction: Since it contains indole alkaloid drugs such as ACE inhibitors and antipsychotic should be avoided (Appendix P, S).

Herbal Interaction: Avoid herbal ramedies that raised blood pressure (Appendix H).

Food Interaction: Interact with caffeinated drinks, alcohol, cheese, and tryptamine.

Future Prospects: Plants contains very interesting chemical constituents. It is highly recommended for pharmacological investigations. Clinical data is on rats only and have been reported by the same investigators and need further confirmation. Human clinical trials are recommended. Cyranoski (2005) reported that Malaysia is trying to export products formulated with Tongkat Ali for erectile dysfunction. Active constituent quassinoids of Eurycoma species have great potential for biological activities.

Ferula hermonis (Zallouh Root)

Product(s): Sexitiva A, Zallinex, Zamoreve.
Zallouh-Known as Syrian Sex Plant.
Common Name(s): Zallouh Root, Herbal Viagra, Shirsh Zallouh.
Latin Name: *Ferulis hermonis*
Synonym(s): *Ferula galbanifula, Ferula rinocaulis, Ferula foetida.*
Family: Umbelliferae (Apiaceae).
Geographical Distribution: It grows in Syria, Lebanon and Israel. The plant grows profusely on Mount Hermon at the border of Syria and Lebanon.
Habitat: It prefers cold climate. The plant grows between 6000 and 10,000 feet around Mount Hermon. The root is harvested from August to October.
Part(s) used: Root.
Description of the Plant: Genus comprises 170 species. *Ferulis hermonis* is a small shrub 1-2 meter tall, with thin leaves and tiny white or yellow flowers produces in large umbels.

Chemical Constituents: Zallouh root contains ferulic acid, feruloside, ferutinine, tenuferidine, and feroline. Ferutinine is a complex ester of phenol, like tenuferidine and feroline, is shown to have oestrogenic activity. Ferutinine has been shown to have a type and level activity comparable to di-ethyl stilbestrol. Sesquiterpenes from *Ferula hermonis* has been reported by Galal (2000). More recently three new daucane sesquiterpenes in *Ferula hermonis* have been isolated and characterized by Lhuillier *et al.* (2005).

Aphrodisiac Myth: The roots of the Zallouh plant are historically known for their powerful aphrodisiac qualities and are used in Middle Eastern countries to treat impotence in men and frigidity in women. Zallouh has also been used by healthy individuals, who want to add extra zip to their love life. It is available at the health food stores in the USA.

Ferula is known as Shilsh-el-Zallouh in South Lebanon that means a hairy root while in North Lebanon it is called Hashishat al-Kattira (herb of abundance). In the Jal Eldib area of Beirut, Pierre Malitchev, runs a pharmacy and promotes Zallout root as a supreme elixir of life and vitality. He has been interviewed on CNN and TV many times. According to Malitschev, Shilsh Zallouh is more than a sex plant that contains antioxidants, and helps to delay the aging process. According to him daily intake of Zallouh root will keep a person strong and youthful. The 77 year old pharmacist gives credit of his vitality and good health to Zallouh root. The most popular Zallouh product is the alcoholic extract for sexual enhancement (Kilham 2004). Dr. Ali Abou Hamman, a physician in the village of Chebea near Mount Hermon claims to have treated hundred of impotence cases with *Ferula hermonis*.

Scientific Evidence: Ingestion of Zallouh root helps to cure ED. In the male ferutinine binds strongly with oestradiol receptors of the pituitary glands causing the hypothalamus to release LH (Luteinizing hormone) that circulates to the testicles where it causes the production and release of testosterone into the blood stream (Ahmedhodjaieva *et al.*1999). There are anecdotal stories of Arabic men in their seventies and eighties who appear to have regained lost youth after using Ferula. Testosterone level decreases with age in both the sexes, so stimulating testosterone level can affect age related impotence. The Lebanese government is supporting research on its precious herb Zallouh root. In Beirut, the Lebanese Urological Society has sponsored clinical trials on men. To date over 7000 men have participated in this research. Clinical studies on Zallouh, showed some encouraging results in erectile dysfunction.

Nutranex Company has refined and standardized the Zallouh root extract under the brand name Pur-Zall(r). Extract is also incorporated with other herbs and dispensed in capsule form that is medically proven to increase sexual performance in men and women. This proprietary blend of ingredients works synergistically to enhance sexual desire, increase proper sexual function, and nutritionally support normal reproductive heath. Zallnex for men and Zamoreve for women are sold worldwide (Kilham 2004).

A comparative study of *Ferula hermonis* root extracts and sildenafil (Viagra) on copulatory behavior of male rats were carried out by Hadidi *et al.* (2003). The results were not encouraging except there was a significant prolongation in IL (Intromission latency). The seed oil from the *Ferula hermonis* revealed encouraging results for enhancing erectile function in rats by El-Thaher *et al.* (2001).

The effects of an aqueous extract of *Ferula hermonis* on social aggression, fertility and some physiological and biochemical parameters were investigated by Khelifat *et al.* (2001) in male mice. The ingestion of 3 mg/kg of aqueous extract of *F. hermonis* for six weeks clearly inhibited social aggression. Body wet weight and other sex accessory organ weights were significantly reduced by this treatment. The ingestion of this extract by male mice resulted in a significant reduction of their fertility. This treatment caused a significant decrease in the number of pregnant females, number of implantations and viable fetuses in females impregnated by males that ingested this extract. Additionally, the numbers of epididymal sperm and their motility were dramatically reduced *in F. hermonis*-treated mice. Concomitant increases in sperm abnormalities were also observed when compared with control. These data indicate that *F. hermonis* exposure during this period puts the exposed animals at significant risk for reduced reproductive capacity in adulthood.

Zanoli *et al.* (2005a) studied the influence of *Ferula hermonis* root extract on sexual behavior in female rats. Sexual receptivity, proceptivity and paced mating behavior were evaluated in ovariectomized females primed with estradiol benzoate (EB) and progesterone (P) and then treated with *F. hermonis* extract acutely (30 and 60 mg/kg) or subchronically (1 and 10 mg/kg daily for 10 consecutive days). The results demonstrate that the acute or repeated ingestion of *F. hermonis* specifically impairs the receptive and proceptive components of female sexual behavior. The effect could be the consequence of an antiestrogenic action. Zanoli *et al.* (2005b) also studied the activity of single components of *Ferula hermonis* on male rat sexual behavior and found that ferutinin was able to stimulate sexual behavior after acute ingestion where as teferdin only improve copulatory performance of sluggish/impotent animals.

Zavitri *et al.* (2006) studied the effects of single components of *Ferula hermonis* extract on female rat sexual behavior. Ovariectomized rats hormonally primed with estradiol benzoate and progesterone were treated by oral gavage with ferutinin, teferin and teferdin. None of the three compounds showed the capacity to alter sexual motivation. However, only ferutinin significantly inhibited female receptivity. These results suggest a primary role of ferutinin in the impairment of sexual behavior elicited by *F. hermonis* extract in hormone primed-female rats.

Additional Benefits: Carminative, used in arthritis, asthma and for culinary purposes.

Dosage: Zallnex and Zamoreve are available in capsule form. One capsule a day is recommended. 0.5 grams to 1.0 grams freeze dried root is recommended.

Toxicity: Large doses can be toxic. Toxicity observed above 50 mg/kg.

Contraindication: Those with hypertension, significant heart disease, or diabetic neuropathy should not take Zallouh root or formulation without the approval of a physician.

Side Effects: Some individuals who take Zallouh experience flushing and headaches as a result of the circulatory effects of the root. For the most part, however, Zallouh root and its various preparations appear safe and effective for a majority of users.

Drug Interaction: Avoid anticoagulant and anti-inflammatory drugs (Appendix O).

Herbal Interaction: Avoid anticoagulant and hypertensive herbal products (Appendix D & H).

Food Interaction: Not known.

Future Prospects: It has a bright future. Owing to contradictory research on the plant, more clinical studies are required to prove its use as an aphrodisiac.

Ginkgo biloba (Maidenhair Tree)

Product(s): Intim X, VP-RX, Libido Fort, VG-RX oil, Enzyte, Vipra and Erect pills.
Common Name(s): Maidenhair Tree, Smart Herb, Yin Sing, Japanese Silver Apricot, Kew Tree, Duck Foot Tree, Ginkgo Balm, Ginkgold, Oriental Palm Tree.
Latin Name: *Ginkgo biloba.*
Synonym (s): *Ginkgo folium* and *Salisburia adiantifolia.*
Family: Ginkgoaceae.
Geographical Distribution: The plant is Native to China but found in Asia, Europe and North America. It was planted in temple grounds in China and later on spread to Japan.
Habitat: Easily grown in average wet soil in full sun. It prefers moist, sandy, well-drained soils.

Part(s) Used: The green-yellow leaves of the plant are used for aphrodisiac activity.

Description of the Plant: It is a unique tree with no living relatives. It is a single species, one of the best examples of a living fossil known inhibited on the earth for 225 million years, when dinosaurs were flourshing on the earth. The *Ginkgo* is a medium-large tree, reaching 20-35 m tall. Some specimens believed to be 2,500 years old. Some specimens are about ten centuries old. Ginkgos are dioecious (separate male and female trees). It is commonly known as Maidenhair Tree in reference to the resemblance of the fan-shaped leaves to maidenhair fern leaflets (pinnae). The male trees have catkin flowers in the spring. The female trees contain small round solitary flowers and plum-shaped fruits.

Chemical Constituents: The leaf extract contains flavoglycosides, quercetin, kaempferol, proanthocyaninidins, kaeterpin lactones, and phenolic acids. Terpenoids such as bilobalide, diterpenes ginkgolides A, B, C, J, and M, dimeric flavones bilobretin, ginkgetin, isoginkgetin, sciadopitysin and sterols have been reported by Hua and Staba (1992). Two coumaroyl flavonol glycosides from the leaves of *Ginkgo biloba* have been found by Tang *et al.* (2001). Stromgaard and Nakanishi (2004) reviewed the chemistry and biology of terpene trilactones from *Ginkgo biloba*.

Recent phytochemical investigation of *Ginkgo biloba* has resulted in the isolation of two new biflavone glucosides, ginkgetin 7"-O-beta-D-glucopyranoside and isoginkgetin 7-O-beta-D-glucopyranoside. The structures were determined on the basis of chemical and spectroscopic evidences. Three new compounds were isolated from *Ginkgo biloba*, together with 27 known compounds. The structures of the new compounds were determined primarily from 1D-and 2D-NMR analysis (Bedir *et al.* 2002; Hyun *et al.* 2005). Lai *et al.* (2005) isolated trilactones including ginkgolides and bilobalide (BB) from the leaves. A new polysaccharide from leaf of *Ginkgo biloba* has been reported by Yang *et al.* (2009). Very recectly, biology and chemistry of *Ginkgo biloba* has been recently reviewed by Singh *et al.* (2008).

Aphrodisiac Myth: Ginkgo is a Chinese word *Ginkyo* meaning *silver apricot* (gin meaning silver, *kyo* meaning apricot. *biloba*: two-lobed; *bi* from Latin *bis* meaning double, *loba* meaning lobes). The leaf is fan shaped with a split in the middle, hence two-lobed. *Ginkgo biloba* is mentioned in Chinese Pharmacopeias. In China, the *Ginkgo* tree is considered sacred. It has been used in the Chinese System of Medicine for more than a millennium. Fruits contain edible ivory-colored kernels. In Asia, the kernels have been used in soups, meat dishes and desserts as an aphrodisiac (Huh and Staba 1992).

It is believed that old age impotence such as arterial erectile impotence can be cured by Ginkgo. Leaves Extract of *ginkgo biloba* increases circulation. It dilates blood vessels. It is believed that the flavoglycosides in the plant dilate the blood vessels and allow more blood to the male and female sex organs, since lack of adequate blood flow to the genital organs is a root cause of impaired performance in both sexes. In fact, women were found to be more responsive to ginkgo's sexual enhancing effects than men (Braquet 1988).

Scientific Evidence: Sohn and Sikora (1991a, b) evaluated the effect of Ginkgo leaf extract in the treatment of erectile dysfunction in fifty patients. The men, diagnosed with arterial erectile impotence, received 240 mg of ginkgo leaf extract daily for a period of nine months. The patients were divided into two groups based on their response to conventional therapies. Twenty of the patients had previously experienced some success with conventional drug therapy, and were placed in the first group. The second group of thirty patients had not experienced erection following conventional therapies. Following treatment with the Ginkgo leaf extract, all patients in the first group (20 men) regained sufficient and spontaneous erections following six months of treatment. Rigidity at both the tip and base of the penis were found to significantly improve after six months. The improvement continued through the nine-month treatment period. Nineteen of the thirty patients in the second group responded positively to the treatment, while eleven remained impotent. No side effects were reported in the study.

Cohen and Bartlik (1998) found that *Ginkgo biloba* could be used to rectify antidepressant-induced sexual dysfunction. Sexual dysfunction in these patients was determined to be secondary to antidepressant medications. The mechanism of antidepressant-induced ED appears to be related to the therapeutic activity of selective serotonin reuptake inhibitors (SSRIs). Wheatly (1999) published an open clinical trial to investigate the effect of Ginkgo on antidepressant-induced sexual dysfunction. Twenty-four patients (12 men and 12 women) reported significant improvement in sexual response after both three and six weeks of use. This trial provides little support to the efficacy of Ginkgo for ED. There was substantial variability among responses, with both ends of the spectrum represented, two patients reported complete restoration of sexual function and two others reported no response at all.

Ashton *et al.* (2000) investigated the use of Ginkgo for sexual dysfunction related to antidepressant use. Improvement from Ginkgo was found in only three of 13 women tested and no improvement was reported in nine men. Kang *et al.* (2002) comprising the first placebo-controlled, double-blind trial of Ginkgo for antidepressant-induced sexual dysfunction reported no statistically significant difference in sexual function between Ginkgo and placebo. Elevated serotonin in the CNS is thought to inhibit the effects of pro-erectile neurotransmitters in the erection-generating center of the spine. This in turn decreases the activity of NO (Nitric oxide) synthase and reduces NO availability to penile smooth muscle cells. Ginkgo is presumed to mitigate the effects of SSRIs on sexual function by increasing NO availability. Marcocci *et al.* (1994) has shown Ginkgo can strengthen the activity of NO synthase, presumably by passing serotonin's ability to block nitric oxide production.

Several studies conducted on isolated rabbit aorta have shown Ginkgo induced a dose-dependent relaxation of vascular smooth muscles (Auguet 1982, 1983; Paick and Lee 1996). Collectively, this evidence suggests Ginkgo has vascular smooth muscle relaxing activity in the heart, brain, and penis and it appears nitric oxid activity provides a unifying association for *Ginkgo's effects*. Increasing nitric oxide bioavailability may not only improve sexual function and vascular health, but may positively impact other age-related chronic diseases.

Waynberg and Brewer (2000) investigated an alternative to synthetic drugs for the treatment of sexual dysfunction in healthy women. The efficacy of a unique herbal formulation of *Muira puama* and *Ginkgo biloba* (Herbal vX) was assessed in 202 healthy women complaining of low sex drive. Various aspects of their sex life were rated before and after 1 month of treatment. Responses to self-assessment questionnaires showed significantly higher average total scores from baseline in 65% of the sample after taking the supplement. Statistically significant improvements occurred in frequency of sexual desires, sexual intercourse, and sexual fantasies, as well as in satisfaction with sex life, intensity of sexual desires, excitement of fantasies, ability to reach orgasm, and intensity of orgasm. Reported compliance and tolerability were good. These initial findings support the strong anecdotal evidence for the benefits of Herbal vX on the female sex drive. A double-blind study was further planned to investigate these results.

Additional Benefits: It improves memory and may be helpful in Alzheimer's disease (Perry *et al.*1999). It reduces depression, slows aging and helps to prevent blood clots and heart attacks. It is also beneficial in macular degeneration of the eye. Lebuisson *et al.* (1986) found potential benefit of *G.biloba* for people with early stage of macular degeneration. Evans (2000) analysed the data available on *G.biloba* for the treatment of MD and reported that the research on this topic was still inconclusive. It is also used as anti-aging supplement. Multifaceted therapeutic benefits of *Ginkgo biloba* L along with chemistry, efficacy and safety has been reviewd by Mahadevan and Park (2008).

Dosage: 120 to 320 mg/d. The most common dose is 40 mg standardized extract of ginkgo leaves three times a day. For impotence, 240 mg once a day is recommended. GBE (*Ginkgo Biloba* Extract) should contain at least 24% flavones glycosides responsible for the herb's antioxidant and anticlotting properties and 6% terpene lactones ginkgolides and bilobalides. *Ginkgo biloba* extract is obtained by drying and milling the leaves and then extracting the active ingredients in a mixture of acetone and water. The solvent is recovered under vacuum.

Toxicity: Ingestion of seeds may be quite toxic, leading to cramps, headaches, indigestion, and overall weakness. It is not recommended to persons who are allergic to mangos or cashew nuts.

Contraindication: Do not use the membrane (pulp) around the nut of the Ginkgo; it will produce allergic reactions, similar to those caused by members of the Sumac family. In Asia, the nut is eaten after being roasted, after the orange-colored membrane is removed from around the nut. Ginkgo can enhance the blood thinning effect if taken with aspirin or prescription blood thinning drugs.

Side Effects: Ingestion of 10 roasted seeds of Ginkgo can cause seizures, difficult breathing, unconsciousness, and even death.

Drug Interaction: *Ginkgo biloba* interacts with MAO inhibitors (antidepressant) and sympathomimetics substances (Appendix S). Ginkgo also helps in preventing clotting by keeping platelets separated. Ginkgo is a natural blood thinner. Avoid the use of aspirin and related platelet inhibitors (Appendix O). These can cause hematoma. *Ginkgo biloba* also interacts with anticonvulsants drugs (Appendix Y).

Herbal Interaction: Some of the common herbs that also inhibit blood clotting such as Danshen, Devil's Claw, Eleuthero, Garlic, Ginger, Horse Chestnut, Panax Ginseng, and *Saw palmetto* as shown in Appendix D should be avoided with Ginkgo. It can lower blood pressure so avoid the herbal products that affect the blood pressure (Appendix G). Ginkgo may lower blood sugar level and can interact with Antidiabetic herbs (Appendix E).

Food Interaction: While taking this product, avoid intake of cheese and red wine. Avoid tyrosine, tryptophan containing food, caviar, champagne, chocolate, sausages, salami, yeast extracts, and yogurt.

Future Prospects: It is the top selling herb in the United States accounting over $140 million of sales in 2001. Sales in Europe are over $280 million per year. The plant has a good prospect not only for an aphrodisiac but also for atherosclerosis, macular degeneration of the eye and Alzheimer disease. It is highly recommended for further investigation.

Glycyrrhiza glabra (Licorice)

Product(s): Roots, root powdered, tablets. Root powdered in capsules.
Common Name(s): Licorice, Yashtimadhu, the Great Harmonizer, Grandfather Herb, Sweetwood.
Latin Name: *Glycyrrhiza glabra.*
Synonym(s): *G. glandulifera, G. glabra.*
Family: Fabaceae (Leguminosae).
Geographical Distribution: Licorice is a perennial herb native to southern Europe, Asia, and the Mediterranean. However, it is extensively cultivated in Russia, Spain, Iran and India.
Habitat: Native to the Mediterranean region and parts of Asia. It is cultivated worldwide.
Part(s) used: Roots.

Description of the Plant: The plant is a very hardy deciduous perennial bush that grows to a height of 1-2 meters. It has spreading pinnate leaves and produces lavender-blue pea-like flowers in a spike-like cluster. It produces small flat pods 2-3 cm in length which turns brown at maturity. The pods contain 1-7 small dark kidney shaped seeds about the size of a pinhead.

Roots are long and cylindrical. Externally they are dark-brown and are wrinkled. Internally they are light-yellow. When fresh the roots are pliable, fibrous, tough, readily tearing into long, fibrous strips. They smell earthy and taste sweetish. The plant becomes dormant during the winter and sheds its leaves. In the spring, the plant produces new shoots from buds on the underground stolon.

Chemical Constituents: Glycyrrhizin, glycyramarin, an acrid resin, flavonoids, saponins, sterols, essential oil, ten different bioflavonoids, starch (up to 30%), various sugars (up to 14%) and amino acids are present in licorice. Li *et al.* (2000) isolated an unusual biflavonoid named licoagrodin from the hairy root cultures of *Glycyrrhiza glabra* along with three prenylated retrochalcones, licoagrochalcones B, C, D, a prenylated aurone, licoagroaurone and four known prenylated flavonoids, licochalcone C, kanzonol Y, glyinflanin B and glycyrdione A. From the glycosidic fraction, an isoflavone glycoside, licoagroside A, and a maltol glycoside, licoagroside B were isolated together with four known isoflavone glycosides, two flavone C-glycosides, and three other glycosides. Their structures were elucidated on the basis of spectroscopic evidence. Recently, two new flavonoids compounds from *Glycyrrhiza glabra* have been isolated by Li *et al.* (2005).

From the air-dried roots of *Glycyrrhiza glabra* L. collected in Xinjiang, China, five new flavonoid compounds named glucoliquiritin apioside (a flavonone bisdesmoside), prenyllicoflavone A (a bisprenylflavone), shinflavone (a prenylated pyranoflavanone), shinpterocarpin and 1-methoxyphaseollin (both pyranopterocarpans), were isolated together with eight known saponins, seven known flavonoid glycosides, and eleven flavonoids by Kitagawa *et al.* (1994).

Aphrodisiac Myth: Glycyrrhiza is derived from the ancient Greek term glykos, meaning sweet, and rhiza, meaning root (Anon 2005). The origin of licorice is uncertain, it is believed to have originated in the East and has been grown since early times in China, Persia, Turkey, Syria, Iraq, and Northern India, Russia, North Africa, and the Mediterranean countries. It was used as folkloric medicine by ancient Egypt, the Roman Empire, the ancient Hindus, and early Buddhists. It was used in Chinese herbal medicines dating back 5000 years and even now enjoys good reputation. Currently, it is grown commercially in Spain, France, Russia, Germany, England, the Middle East, and Asia. Warriors used it for its ability to quench thirst while on the march, while others including Indian prophets and Chinese Buddhist sages recognized Licorice's valuable healing properties long time back (Fiore *et al.* 2005; Davis and Morris 1991).

The rejuvenating and nutritive properties have made licorice one of the most versatile herbs widely used by herbalist of East and West. Licorice has been used as aphrodisiac for vitality, and longevity, often known as an elixir of life. The earliest clay tablets found in Mesopotamia signifies licorice as a panacea potion. In India, licorice has an ancient reputation as an aphrodisiac in the *Kama Sutra* and *Ananga-Ranga* that comprise numerous recipes containing licorice for increasing sexual vigor.

Scientific Evidence: Though it is used as an aphrodisiac yet we have contradictory studies. Licorice may be a hidden cause of erectile dysfunction. Recent research suggests that licorice may reduce testosterone levels in men (Armanini *et al.* 1999, 2003). After one week of treatment, mean testosterone levels decreased 26%. Testosterone has a key role in erectile dysfunction. Men with erectile dysfunction, infertility, or decreased libido may wish to avoid this herb. Interestingly, licorice is an ingredient in some aphrodisiac and sexual enhancement formulations. Very recently a review of pharmacological effects of Glycyrrhiza species and its bioactive comounds have been described by Nassiri and Hosseinzadeh (2008).

Additional Benefits: It is used in the formulations of expectorants and cough syrups. It is also used for mouth ulcers, peptic ulcers, irritable bowl syndrome, and Crohn's disease.

Dosage: 1-4 grams of root powdered and Standard Extract 380 to 1000 milligrams have been recommended.

Toxicity: Licorice is not a candy. It can be toxic. Avoid during pregnancy, diabetes, obesity and liver and kidney disorder. Long-term (more than a month) intake of products containing more than 1 gram of glycyrrhizin may increase blood pressure.

Contraindication: Not recommended during pregnancy and breast feeding.

Side Effects: Prolong or excessive use can lead to excessive thirst, high blood pressure, increased urination, low potassium levels, muscle weakness, and erectile dysfunction.

Drug Interaction: Avoid hypertensive drugs, antiplatelet, and diuretic that decrease potassium (Appendix Q, R, O). Avoid the use along with oral contraceptive and corticosteroids. Cardiotonic drugs such as digoxin and strophanthus should be avoided.

Herbal Interaction: Avoid the use of herbs that cause hypertension (Appendix H) and likewise, prevent anticoagulant herbs such as Danshen, Devil's Claw, Eleuthera, and Garlic as shown in Appendix D. Also avoid the use of herbs such as Senna, Rhamnus, and Aloe that decrease potassium.

Food Interaction: Avoid alcohol and grapefruit juice.

Future Prospects: Contradictory reports on Glycyrrhiza regarding its use in sexual dysfunction are confusing. Myth recommends its use as an aphrodisiac and some clinical studies indicate that it reduced testosterone level in the body leading towards erectile dysfunction. Systematic clinical evidences are required to confirm the validity of contradictory reports for its aphrodisiac qualities. At present, it will remain as myth.

"We need to find God, and he cannot be found in noise and restlessness. God is a friend of silence. See how the nature- trees, flowers, grass grow in silence; See the stars, the moon, and the sun, how they move in silence…….we need silence to be able to touch souls."

—Mother Teresa

Hypericum perforatum (St. John's Wort)

Product(s): Prolast (extract, capsules and tablets).
Common Name(s): Amber Touch-and-Heal, Hardhay, Hypericum, Klamath Weed, Millepertuis, Rosin Rose, SJW, Tipton Weed.
Latin Name: *Hypericum perforatum.*
Synonym(s): *Hypericum vulgare.*
Family: Clusiaceae.
Geographical Distribution: *Hypericum* grows wild in Europe, Western Asia, North Africa, Madeira, and the Azores. It has also been naturalized in many parts of the world, especially in regions of North America and Australia.
Habitat: This perennial shrub grows 12-36 inches tall and is covered with, fragrant flowers from mid to late summer. It prefers sandy loam soil and warm weather. St. John's Wort is usually propagated from runners in the autumn or by seed sown early in the spring. The plants grow rapidly but last 4-5 years. New plants can be started from softwood cuttings of the young plant. Flowers are cut when fully open. Harvesting is done before the heat of the day to prevent loss of volatile oils.

Part(s) Used: Extracts of flowers and leaves are used.

Description of the Plant: *H. perforatum* is a small plant. The yellow flowers have 5 petals and clusters of feathery gold stamens. The flowers are covered with black dots. When these dots are rubbed between the fingers, the fingers turn red. It is believed that the black-red dots and the translucent 'perforations' contain the active constituents. The oval shape leaves of Hypericum, when held to light, reveal translucent dots, giving an impression of perforations on the leaves. The dots contain colorless essential oils and resin.

Chemical Constituents: Plant contains about ten active constituents that may contribute to its effects to Hypericin complex, including hypericin, pseudo-hypericin, isohypericin, and protohypericin. Flavonoids include hyperoside, quercitin, rutin bi flavones and hyperforin-a phloroglucinol derivative as found by Smelcerovic and Spiteller (2006). Volatile oils and hydrocarbons include n-alkanes, a-pinene and other monoterpenes, tannins and procyanidins. Tanaka (2006) and Sleno *et al.* (2006) isolated six new xanthones and 21 known xanthones from the leaves and stems of Chinese Hypericum. Their structures were established by spectroscopic studies.

Aphrodisiac Myth: *Hypericum perforatum* has long been used as an ancient folk cure for depression (Stevenson and Ernst 1999; Gaster and Holroyd 2000). Hyperforin is considered the active constituent to treat depression (Chatterjee *et al.*1998). *H. perforatum* was even prescribed by Hippocrates, the father of medicine, Dioscorides, the foremost Greek physician, and Pliny, a well-known Roman physician. *H. perforatum*, derived from Greek, meaning *over the apparition*, a name based on the belief that the odor of the herb was so potent that evil spirits would depart when they smelled it. Hypericum is also known as St. John's Wort.

In Germany, it is the most popular prescription drug for the treatment of mild depression. Concentrated extract of the flowers and leaves of St. John's Wort, are sold as Hypericum. About 200,000 prescriptions per month are filled for a single brand (Jarsin) in Germany, compared with about 30,000 per month for fluoxetine (Prozac). This figure does not include sales of other hypericum products. Approximately 80% of the sales are prescriptions, since their cost can be reimbursed by the German health-insurance system.

Scientific Evidence: It reduces the stress and work as an aphrodisiac for those persons who have low libido due to stress. More recently, Cannon-smith and Kaufman (2007) performed a pilot study on *Hypericum perforatum* and demonstrated that hyperforin extract of hypericum increased the duration of sexual intercourse before ejaculation for men with or without complaints of premature ejaculation. Sexual satisfaction improved after hyperforin use an attractive option for men who wish to increase time to ejaculation during sexual intercourse.

Additional Benefits: In some European countries, St. John's Wort is registered for the treatment of mild to moderate depression; in the United States, it is marketed as a dietary supplement. It strengthens the immune system. It may help to resist viral infections such as herpes and AIDs.

Dosage: Dry extracts or freeze-dried standardized for hyperforin content, 300 mg one to three times a day. It may take 4-8 weeks to get optimal effects. It is recommended with food to minimize possible gastrointestinal upset. Nonstandardized extracts should be discouraged since it can be toxic. For premature ejaculation 20 mg of hyperforin is recommended at least 15-30 minutes before the intercourse.

Toxicity: It may cause a photosensitivity reaction in some people. Use sunscreen and avoid sun exposure while taking St. John's Wort.

Contraindication: Severe depression with suicidal, psychotic, or severe melancholic symptoms has been reported such as bipolar depression (Ernst *et al.* 1998). Avoid its use in pregnancy or lactation. It is known that St. John's Wort inhibits pituitary secretion of prolactin, so the possibility exists that there could be problems with breast milk production. People suffering from bipolar depression should avoid it.

Side Effects: Gastrointestinal upset, allergic reactions, dizziness, tiredness, fatigue, and loss of hair. All side effects are completely reversible-stop taking the St. John's Wort and side effects go away (Parker *et al.* 2001).

Drug Interaction: It interacts with anticonvulsants (Appendix Y) and Drugs for insomnia such as zaleplon and zolpidem, Tricyclic antidepressants such as amitriptyline, amoxapine, doxepin, and nortriptyline. Don't combine with prescription antidepressants (Lantz *et al.*1999; Miller 1998). It may interfere with iron absorption (Miller 1998). Also avoid amphetamines, narcotics, tryptophan, and tyrosine. Diphenhydramine should not be taken along St John's Wort. It causes excessive drowsiness. It can interact with Cyclosporine, Indinavir, and oral contraceptives. Dextromethorphan is an anti-coughing ingredient in many non-prescription cough and cold products such as Nyquil and Robitussin DM. It may have an increasing effect on serotonin levels, taking dextromethorphan with St. John's Wort may result in a higher risk of side effects. People suffering with migraine or headache usually take triptans. Avoid the use of St John's Wort with triptans group of drugs (Appendix W).

Herbal Interaction: St. John's Wort may cause synergistic effect, if it is taken with other sedating herbs such as Catnip, Hops, Kava, Valerian, and Withania.

Food Interaction: Do not consume along with cheese, wine, alcohol and pickled foods.

Future Prospects: It can be used as one of the ingredients in the formulation of antidepressant female sexual dysfunction. A recent finding by Cannon-smith and Kaufman (2007) that hypericum extract can help premature ejaculation has opened a new door for research. It has a great potential since premature ejaculation is very common in all ages. SSRI's drugs are given for premature ejaculation but their use is associated with side-effects that include nausea, vomiting, cognition impairment anticholinergic and sexual side effects such as genital anesthesia and decreased libido. Rapid action, ease of use and safety make hyperforin extract an attractive option for men suffering from premature ejaculation.

> *"One touch of nature makes the whole world kin."*
> —*William Shakespear*

Lepidium meyenii (Maca Root)

Product(s): IntimX, MagnaRx, male Boost, Enzyte.
Common Name(s): Maca, Peruvian Ginseng, Maka, Mace, Maca-Maca, Maino, Ayak Chichira, Ayuk Willku, Pepperweed, Andean Ginseng, Peru Ginseng, Vegetal Viagra.
Latin Name: *Lepidium meyenii.*
Synonym(s): *Lepidium peruvianum, L. weddellii, L. affine, L. gelidum.*
Family: Brassicaceae.
Geographical Distribution: Maca is found in the central Andes of Peru.
Habitat: The plant cultivated high in the Andean Mountains at altitudes from 11,000-14,500 feet. The plant prefers sandy loamy and heavy soils. It can grow in acid, neutral and basic soils and requires semi-shade area. It can survive under extreme sunlight and cold weather.
Part(s) Used: Tuberous root
Description of the Plant: The genus *Lepidium* comprises about 175 species. *Lepidium*
meyenii is the one known as an aphrodisiac. Maca Root is a pear shaped vegetable root or tuber. Unlike many tuber plants, seeds propagate Maca plants. Its scalloped leaves lie close to the ground and it produces self fertile off white flowers. The Maca root grows wild in the Peruvian Andes just below the glacial icecap. It only grows in areas of extreme weather conditions such as freezing, high winds, and intensive sunlight. The dried root tuber can be stored under suitable conditions for seven years. The rich soil located at these high plateaus of Peru where it is very cold and poor oxygen may account for the high levels of trace minerals found in Maca Root. No other food plant exists in the world that can grow at so high altitude and will survive.

Chemical Constituents: Chemical research shows maca root contains a chemical called *p-methoxy benzyl isothiocyanate*, which has reputed aphrodisiac properties. Two new imidazole alkaloids, lepidiline A and lepidiline B have been isolated by Cui *et al.* (2003). The tubers of *Lepidium meyenii* contain the benzylate derivative of 1, 2-dihydro-N-hydroxypyridine, named macaridine, together with the benzylated alkamides (macamides), N-benzyl-5-oxo-6E, 8E-octadecadienamide and N-Benzyl hexadecanamide, as well as the acyclic keto acid, 5-oxo-6E, 8E-octadecadienoic acid (Muhammad *et al.*2002). Five new alkamides were recently isolated and characterized by Muhammad *et al.* (2005). Very recently Gonzalez *et al.* (2008) found the presence of methyltetrahydro-{β}-Carbolines in *Lepidium meyenii*.
The essential oil profile of *Lepidium meyenii* obtained from Lima, Peru, was examined by Tellez *et al.* (2002). 53 oil components were identified. Phenyl acetonitrile (85.9%), benzaldehyde (3.1%), and 3-methoxyphenylacetonitrile (2.1%) were the major components of the steam distilled oil. Significant concentrations of calcium,

magnesium, potassium, zinc, and copper are also present in the root. Vitamins in maca are thiamine, riboflavin, and ascorbic acid. Piacente (2005) identified the amino acids in Maca such as aspartic acid, glutamic acid, serine, histidine, glycine threolline, cystine, alanine, arginine, tyrosine, valine, methionine, isoleucine, lysine, proline, hoproline and sarcosine. Zhao *et al.* (2005) isolated and characterized five new alkamides from Maca root. Their structures were established by spectrometric and spectroscopic methods including ESI-HRMS, EI-MS, (1) H, (13) C, and 2D NMR, as well as (1) H-(15) N 2D HMBC experiments.

Aphrodisiac Myth: Maca is the edible root of the Peruvian plant *Lepidium meyenii*, traditionally employed for its aphrodisiac and fertility-enhancing properties. Maca was recognized about 2,000 years ago by the Incas. Primitive cultivators of maca have been found in archaeological sites dating back as 1600 B.C. Native Peruvians have traditionally utilized Maca for both nutritional and medicinal purposes even before the time of the Incas. The native for sexual disorder has used it in fertility and impotence (Rowland and Tai 2003). The Peruvian Indian considered it as a gift along with corn and potatoes from God. It was also used as hallucinogen and often consumed during dances and religious ceremonies by the Incas. Maca has been used medicinally for centuries to enhance fertility and low libido in humans and animals (Popeno *et al.* 1990). Native shepherds learned quickly that at higher elevations where the Maca grew naturally, their grazing herds became healthier, their stamina increased, and they became much more sexually active. Natives used it for sexual disorder particularly for impotence.

Scientific Evidence: Zheng (2000) reported for the first time that Maca has beneficial affects on the sexual function of impotent mice and rats. Cicero *et al.* (2001) found that maca stimulate male sperm production and improvement of sperm count and motility in nine healthy adult men. In the second study Cicero *et al.* (2002) reported improved sexual performance in inexperienced male rats and another "self-perception on sexual desire" test in healthy men reported aphrodisiac or libido enhancement effects. None of these studies, however, indicated a possible mechanism of action or could pinpoint the active constituents contained in maca root. Comhair and Mahmoud (2003) reported that the extracts of the Peruvian plant *Lepidium meyenii* were shown to improve sperm morphology and concentration in men. Gonzales *et al.* (2001, 2002) studied the effect of *Lepidium meyenii* (Maca) root on serum reproductive hormone levels in adult healthy men. The study aimed to test the hypothesis that Maca has no effect on serum reproductive hormone levels in healthy men when administered in doses used for aphrodisiac and/or fertility-enhancing properties. In conclusion, treatment with Maca does not affect serum reproductive hormone levels. Cicero *et al.* (2002) in their investigations on mice noted that male mice triple the frequency of sexual activity when fed high doses of maca.

A study conducted by Gonzalez *et al.* (2003) with humans in Peru confirmed the earlier investigation on mice. In a double blind, placebo-controlled study, men between the ages of 21 and 56 were given 3 g of maca or placebo for a period of 12 weeks. The men who took Maca reported increased sexual desire after about 8 weeks of treatment, even though the supplements did not appear to affect levels of testosterone in the men's blood.

In a recent widely publicized clinical trial at La Molina National Agrarian University in Peru, maca produced an increase in sex drive among men. The effect was observed within two weeks. Gonzales *et al.* (2004) studied the effect of *lepidium meyenii* on spermatogenesis in male rats. The aim of the study was to prevent high altitude induced testicular disturbance. Treatment of rats with Maca at high altitude prevented high altitude induced spermatogenic disruption. A double-blind, randomized, pilot dose-finding study of maca root (*L. meyenii*) for the management of SSRI-induced sexual dysfunction was carried out by Dording *et al.* (2008) and found that Maca has a beneficial effect on low libido.

More recently Bogani *et al.* (2006) found that *Lepidium meyenii* (Maca) did not exert direct androgenic activities. This study aimed at testing the hypothesis that Maca contains testosterone-like compounds, able to bind the human androgen receptor and promote transcription pathways regulated by steroid hormone signaling. Maca extracts obtained with different solvents: methanol, ethanol, hexane, and chloroform are not able to regulate GRE (glucocorticoid response element) activation. Very recently effect of Black Maca (*Lepidium meyenii*) on one spermatogenic cycle in rats was studied by Gonzales *et al.* (2006). It increased the sperm count. Beneficial effects of Lepidium meyenii (Maca) on psychological symptoms and measures of sexual dysfunction in postmenopausal women related to estrogen or androgen content were determined by Brooks *et al.* (2008). Preliminary findings showed that *Lepidium meyenii* (Maca) reduces psychological symptoms, including anxiety and depression, and lowers measures of sexual dysfunction in postmenopausal women independent of estrogenic and androgenic activity. More recently, Zenico *et al.* (2009) studied the effects of *Lepidium meyenii* extract on well-being and sexual performance in patients with mild erectile dysfunction. Their data support a small but significant effect of Maca supplementation on subjective perception of general and sexual well-being in adult patients with mild ED.

Another admirer of Maca is Dr. Garry F. Gordon, President of the International College of Advanced Longevity Medicine in Chicago, Illinois. He said: "Using maca myself, I experienced a significant improvement in erectile tissue response. I call it nature's answer to Viagra". (*http://www.macaroot.com/science/curative.html*).

Additional Benefits: It helps in menopausal symptoms: hot flashes, tender breasts, sleeplessness, emotional upsets, and vaginal dryness. Beneficial in osteoporosis: significant bone rebuilding, improvement in bone density. Maca is believed to be an energy booster that balances the endocrine system such as thyroid, pituitary and adrenal glands. Dr Jorge Aguila Calderon, Dean of the Faculty of Human Medicine at the National University of Federico Villareal in Lima, prescribes Maca for a wide variety of conditions, including osteoporosis and the healing of bone fractures in the very elderly.

Dosage: In herbal medicine in the U.S., dried Maca root tablets, capsules and powders are generally recommended at dosages of 5-20 grams daily. The dried root powder can be stirred into juice, water, or smoothies.

Toxicity: Not known, it can be toxic due to the presence of alkaloids.

Contraindication: Large amounts may cause intestinal gas.

Side Effets: Glucosinolates present in Maca can cause goiter if taken in excess combined with a low-iodine diet.

Drug Interaction: At the time of writing this book no significant drug interaction was found in the literature.

Herbal Interaction: Not known.

Food Interaction: No record was found.

Future Prospects: Maca's traditional aphrodisiac and fertility-enhancing myth is now being turning into reality by modern scientific research and encouraging results are coming to light. Maca has a bright future. More research is needed to confirm the contradictory clinical studies as described above.

> *"All my life through, the new sights of Nature made me rejoice like a child."*
>
> —*Marie Curie*

Mandragora officinarum (Mandrake Root)

Product(s): Roots only.

Common Name(s): Love Apple, Satan's Apple, Mandragora, Autumn Mandrake, Ladykins, Manikin, Raccoon and Bryony Root. Crazy Apple, Dudaim, Herb of Circe, Majnoon, Pome Di Tchin, Sorcerer's Root, True Mandrake, Witch's Manikin.

Latin Name: *Mandragora officinarum.*

Synonym(s): *Atropa mandragora, Mandragora vernalis, M. Acaulis.*

Family: Solanaceae.

Geographical Distribution: It is a native of Southern Europe, especially around Mediterranean region of Greece and Rome.

Habitat: It grows in uncultivated fields and stony wastelands. It cannot survive severe winter. It likes a light, deep soil, as the roots grow far down. Roots grow poorly in a soil that is chalky or excessively gravel and rot in winter if the soil is wet. It is propagated from seeds which should be sown in deep flats or, better, singly in pots.

Part(s) Used: Root.

Description of the Plant: Leaves are big almost 12 inches by 6 inches. Flowers are purple bell-shaped. Fruits are small, orange colored fleshy berries with a strong apple like smell. Thus, it is called Satan's Apples. Large brown roots running 3 to 4 ft into the ground. The root is thick and forked like two legs resembling the human body. Beneath a crown of leave, berries spread out like an American Indian headdress, The root is depicted as having legs, arms, knees and breast, a face and long brown hair for female and male represented a long beard.. The fingers of the toes are long, fibrous and root like. It was considered there was a male and female mandrake. It is the oldest herb surviving illustrated botanical work written by Dioscorides in the first century (Bare 2004).

Chemical Constituents: The principle active components of mandrake are tropane alkaloids such as scopolamine (hyoscine), atropine, hyoscyamine, and mandragorine (Thomson 1968). The concentrations of alkaloids vary 0.2 to 0.6% in dry roots. These tropanes are all parasympathetic depressants. The chemistry of mandrake is similar to that of *Datura* species and is exceptionally rich in mandragorine, a powerful narcotic and hypnotic (Carter 1996).

Aphrodisiac Myth: The name *Mandragora* is derived from two Greek words indicating hurtful to cattle. There are 6 species of the genus *Mandragora;* it is *M. officinarum* of Europe and the Near East that has played the most important role as a hallucinogen in magic and witchcraft. Several members of the Nightshade family were favorite tools of medieval witches of Europe. Among these plants, *Hyoscyamus Niger*, *Atropa belladonn, Datura stramonium* and *Mandragora officinarum* have long history of use as hallucinogens and magic plants linked with alchemy, witchcraft, and superstition.

The most interesting and well-known of all the aphrodisiacs of the ancient world is mandrake. The significance of the plant as an aphrodisiac is supported in Genesis 30:14-17. Some aphrodisiacs have been popular since antiquity and mandrake is one of them. The mandrake plant is mentioned in the *Old Testament* (Fleisher and Fleisher 1994).

The roots of mandrake bear a resemblance to the human form. The female form is the most desired one. The female ones are called woman drakes and are shaped like the body of a female. The male, on the other hand, has a single root and shape of a man. It was the female form that was carved into mannequins in the Middle Ages (in Germany and France). It was believed that they brought good luck and wealth.

The root of the mandrake resembles a phallus, and for this reason was believed to have occult powers. In some areas of Europe, possession of the root was punishable by death (Mendelson and Mello 1986).

A mandrake root that was soaked every Friday in a bottle of wine and carried in a charm bag made of silk and velvet would give its possessor great sexual power and make him or her attractive to the other. Once in family's possession, the magic root became a familiar spirit and was enshrined on the mantle over the fireplace. If properly addressed the root would speak "in Oracles" and bring good luck to the members of the household. (Vandaveer 2003) A mandrake root placed underneath a bed pillow would excite passion between two lovers. Both male and female fertility was promoted by eating mandrake roots also.

There were many superstitions about *Mandrake*. A legend from second-century Rome warns that whoever pulls a mandrake out of the ground would suffer dire consequences. The demon inhabiting the root would be aroused and the sounds of its piercing shrieks of agony would be so horrible that the harvester would die on the spot. But if one drew three circles around the plant with the tip of a willow and tied a black thread from the plant to the collar of a white dog, one would be safe from the demon's spell when the *Mandrake* root was pulled from the ground.

The herbalists began to doubt many of these superstitions about *Mandrake* during the sixteen century. In 1526, the English herbalist Turner had denied that all Mandrake roots had a human form and protested against the beliefs connected with its folk tale. Another English herbalist, Gerard, for example, wrote in 1597: "*All which dreams and old wives tales you shall henceforth cast out of your books and memory; knowing this, that they are all and every part of them false and most untrue. I myself and my servants also have dig up, planted and replanted very many . . . ."* But many superstitions surrounding Mandrake persisted in European folklore even into the nineteenth century. However, the herb is legally banned in Finland, Norway, and Netherlands.

Scientific Evidence: Survey of the literature does not reveal any evidence about its use as an aphrodisiac. It is all myth.

Additional Benefits: In folkloric medicine it is also used as an anaesthetic, in chronic rheumatism and in scrofulous tumors (Grieve 2003).

Dosage: It is very poisonous; 30 grain of powdered root is recommended, macerated overnight in red wine. It is extremely toxic and causes insanity. Fresh tincture extracts are also prepared.

Toxicity: It can be very toxic owing to its alkaloidal contents. Root contains hyoscine a powerful alkaloid with the ability to cause hallucinations, delirium and, in larger doses, coma.

Contraindication: Do not take during pregnancy and breast-feeding.

Side Effects: Mandrake causes extreme sleepiness, seizures, hallucinations, breathing difficulty, heart stoppage, and death.

Drug Interaction: Mandragora interacts with anti allergic drugs, antidepressants, drugs used in heart rhythm disorders, motion sickness, Parkinson's disease, and psychiatric illnesses.

Herbal Interaction: Herbal products containing tropane alkaloid such as Datura, Henbane etc should not be taken. Valerian, Damiana, Withania, and St. John's Wort should also be avoided.

Food Interaction: Avoid the use of alcohol.

Future Prospects: Survey of the literature does not show any scientifc data in support of its aphrodisiac effect. It will remain as myth. Mandrake contains extremely toxic constituents and its use as an aphrodisiac is very risky.

> *"Study nature, love nature and stay close to nature. It will never fail you."*
> — *Frank Lloyd Wright*

Mucuna pruriens (Velvet Beans)

Product(s): Avenavin, Kama Raja, Kama Rani, Nirvana and Ziozpher.
Common Name(s): Cow-itch, Cowhage, Velvet Bean, Kapikachu, Kawanen, Atmagupt-Naik, Sea Beans.
Latin Name: *Mucuna pruriens.*
Family: Fabaceae (Leguminosae).
Synonym(s): *Dolichos pruriens, Stizolobium pruriens.*
Geographical Distribution: It is found in South America, Africa, India, Pacific Islands, Caribbean, and West Indies.
Habitat: It prefers moist soil. It grows in grasslands, bushland, and river forest in tropical and subtropical region.

Part(s) Used: Seeds, legumes and roots of the plants are used.

Plant Description: The plant is an annual shrub with long vines that can grow up to 10-15 meter height. Leaves are alternate with three rhomboid-ovate leaflets. Terminal leaflets are smaller than the lateral. Large, dark purple flowers are in drooping racemes inflorescence. Fruits are curved pod with 4-6 seeds. Pods are 3 inches long and are densely covered with loose pale brown trichomes that cause itching and blisters to skin. Seeds are black in color, oblong, variegated, and show distinct white hilum.

Chemical Constituents: 3:4 dihydroxyphenylalanine was isolated from the seeds of *Mucuna prureiens* by Damodaran and Ramaswamy (1937). Goshal *et al.* (1971) reviewed the chemistry and pharmacology of *Macuna pruriens*. Four indole-3-alkylamines, viz., N, N-dimethyltryptamine, its N_b-oxide, bufo-tenine, and 5-methoxy-N, N-dimethyltryptamine along with two uncharacterized 5-oxy-indole-3-alkylamines and a β-carboline were isolated from the various parts, except the trichomes of pods, of *Mucuna pruriens* DC. The trichomes of pods afforded only serotonin. Besides these, choline was found in all parts of the plant. The seeds contain L-dopa in fairly significant amount as found by Modi *et al.* (2008), besides compounds such as glycosides, nicotine, prurenine, prurenidine, mucunine, mucuadinine, β-sitosterol, venolic and gallic acid. The trichomes of the pods comprise serotonin and mucunine. Recently, Misra and Wagner (2004) reported the presence of four tetrahydroisoquinoline alkaloids from the seeds of *Mucuna pruriens*. Their structures were confirmed by spectroscopic analysis. Two of them were found as novel compounds. More recently, Misra and Wagner (2007) developed extraction techniques using different solvents for the bioactive principles from *Mucuna pruriens* seeds.

Aphrodisiac Myth: Mucuna seeds have been used as an aphrodisiac for a long time in the Indian system of Ayurvedic Medicine. The plant has been cited in ancient Indian text such as *Charaka Samhita* and the *Susruta Samhita*. It is still used to increase low libido in men and women.

Scientific Evidence: Amine *et al.* (1996) found some interesting results of *Mucuna pruriens* by experimenting on rats. They found it stimulates the sexual activity of rats. *Mucuna* contain L-dopamine that definitely works as neurotransmitters to stimulate low libido as shown by Giuliano and Allard (2001a, b). Shukla *et al.* (2008) found that *Mucuna pruriens* improved male fertility by its action on the hypothalamus-pituitary-gonadal axis. Seventy-five normal healthy fertile men (controls) and 75 men undergoing infertility screening were selected. High-performance liquid chromatography assay for quantitation of dopa, adrenaline, and noradrenaline in seminal plasma and blood was used. Treatment with *M. pruriens* significantly improved LH, dopamine, adrenaline, and noradrenaline levels in infertile men and reduced levels of FSH and PRL. Sperm count and motility were significantly recovered in infertile men after treatment. Treatment with *M. pruriens* regulated steroidogenesis and improved semen quality in infertile men.

Additional Benefits: It is beneficial in Parkinsonism due to the presence of dopamine (Manyam *et al.* 2004; Modi *et al.* 2008). It acts as an anthelmintic. It also possesses antioxidant and hypoglycemic properties. The protective effect of *Mucuna pruriens* seeds against snake venom poisoning of Asiatic cobra has been recently shown by Tan et al. (2009).

Dosage: 1-2 grams seed powdered in the form of tablets or capsules. 100 mg of capsule comprising 60% of L-dopa, 1-2 capsules per day.

Contraindication: The seeds may cause birth defects, thus these are not recommended during pregnancy.

Toxicity: It can be toxic due to the presence of alkaloida contents. Person suffering from Parkinson's disease should only use under the supervision of his physician.

Side Effects: High doses can cause overstimulation, restlessness, and insomnia.

Drugs Interaction: It possesses synergistic effect with l-dopa and antidiabetic medicines.

Herbal Interaction: Avoid herbs comprising alkaloids such as Yohimbe, Mandrake, and Henbane. It has hypoglycemic effect so use with caution along antidiabetic herbs (Appendis E).

Food Interaction: Wine and alcohol can be worse.

Future Prospects: The plant contains interesting alkaloids and further clinical studies are recommended to investigate its potential as an aphrodisiac. It contains dopamine that is one of the important neurochemicals to stimulate low libido.

Muira puama (Potency wood)
(Sex Stimulant of Brazil)

Product(s): Intim X, Cobra, Chinese Virility pills, VF-RX, Sextiva A, Male Boost, Viacyn and Erect Pills. These preparations are sold in the form of pills, tablets, capsules and extracts.

Common Name(s): Potency Wood, Marapuama, Muirata, Muiratum, Pau-homen, and Brazilian Jungle Passion.

Latin Name: *Ptychopetalum olacoides.*

Synonym(s): *Liriosma ovata.*

Family: Oleaceae or Olacaceae.

Geographical Distribution: Brazil, Amazon rainforest.

Habitat: Forests, prefers sandy loamy soil.

Part(s) Used: The bark and roots are used for erectile dysfunction.

Description of the Plant: *Muira puama* is a small tree that grows to 5-8 meter high. The small white flowers smell like jasmine. The roots and bark are slightly pink.

Chemical Constituents: *Muira puama* contains, alkaloids muirapuamine, phlobaphene, alpha-resinic acid, beta-resinic acid, a mixture of esters including behenic acid, lupeol and sterols, as well as tannin, volatile oils, monoterpenes, triterpenoids, coumarins, fatty acids and trace elements chromium (Steinmetz 1971; Duke 1985; Bucek *et al.*1987). Very recently, four new clerodane-type diterpenoids, ptychonolide, 20-O-methylptychonal acetal and an equilibrium mixture of ptychonal hemiacetal and ptychonal were isolated from the MeOH extract of the bark of a Brazilian plant, *Ptychopetalum olacoides* by Tang *et al.* (2008). The structures were established by nuclear magnetic resonance and other spectroscopic techniques.

Aphrodisiac Myth: Amazon natives have been using *Muira puama* as an aphrodisiac since the time immemorial (Bernardes 1984). Tribes in Brazil have used the roots and bark for years treating sexual debility and impotence. It has been used in Brazil since1950 and is still in great demand as a powerful aphrodisiac. *Muira puama* was used as a natural aphrodisiac by South American shamens for generations. They drank a beverage of the plant prior to lovemaking to enhance low libido (Schultes and Raffauf, 1990). Early European herbalist realized the aphrodisiac properties of *Muira puama* and brought it to Europe. It was included in herbal medicine in England (Mowrey 1993, *British Herbal Pharmacopoeia* 1983**).

Scientific Evidence: The clinical studies on impotence with *Muira puama* have been carried out in two human trials in France (Waynberg 1990, 95; Werbach *et al.* 1994). It was revealed that *Muira puama* improved libido and erectile dysfunction. The first study comprised of 262 male patients with lack of sexual desire and erectile dysfunction. 62% of the patients reported that the extract of *Muira puama* had a significant effect. 51% of patients with erectile dysfunction felt that *Muira puama* was useful. The second study (Waynberg1995) indicated psychological benefits of *Muira puama* in 100 men with male sexual asthenia. They found that *Muira puama* could enhance libido in 85% of the test group, increase the frequency of intercourse in 100 % and improve the ability to maintain an erection in 90%. Waynberg and Brewer (2000) investigated the possibility of an alternative to synthetic medication in the treatment of sexual dysfunction in healthy women. The efficacy of unique formulation of *Muira puama* and *Ginkgo biloba* was assessed in 202 healthy women complaining of low sex drive. Various aspects of their sex life were rated before and after one month treatment. Statistically significant improvements occurred in frequency of sexual desire, sexual intercourse and sexual fantasies, ability to reach orgasm and intensity of orgasm. A double blind study is planned to further investigate these findings.

Muira puama is widely used around the world even today in herbal medicine. *Muira puama* has been considered the most effective natural treatment for erectile dysfunctions and libido enhancement in the herbal world (Rowland and Tai 2003). In the United States, *Muira puama* has also been gaining popularity. *Muira puama* is considered to be safe as sex arousal. It has synergistic effect with arginine and yohimbine and increase blood flow to both male and female reproductive organs and thus causes stimulation and sexual response. Due to the side effects of yohimbine, precautions are required.

Additional Benefits: Herbalists are using it for depression, menstrual cramps, PMS, neurasthenia, and central nervous system disorders and an anticholestric agent (Bartram 1995; Schwontkowski 1993; Jayasuriya *et al.* 2005). Piato *et al.* (2008) studied antidepressant properties of *Muira puama* and results were encouraging. Antioxidant properties of *Muira puama* have been shown by Siqueira *et al.* (2007).

Dosage: *Muira puama* tea is taken an hour prior to lovemaking. Tea can be made from tincture extract. It works well in combination with other herbal aphrodisiacs. In combination with ginseng, *Saw palmetto*, yohimbine, ashwagandha, tribulus and maca work well. 1.5 grams therapeutic, if given alone and 50-200 mg in combination with other herbs is the recommended dose. It is recommended to be taken an hour before the sex. *Muira puama* is also used in combination with Catuaba.

Tincture extracts, tablets, and capsules are available. Powdered is capsuled, however, the resin is poorly absorbed in the stomach. *Since the constituents of the plant are not soluble in water, the crude powder may not be effective.*

Precautions: *Muira puama* active constituents are water insoluble. Water decoction and infusion won't work. The constituents are not broken down in the digestive process either; therefore taking a powder bark or root in a capsule or tablet will not be very effective. Tincture extracts are fine. To achieve the beneficial effects of the plant, proper preparation methods must be employed. Alcoholic extracts are specially prepared to dissolve the active constituents (Cherksey 1996).

Toxicity: None reported.

Contraindication: Use is not recommended during pregnancy and lactation.

Side Effects: Insomnia, restlessness, breathing problem, skin hives, and rash have been observed.

Drug Interaction: Caffeine, aspirin, and central nervous system stimulants should be avoided.

Herbal Interaction: Anticoagulant herbs (Appendix D) should be avoided.

Food Interaction: None known.

Future Prospects: The herb seems to be promising for further investigation after the report of Jacques Waynberg from the Institute of Sexology, Paris, France. More clinical trials are required to prove its use as an aphrodisiac.

"What is a weed? A plant whose virtues have not yet been discovered."
—Ralph Waldo Emerson

Panax species (Ginseng)

Ginseng is widely used to strengthen the immune system, and increase strength and vigor. It is considered panacea or elixir of life in the Traditional Chinese System of Medicine and has been used in various ailments such as diabetes, stress, cancer, and sexual dysfunction. The most popular species of ginseng are as follows:

Panax quinquefolius-American Ginseng
Panax ginseng C.A. Meyer-Oriental Ginseng (Korean and Chinese)
Panax notoginseng-Sanchi Ginseng, Tenchi ginseng
Panax japonicus-Japanese Ginseng
Panax pseudoginseng-Himalayan Ginseng
Panax trifolius-Dwarf Ginseng, Groundnut
Panax elegantior-Pearl Ginseng
Panax zingiberensis-Ginger Ginseng
Panax bipinnatifidum-Double cut-leaved Ginseng
Panax vietnamensis-Vietnamese Ginseng
Eleutherococcus senticosus-Siberian Ginseng or Russian Ginseng

Among these species of ginseng, three species such as American, Asian and Siberian ginseng have been used extensively all over the world. Complex terminology and lack of clear distinction between Siberian or Russian ginseng (*Eleutherococcus senticosus*) and Asian (Chinese or Korean) ginseng (*Panax ginseng*) have generated confusion. The genus *Panax* includes other species, such as American ginseng (*P. quinquefolius*) and Japanese ginseng (*P. japonicus*). We will discuss only the most popular three species of ginseng.

American ginseng, *Panax quinquefolium* belongs to *Araliaceae* family. The *Araliaceae* family is subdivided into seventy genera, one of which is Panax or the ginseng family. There are five species of ginseng; two species are native to North America and three species are native to Asia. Of these five species, only two are known for their exceptionally curative properties, *Panax quinquefolius*, known as American ginseng, and *Panax ginseng*, known as Oriental or Asian ginseng. These two species of ginseng look very much alike and have similar chemical compositions as shown in Figure 9. American ginseng is most likely a distant cousin of Asian ginseng, which found its way to North America over the ancient land bridge between Siberia and Alaska.

Both American and Asian ginsengs belong to the genus *Panax* and are similar in their chemical composition. Siberian ginseng (*Eleutherococcus senticosus*) on the other hand, although part of the same plant family *Araliaceae*, is an entirely different plant and does not contain dammarane type ginsenosides, the active ingredients found in both Asian and American ginseng. Siberian ginseng contains eleuthrosides.

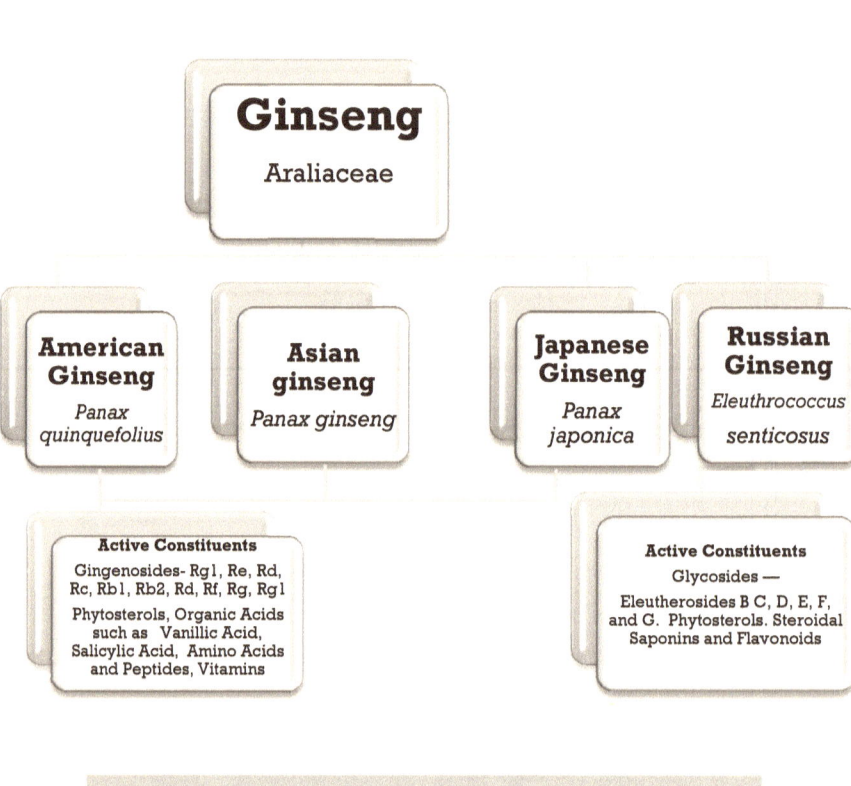

Figure 9: Schematic of Ginseng Family

Panax quinquefolius (American Ginseng)

Product(s): Cobra, Sexstacy, VP-RX, Inferno, Libido fort, VigRX oil, Sextivia A, Enzyte, Varomax.
Common Name(s): American Ginseng, Sang, Redberry, Fivefingers, Divine Root, Root of Life, Manroot.
Latin Name: *Panax quinquefolius.*
Synonym(s)*: Panax quinquefolium, Aralia quinquefolia.*
Family: Araliaceae.
Geographical Distribution: The plant grows wild in shady forests of the northern and central United States, as well as in parts of Canada. It is cultivated in the United States, China, and France.
Habitat: It grows in rich moist soil of hardwood forests.
Part(s) Used: Root is used as an aphrodisiac. The root is branched and forms a shape resembling human form and thus considered as divine root.

Description of the Plant: It grows up to the height of 12-15 inches, bears three leaves at the apex. Each leaf contains five thin stalked leaflets. Three upper one is larger than the lower one. Six to twenty flowers are produced in small clusters during July and August. Fruits are bright crimson berries. The word *Panax* is derived from the Greek word *panakos*, or panacea. American ginseng is considered to possess milder properties than Asian ginseng. It is only used after the roots are at least four years old. Eighty percent of U.S. ginseng is grown in Marathon County, Wisconsin. The commercial harvesting of American ginseng began in Canada in 1716. The environment of Canada closely resembled that of Manchuria. American ginseng was also found growing near Montreal, Canada; thus began a vigorous export of ginseng from Canada to China. Ginseng was discovered growing in the wild in New England, New York, Massachusetts, and Vermont. American ginseng became a lucrative crop while the wild ginseng was almost wiped out along the East Coast due to over harvesting.

Chemical Constituents: Shibata *et al.* (1985) reviewed the chemistry and pharmacology of panax species. American ginseng contains ginsenosides. The type and ratio of ginsenosides are different in American and Asian ginseng. Zhou *et al.* (1998) isolated water-soluble ginsenosides from the water-soluble extract of American Ginseng suggesting that malonyl ginsenosides exist both in American Ginseng and Asian Ginseng. Ten steroidal saponins were isolated from the roots of *P. quinquefolium* by Su *et al.* (2003). These were identified as ginsenosides Rg1, Re, Rd, Rc, Rb1, Rb2,

24(R)-ginsenosides Rg3, 24(R)-pseudoginsenoside RT5, F11, and notoginsenoside K, respectively. Phytosterol such as campesterol, stigmasterol, and clerosterol, and triterpenoid β-amyrin, lupeol were identified in the seed oil of American ginseng by Beveridge *et al.* (2002). Isolation and characterization of ginsenosides in American ginseng leaves and roots were carried out by Ligor *et al.* (2005). Five new triterpenoides floralquinquenosides A, B, C, D, and E were isolated from the flower buds of *Panax quinquefolium* together with 18 known dammarane type terpenoids and 3 flavone glycosides by Nakamura *et al.* (2007) and their structures were determined by physicochemical techniques.

Aphrodisiac Myth: The Cherokee Indians referred to the roots of American ginseng as little man and used the plant for medicinal purposes (Duke 1989). They used the roots of ginseng as tonic to strengthen the whole body. The use of ginseng by the natives in sexual dysfunctions of male and female has also been cited. American ginseng was considered inferior to Korean ginseng. The Chinese began to use American ginseng after it was imported during the 1700s (Bensky *et al.* 1993). The traditional applications of American ginseng in China are significantly different from those for *Panax ginseng* (Asian ginseng). It has been known to fight impotence in men and frigidity in women. American ginseng is recommended for females and the elderly people.

Scientific Evidence: Contradictory studies are reported in the literature. Morris *et al.* (1996) could not find estrogenic effect in their studies. American ginseng (*Panax quinquefolium*) has been shown to enhance libido and arousal in male rats and mice. Results from humans' studies are also promising. A double-blind crossover study by Hong *et al.* (2002) in men with erectile dysfunction revealed that after eight weeks of treatment, the group that was treated with ginseng showed significant improvement in erectile function, sexual desire, and intercourse satisfaction, as compared to the placebo group. American ginseng is mostly cited as an item of commodity. It has been cultivated, harvested, and exported to Asia.

Additional Benefits: Tonic, adaptogen, strengthens immune system, and possesses hypoglycemic and anticancer activity (Wang *et al.* 2006).

Dosage: Standardized extracts of American ginseng should be used. However, dried root powder, 1-2 grams per day in capsule or tablet form are used (Foster 1996).

Toxicity: Hypertonia with muscle twitching and edema. Males may experience gynecomastia (enlargement of breast).

Contraindication: Some of the adverse effects of ginseng include hypertension, nervousness, insomnia, and headache. Though not an absolute contraindication, there is a theoretical risk in using ginseng for patients with certain cardiac problems, so they should check with their primary care physician before starting therapy with ginseng (Drewes *et al.* 2003).

Side Effects: Ginseng causes *Insomnia*. These symptoms have been observed when *caffeine*-containing foods and beverages are consumed along with ginseng. Adverse effects such as hypertension, insomnia, vomiting, headache, and epistaxis have been reported. Isolated case reports have noted post-menopausal vaginal bleeding and breast nodularity with prolonged usage, suggesting a modest estrogen-like effect. A two-year study by Siegel (1979) comprising 133 people using ginseng noted a central nervous system stimulant effect. Nervousness and insomnia were noted in 11% of subjects. Thus ginseng is not recommended for patients with bipolar disorder.

Drug Interaction: Aspirin and warfarin interact with American ginseng. There have been reports that ginseng may decrease the effectiveness of the blood-thinning medication, warfarin. It may inhibit platelet activity and should not be used with aspirin and related drugs (Appendix O). It may stimulate the effects of the anti-psychotic drugs so should be avoided while taking Haloperidol. Ginseng may block the pain killing effects of Morphine. Phenelzine and MAOIs used for depression also interact with ginseng. Avoid use of ginseng with antidiabetics, antipsychotics, and antidepressants (Appendix U; S). Avoid with Central Nervous System stimulants such as amphetamine (Adderall), dextroamphetamine (Dexedrine), methylphenidate (Concerta, Ritalin), phentermine (Adipex-P, Ionamin). Use of estrogens or oral contraceptives may be avoided.

Herbal Interaction: Avoid the ue of Eleuthero, Fenugreek, Ginger, Kudzu, Panax Ginseng, and hypoglycemic herbs (Appendix E).

Food Interaction: Foods and beverages containing caffeine cause synergistic effect.

Future Prospects: Most of New York State has the potential for growing ginseng. It is estimated that ginseng harvesting in New York is over $3 million annually, and another $50 million of out-of-State harvested ginseng is bought and sold by New York dealers each year. Ginseng is an important source of income for many New Yorker's. The cultivation of ginseng is a rapidly expanding business in the State with an unlimited potential. More humans' clinical studies are required to confirm the contradictory findings.

> *"How on earth are you ever going to explain in terms of chemistry and physics so important a biological phenomenon as first love?"*
>
> — *Albert Einstein*

Panax ginseng (Korean Ginseng)

Product(s): *Cobra, VP-Rx, Magna Rx, Boom, Male boost, Enzyte.*

Common Name(s): Korean Ginseng, Ginseng, Asian ginseng, Asiatic Ginger, Chinese Ginseng, Shen Tsao.

Latin Name: *Panax ginseng.*

Synonym(s): *P. schinseng, Aralia ginseng*

Family: *Araliaceae.*

Geographical Distribution: Asian ginseng has been a part of Chinese medicine for over 2,000 years. The first reference to the use of Asian ginseng dates to the 1st century.

Habitat: Asian ginseng commonly grows on mountain slopes in moist and shaded soil and is usually harvested in the fall. The root is used, preferably from plants older than six years of age. The root grows only 0.4 inch per year

Part (s) Used: Root. It is sweet in taste, but aftertaste is bitter.

Description of the Plant: Korean ginseng is a deciduous perennial herb. It grows 4-5 ft in height. The leaves are dark green and oval shaped. It is a compound leaf and consists of five leaflets; the three terminal leaflets are larger than the two lateral ones. The stem is erect and dark red. Fruits are red berries In China, the Korean ginseng roots are called *Jin-chen*, meaning *like a man* as they resemble the same shape of a human body. The ginseng plant can live for over 100 years. *Panax ginseng* was discovered over 7000 years ago in the mountains of Manchuria, China. Initially, it was used as food but soon it was known for its rejuvenating properties.

Chemical Constituents: The active group is the ginsenosides which are saponin glycosides. The total ginsenosides content of a 6 year old root varies between 0.7 and 3% as described by Shibata *et al.* (1985). Thirteen ginsenosides have been identified in Asian ginseng. Two of them, ginsenosides Rg1 and Rb1, have been studied. The other constituents are phytosterols, polysaccharides organic acids such as vanillic acid, salicylic acid, nitrogenous substances, amino acids, peptides, vitamins and enzymes have been isolated and characterized (Hou 1977; Tomodo *et al.* 1993). Six new dammarane-type triterpene saponins from the leaves of *Panax ginseng* were isolated by Duo *et al.* (2001). Korean ginseng contains steroids such as panaxtriol. The steroids are remarkably similar in structure to anabolic steroids found naturally in our body. Volatile constituents of *Panax ginseng* were studied by comprehensive two-dimensional gas chromatography/time-of-flight mass spectrometry by Qui *et al.* (2008). Thirty-six terpenoids were tentatively identified based on the MS library search and retention index in a ginseng sample at the age of 3 years. An obvious group-type separation was obtained in the GC x GC-TOF MS chromatogram. The data collected by GC x GC-FID were processed using a principal component analysis (PCA) method to classify the samples at different ages. The compounds responsible for the significant differentiation among samples were defined. It was found that the relative abundances of alpha-cadinol, alpha-bisabolol, thujopsene, and n-hexadecanoic acid significantly rise with the increase in age.

The number of ginsenoside types contained in Korean ginseng (38 ginsenosides) is substantially more than that of ginsenoside types contained in American ginseng (19 ginsenosides). Furthermore, Korean ginseng has been identified to contain more main non-saponin compounds, phenol compounds, acid polysaccharides and polyethylene compounds than American ginseng and Sanchi ginseng (Choi 2008).

Aphrodisiac Myth: It is called "root of heaven" in the Traditional Chinese System of Medicines and can provide harmony in both ying and yang energy since impotence and frigidity are considered imbalance of yin and yang energy. Early legends believed that consuming ginseng root could lead to eternal life, being an elixir for all ailments. The wild roots of ginseng from Korea still receive exuberant prices when they have humanlike and phallic form. The root of ginseng is combined with deer antlers, deer phalli, tiger bones, and toad meat and snake venom for aphrodisiac activities.

Scientific Evidence: It has been found useful in infertility in men. A double-blind trial with a large group of infertile men found that 4 grams of Asian ginseng per day for three months improved sperm count and sperm motility (Salvati *et al.* 1996). A double-blind trial in Korea found that 1.8 g daily dose of Asian ginseng extract for three months improved libido and an erection in men with erectile dysfunction (Choi *et al.* 1995). This finding was confirmed in another double-blind study, in which 900 mg extract was given three times a day for eight weeks (Hong *et al.* 2002). Korean red ginseng is effective for erectile dysfunction has been shown by Price and Gazewood (2003).

Laboratory studies of *Panax ginseng* have shown, it may improve vascular endothelial abnormalities by increasing the production of NO. Several studies suggest *Panax ginseng* possesses antioxidant and organ-protective action associated with enhanced NO synthesis in the endothelium of the *corpus cavernosum* of the penis (Gillis 1997).

The principal active constituents of *Panax ginseng* ginsenosides have been shown to cause a dose-dependent relaxation of the *corpus cavernosum* smooth muscle in rabbits by increasing release of nitric oxide as suggested by Tamaoki *et al.* (2000). Choi *et al.* (1995, 1998, 1999) demonstrated *Panax ginseng* was superior to placebo for the treatments of ED. Ninety patients were divided into three groups and given *Panax ginseng*, placebo, or trazodone orally. Frequency of intercourse, premature ejaculation, and morning erections after treatment were unchanged in all three groups. However, in the *Panax ginseng*-treated group a significant improvement in erectile parameters such as penile rigidity, girth, duration of erection, improved libido, and patient satisfaction were reported. No changes in serum testosterone were observed. The overall therapeutic efficacy on erectile dysfunction was 60 percent for the *Panax ginseng* group and 30 percent for the trazodone and placebo groups.

A more recent, double-blind, placebo-controlled, crossover study by Hong *et al.* (2002) revealed that *Panax ginseng* is an effective alternative for treating ED. Forty-five men diagnosed with ED were randomized and received either 900 mg *Panax ginseng* or placebo (starch capsule with ginseng flavor) three times daily for eight weeks. The first eight weeks of treatment were followed by a two-week washout period, after which the patients received crossover treatment of placebo or *Panax ginseng* for another eight weeks. Treatment efficacy was determined based on changes observed in indexes of erectile function, including the International Index of Erectile Function (IIEF), RigiScan, serum testosterone levels, and penile duplex ultrasonography with audiovisual sexual stimulation. Mean scores on the IIEF for *Panax ginseng* were significantly higher than for placebo after eight weeks of each treatment. Although the number of patients examined was small, the significant increases in IIEF data have been suggested to represent clinically relevant success. Penile tip rigidity, based on RigiScan parameters, was significantly better after eight weeks of *Panax ginseng* compared to placebo. No significant changes in hemodynamics on duplex ultrasonography with audiovisual stimulation were recorded. No changes in serum testosterone were observed in this or the previously mentioned humans' trial of ginseng for ED.

More recently, the efficacy of the Korean Red Ginseng in the treatment of erectile dysfunction was studied by de Andrade *et al.* (2007). A total of 60 patients presenting mild or mild to moderate erectile dysfunction were enrolled in a double-blind, placebo controlled study in which the effects of Korean Red Ginseng and a placebo were compared. The patients received either 1000 mg, 3 times daily, of ginseng or a placebo. The five-item version of the International Index of Erectile Function score after treatment was significantly higher in the ginseng group. In contrast, there was no difference before and after the treatment in the placebo group. In the ginseng group, 20 patients (66.6 %), reported improved erection, significant in the global efficacy question ($P < 0.01$); in the placebo group this had no significance. Scores on questions 2 (rigidity), 3 (penetration), 4 and 5 (maintenance), were significantly higher for ginseng than for the placebo after 12 weeks of each treatment. The score in the ginseng group was better than the placebo group. Data showed that Korean Red Ginseng can be an effective alternative for treating male erectile dysfunction.

Additional Benefits: It possesses adaptogenic, anticarcinogenic, antioxidative and antiinflamatory properties

Dosage: Standardized herbal extracts, supplies approximately 5-7% ginsenosides (Brown 1996). Standard ginseng root extracts are sometimes recommended at 200-500 mg per day. Non-standardized extracts require a higher intake, generally 1-4 grams per day for tablets or 2-3 ml for dried root tincture three times per day. Ginseng is traditionally used for two to three weeks continuously, followed by a one to two-week rest period before resuming.

Toxicity: Long-term use of ginseng may cause menstrual abnormalities and breast tenderness in some women. Ginseng may produce manic symptoms a case has been reported recently. A special risk situation seems to be effective in patients under antidepressant medication as suggested by Vazquez (2002).

Contraindication: Ginseng is not recommended for pregnant or breast-feeding women. Nervousness and insomnia were noted in 11% of subjects, thus ginseng is not recommended for patients with bipolar disorder.

Side Effects: Used in the recommended amounts, ginseng is generally safe. In rare instances, it may cause over-stimulation and possibly *insomnia* as described by Newall *et al*. (1996). People with uncontrolled *high blood pressure* should use ginseng cautiously.

Drug Interaction: Certain medicines may interact with Asian ginseng such as warfarin and ticlopidine. When it is taken with antiplatelet or anticoagulant drugs, the effect of the drug may be increased, possibly resulting in uncontrolled bleeding. Some drugs used for asthma, heart problems, or other reasons can affect heart rhythm. All central nervous stimulants such as theophylline and related drugs for asthma, albuterol, and clonidine may be avoided with ginseng. *Panax ginseng* can affect the force and rate of heart beat, viagra should be avoided with ginseng.

Herbal Interaction: If *Panax ginseng* is used with other herbs that affect blood clotting, bleeding may occur. Some of the most common herbal products that might inhibit blood clotting are shown in Appendix D. If *Panax ginseng* is taken at the same time as other herbs that also affect the heart, potentially dangerous changes in heart function may result. Some herbal products with heart effects are European Mistletoe, Digitalis, Hawthorn, Motherwort, Pleurisy Root and Squill should be avoided. *Panax ginseng* may decrease blood sugar levels, taking it with other blood sugar-lowering herbal products may result in hypoglycemia-blood sugar that is too low. Herbs that may reduce blood sugar are shown in Appendix E.

Food Interaction: Tea, coffee and caffeine containing products with ginseng increases the risk of over-stimulation and gastrointestinal upset.

Future Prospects: Worldwide consumption of this herb is quite significant. The product needs further clinical investigations since the data is contradictory. Jang *et al.* (2008) reviewed the scientific data available on *Panax ginseng* for ED and concluded that the methodological quality of the primary studies were too low to draw definitive conclusions. Thus more rigorous studies are necessary.

Eleutherococcus senticosus (Siberian Ginseng)

Product(s): Male Boost, Magna Rx, Varomax.
Common Name(s): Siberian Ginseng, Ci wu jia, Touch-me-not, Devil's Shrub, Russian Ginseng.
Latin Name: *Eleutherococcus senticosus.*
Synonym(s): *Acanthopanax senticosus.*
Family: Araliaceae.
Geographical Distribution: Eleuthero is native to the Taiga region of the Far East (southeastern part of Russia, northern China, Korea, and Japan).
Habitat: In Russia, it occurs in forest undergrowth and margins
Part(s) Used: The root and the rhizomes are used.

Plant Description: The plant is deciduous shrub can attain a height of 10-15 ft. Stem is hard and bears 5-7 leaflets. Stem and branches have thorns. Compound leaves 5 leaflets, three are large and two are small. Flowers grow in umbrella-shaped clusters and black berries fruits develop in late summer. Root is brownish, woody, and twisted.

Chemical Constituents: The constituents in eleuthero are eleutherosides (Collisson 1991). Primary eleutherosides have been identified, along with eleutherosides B and E by Farnsworth *et al.* (1985). Eleuthero root also contains complex polysaccharides such as elutherans A, B, C, D, E, F, and G as shown by Hikino *et al.* (1986). *E. senticosus* root contains a number of glucosides, including the glucoside of β-sitosterol, eleutheroside B1, which is a coumarin derivative, and eleutherosides C, D, E, F, and G. Non-glucoside constituents include l-sesamen and syringaresinol. Other ingredients of eleuthero root include saponins, flavonoids, and polysaccharides. At least thirty-five compounds have been identified in the root, and while the constituents of the leaves differ significantly; the leaves are not used medicinally.

Aphrodisiac Myth: Use of Eleuthero for vitality is date back to 2,000 years in China. In the Traditional Chinese Medicine, this herb is considered to improve qi, treat deficiencies of yang in the spleen and the kidney, and bring bodily functions back to normal. Like ginseng, eleuthero has been considered an "adaptogen." In the Soviet Union it was widely available than *Panax ginseng*. In Russia, people in the Siberian Taiga region consumed it to increase performance and quality of life.

Scientific Evidence: No scientific evidence or clinical data was found regarding its evidence as an aphrodisiac or use in any kind of sexual dysfunction in males and females. It has been cited as a tonic to improve the general weakness. That is why it has been used in combination with other herbs. Eleuthero is an *adaptogen* (an agent that helps the body adapt to stress). Eleuthero has been shown to enhance mental acuity and physical endurance. Wagner *et al.* (1994) and Dowling *et al.* (1996) have shown that eleuthero improves the use of oxygen by exercising the muscle. This means that a person is able to maintain aerobic *exercise* longer and recover from workouts more quickly has been confirmed by Kelly (1997); McNaughton (1989) and Asano *et al.* (1986).

Additional Benefits: It has shown a protective action in animal studies against chemicals such as ethanol, sodium barbital, tetanus toxoid, and chemotherapeutic agents. (Collisson 1991) According to a test tube study by Ben-Hur (1981) eleuthero also helps protect the body during radiation exposure. Preliminary research in Russia by Kupin (1986) has suggested that eleuthero may help alleviate side effects and help the bone marrow recover more quickly in people undergoing *chemotherapy* and radiation therapy for *cancer*. Eleuthero may be useful as a preventive measure during the *cold* and *flu* season. Eleuthero may also support the body by helping the liver detoxify harmful toxins. Preliminary evidence also suggests that eleuthero may prove valuable in the long-term management of various diseases of the immune system, including *HIV* infection and *chronic fatigue syndrome* (Bohn 1987).

Dosage: Dried, powdered root and rhizomes, 2-3 grams per day, are commonly used (Brown 1996). Alternatively, 300-400 mg per day of concentrated solid extract standardized on eleutherosides B and E can be used, as can alcohol-based extracts, 8-10 ml in two to three divided dosages. Historically, eleuthero is taken continuously for six to eight weeks, followed by a one-to two-week break before resuming.

Toxicity: Elevated serum digoxin levels in a patient taking digoxin with Siberian ginseng have been found by McRae (1996).

Contraindication: Eleuthero is not recommended for people with uncontrolled high blood pressure. Avoid Eleuthero during pregnancy and breast-feeding.

Side Effects: Reported side effects have been minimal with use of eleuthero (McGuffin *et al.* 1997). Mild, transient diarrhea and vomiting have been reported. Eleuthero may cause insomnia, confusion, high blood pressure, and irregular heart rhythm.

Drug Interaction: Certain medicines such as caffeine, theophylline, aspirin, and warfarin may interact with eleuthero. In one case report, a person taking Eleuthero with digoxin developed dangerously high serum digoxin levels (McRae 1996). Although a clear relationship could not be established, it is wise for someone taking digoxin or any prescription drug to seek the advice of a doctor before taking Eleuthero or any other herbal product. Avoid using with antipsychotic, antidepressants (Appendix S), anticonvulsants, and antiallergic drugs. Antifungal drugs such as ketoconazole (Nizoral) and Sporanox and anticancer drugs such as etoposide, paclitaxel, vinblastine, or vincristine, cyclobenzaprine (Flexeril) should be avoided. Drugs for high cholesterol such as lovastatin, fluvoxamine, haloperidol (Haldol) and Oral contraceptives also interact with Siberian ginseng.

Herbal Interaction: Anicoagulant (Appendix D), antidiabetic (Appendix E) and antidepressant herbal products need to be avoided along with the consumption of siberian ginseng.

Food Interaction: Alcohol, pickles, spicy food and bitters should be avoided.

Future Prospects: It can prove to be a good adaptogen to improve the physical endurance of the athletes. As an aphrodisiac it can be used for general weakness to improve the quality of life. It also needs significant clinical data for biological activity.

> *"The Earth laughs in flowers."*
> —*Ralph Waldo Emerson*

Pausinystalia yohimbe (Yohimbe Bark)

Product(s): IntimX, Cobra, Libidoex, VP-RX, Magna Rx, Boom, Viaeyn and Maxoderm, Afrodex, Aphrodyne, Yocon, Yohimex.
Common Name(s): Johimbe, Yohimbehe, Johimbi, Pausinystalia.
Latin Name: *Pausinystalia yohimbe.*
Synonym(s): *Corynanthe yohimbe.*
Family: Rubiaceae.
Geographical Distribution: Native to Southwest Nigeria, Cameroon, Gabon and Congo.
Habitat: it is found in low altitude forests.

Part(s) Used: Bark
Plant Description: A tall evergreen forest tree grows up to 90 ft high with a width of 40 feet. The leaves are 3-5 inches in length, oblong and oval. The seeds are winged.

Chemical Constituents: The plant contains indole alkaloids, presence of yohimbine, isoyohimbine, allo-yohimbine, corynantheine and ajmaline was confirmed by Chan *et al.* (2008) by modern chromatographic and spectroscopy techniques. Yohimbine is the primary active constituent in yohimbe, although similar alkaloids may also play a role. Yohimbine is also found in related trees of the genus *Pausinystalia* as well as Indian snakeroot *Rauwolfia serpentina*, *Aspidosperma quebracho-blanco* and *Alchornea floribunda* (Betz *et al.* 1995; Bruneton 1995; Budavari 1996).

Aphrodisiac Myth: Yohimbe has been used for centuries as a sex booster and rejuvenator. The pure extract is made from the bark of the Yohimbe. It is known as "love bark". Bantu speaking tribes in Africa used yohimbe bark maceration for pagan matrimony as an aphrodisiac and stimulant. It is only used when mating ritual occurs with dancing and drumming. These orgy rituals have been known to continue up to 10-15 days (Bown 1995; Duke; 1997; Grasing *et al.* 1996).

Scientific Evidence: The majority of pharmacological data is on one of its isolated constituents, the indole alkaloid *yohimbine,* rather than on whole bark preparations. Yohimbine blocks alpha-2 adrenergic receptors, part of the sympathetic nervous system. It also dilates blood vessels. Yohimbine has been shown in double-blind trials to help treat men with *erectile dysfunction* by Vogt *et al.* (1997). Some pharmacokinetic studies have been performed in humans (Grasing *et al.* 1996; Owen *et al.* 1987; Reichart 1997) and human clinical studies have investigated its use for erectile dysfunction or male impotence (Morales *et al.* 1987; Reid *et al.* 1987; Riley 1994; Susset *et al.* 1989; Guirguis 1998; Ernest and Pittler 1998). One study by Grasing *et al.* (1996) indicated that lower doses of yohimbine, given to patients who are fasting or eating a low-fat diet, may be more effective in controlling ED.

Additional Benefits: Small doses of Yohimbe increased fat metabolism by blocking alpha-2 adrenoreceptor that stimuled sympathetic nervous system to get more adrenaline (Galitzky *et al.* 1988; Goldberg 1983). Thus helps in obesity. Athletes and fitness promoters may claim fat reduction benefits for yohimbe. These are individual opinions and may not be supported by controlled clinical studies or additional published scientific data on yohimbine.

Dosage: Yohimbe bark extract standardized to the maximum allowable concentration of yohimbine HCL, dose is 500 mg per tablet-3 tablets per day or as recommended by the physician. Yohimbine hydrochloride 5.4 mg three times a day. The most commonly prescribed drugs for impotence are Afrodex, Aphrodyne, Yocon, Yohimex, and Yohydrol.

Toxicity: Use is not recommended in liver and kidney or heart diseases and in chronic inflammation of the sexual organs or prostate glands. An ingested dose of 1.8 grams resulted in unconsciousness with priapism (Duke 1997). Symptoms of toxicity include paralysis, low blood pressure, heart rhythmic abnormalities, heart failure, and death as described by Mcguffin *et al.* (1997). Agranulocytosis associated with yohimbine use has been reported earlier by Siddiqui et al. (1996).

Contraindication: Non-prescription cough and cold remedies often contain pseudoephedrine (PSE) or phenylephrine, drugs that may increase the risk of side effects. Increase blood pressure; produce unpleasant digestive and nervous systems symptoms. Long term use may potentiate MAO inhibitors. People with heart or kidney trouble, psychological disorders, low blood pressure, diabetes, or ulcers, as well as pregnant women and the elderly should not use it.

Side Effects: Dizziness, nervousness, and anxiety. It can also raise blood pressure.

Drug Interaction: The herb interacts with ACE inhibitors (Appendix P) Beta blockers (Appendix Q) and calcium channel blockers (Appendix X) with yohimbine formulations. Diuretics such as Dyazide, furosemide, and hydrochlorothiazide and Alpha blockers should be avoided (Appendix R). Asthma drugs such as albuterol, metaproterenol, clonidine (Catapres) and guanabenz (Wytensin) also interfere with yohimbine. Antipsychotic drugs such as chlorpromazine, fluphenazine, and prochlorperazine also interact with Yohimbine.

Herbal Interaction: Ephedra and other herbal products that raise blood pressure as shown in Appendix H should be avoided.

Food Interaction: It should not be taken with the amino acid tyramine found in cheese, liver, red wine, and various medications or alcohol. Moreover, it causes an allergic reaction in some individuals. Caffeinated beverages such as coffee, soft drinks, and tea should not be consumed when taking yohimbe or yohimbine. Caffeine enhances the central nervous system stimulation effect of yohimbe and yohimbine. The combination may cause excessive nervousness and irritability. Yohimbe and yohimbine may stimulate the effects of alcohol, leading to a feeling of intoxication with only a small amount of alcohol ingestion.

Future Prospects: The Commission E of Germany did not find adequate documents to support its use for sexual disorders, hence categorized it as an unapproved herb. Contradictory studies are available. It needs systematic clinical studies to prove as a viable aphrodisiac. Though yohimbe bark and its formulations are freely available in the United States by mail order, it should be used with caution and with the consultation of a physician on account of its toxicity.

> *"To be overcome by the fragrance of flowers is a delectable form of defeat."*
> —*Beverley Nichole*

Piper methysticum (Kava-Kava)

Product(s): Capsules by Arcadia, root in powdered form, tablets and extracts of roots.
Common Name(s): Ava, Kava-Kava, Intoxicating Long Pepper, Ava Pepper Shrub, Wati, Yogona and Waka.
Latin Name: *Piper methysticum.*
Synonym(s) *Piper cubeba.*
Family: Piperaceae.
Geographical Distribution: Polynesia, and South Sea Islands.
Habitat: It grows well in forests, cool, moist highland 1000 ft above sea level.

Part(s) Used: Root of the plant is used. Root has a pleasant, lilac odor, and a slightly pungent and astringent taste

Description of the Plant: This is a shrub about 6 feet high, a native of South Sea Islands. Captain James Cook, discovered it in 1769, in the Tahiti Islands and J.G.A Foster, a botanist on Cook's ship gave the name as "Intoxicating Long Pepper." The leaves are alternate, cordate, with a wavy, entire margin, and acute apex. The flowers are small, apetalous, and arranged on slender spikes. Those bearing male flowers are axillary and solitary. The female spikes are numerous. There are 200 varieties of the kava-kava plant in the Pacific, each differing in the potency of psychoactive components as mentioned by Norton (1998). Hawaiian species are the most potent.

Chemical Constituents: It contains kavahin or methysticin. The active principle of kava-kava consists of an acrid resin that was differentiated into *alpha-resin and beta-resin*. Alpha form is more active than beta. Seventeen kava lactones have been isolated and chemically characterized by *Dharmaratne et al.* (2002). Novel compounds such as Pinostrobin flavokawain B and 5, 7-dimethoxyflavanon were isolated and characterized by Wu *et al.* (2002). Six lactones are considered major constituents: (+)-methysticin, (+)-kavain, 7,8-dihydro-(+)-kavain (marindinine), yangonin, 7,8-dihydro-(+)-methysticin and desmethoxyyangonin (Chienthavorn 2005). Piperidine alkaloids kavain and methystine have been isolated *by* Dragull *et al.* (2003). The chemical composition of *Piper methysticum* lactones and related compounds obtained following the sonication of ground kava roots extracted in the solvents hexane, chloroform, acetone, ethanol, methanol and water, respectively, was analyzed. Eighteen kava lactones, cinnamic acid bornyl ester and 5,7-dimethoxy-flavanone, known to be present in kava roots, were identified, and seven compounds, including 2,5,8-trimethyl-1-naphthol, 5-methyl-1-phenylhexen-3-yn-5-ol, 8,11-octadecadienoic acid-methyl ester,

5,7-(OH)(2)-4'-one-6,8-dimethylflavanone, pinostrobin chalcone and 7-dimethoxyflavanone-5-hydroxy-4', were identified for the first time by Xuan *et al.* (2008). Glutathione (26.3 mg/g) was found in the water extract. Dihydro-5, 6-dehydrokavain (DDK) was present at a higher level than methysticin and desmethoxy-yangonin, indicating that DDK is also a major constituent of kava roots. Acetone was the most effective solvent in terms of maximum yield and types of kava lactones isolated, followed by water and chloroform, whereas hexane, methanol, and ethanol were less effective as solvents. Total phenolic and antioxidant activity varied among the extracting solvents, with acetone and chloroform producing the highest effects, followed by water, while methanol, ethanol, and hexane were less effective,

Aphrodisiac Myth: Kava-Kava, the term Kava means bitter or sour or sharp in South Pacific societies. Kava can be traced back in 1886. It is still popular in the US for its relaxing and euphoric effect. More recently, it has also gained popularity with the natives of Hawaii, Australia, and New Guinea. The natives of these islands prepare a drink in water and coconut milk (Whiton *et al.* 2003). This drink is made by converting the root of the plant to a pulp, macerating a short time in water and coconut milk, and then straining it through "fow," a fibrous material obtained from the bark of a certain native tree. The leaf is chewed with the betel-nut, and the dried root, under the name *pipula moola* in India. For centuries, it has been used as relaxant to the body and mind at weddings and other festivities. During South Pacific welcoming ceremonies, such dignitaries as Queen Elizabeth II, Pope John Paul II, Lyndon, and Lady Bird Johnson, and Hillary Rodham Clinton all drank Kava-Kava (Reader's Digest, 1999). Today, Kava-Kava is called the official drink of the Pacific and there are kava bars in Fiji and Polynesia.

Scientific Evidence: Aphrodisiac use is psychotic. If there is a problem due to stress, Kava-Kava may help. The pharmacological effects of Kava-Kava come from root's active constituent kava lactones. Electroencephalogram studies of anxious subjects have indicated that kava-kava is as effective as the benzodiazepines (Keville and Korn, 1996). In addition, it produces anxiolytic changes similar to those seen with diazepam as observed by Gessner *et al.* (1994) and Linderberg (1990). In a placebo-controlled, double-blind study by Lehmann *et al.* (1996), one group of 29 patients received three daily doses of 100 mg of kava extract WS 1490, while a second group received a placebo. The research showed that kava-kava significantly reduced anxiety syndromes not caused by mental disorders. Kava-Kava is an aphrodisiac herb and helps to remember lucid dreams (Chevallier 1996). It is macerated in the oil and used as a lubricant by the natives in their ritual ceremony. The natives used its infusion as an intoxicating beverage for a considerable length of time. It has been used for traditional ceremonies for at least three thousands years. The drink causes euphoria and they claim the ability to attain higher consciousness and communicate with God. The Kava-Kava roots used as aphrodisiacs are of least twenty years old. The roots are minced and mashed in water or fermented to make alcoholic beverages (Head *et al.* 1998; Mayell 1998). It induces hallucination and stimulates sex. As an aphrodisiac, tea is taken before having sex; large doses may induce sleep. It does not alter mental clarity, nor does it interfere with reaction times, alertness, or other cognitive abilities. In fact, there is no clinical or scientific evidence has been reported for its aphrodisiac use.

Additional Benefits: Kava-Kava relieves pain, acts as muscle relaxant, and prevents seizures and CNS disorder (Kumar 2006). It is also effective in *acute vaginitis* or *urethritis* or leucorrhoea. It is a remedy for nocturnal incontinence of urine in the young and old.

Dosage: The adult daily dosage of Kava-Kava extract (standardized to contain 30% kava lactones) is 60 mg/day to 200 mg/day in either divided doses for general anxiety or a single dose before sleep for insomnia (Bone 1993/1994).

Toxicity: Acute hepatitis and liver failure have been reported (Humberston *et al.* 2003; Stickel 2003; MacGregor *et al.* 1989).

Contraindication: Germany's Commission E acts as an adviser to the German equivalent of the U.S. Food and Drug Administration. This group sets the standard for approval of all herbs sold in German pharmacies. According to their monograph, Kava-Kava should not be used during pregnancy, by nursing mothers or by those with endogenous depression (Newell *et al.* 1996). Its use is contraindicated in Parkinson's disease, where it may make symptoms worse. Since it acts as a sedative, it should not be taken when driving or operating machinery.

Side Effects: In recent years several reports in the USA and abroad have linked kava ingestion with liver problems such as hepatitis, cirrhosis, and liver failure. There have been 68 cases of liver problems thought to be linked to Kava-Kava, including 6 cases that resulted in liver transplantation. Regulatory agencies in Germany, Switzerland, France, Australia, Canada, and United Kingdom warned consumers about the potential risk associated with Kava use. In the USA, the FDA issued an advisory in March 2002 the potential risk of liver failure with the consumption of kava containing products. Prolong use can cause dry, scaly, cracked, and ulcerated skin, and vision becomes more or less obscured. Leprous ulcerations may be produced by its habitual use.

Drug Interaction: It should not be used with alcohol, sedatives/hypnotics, barbiturates, antidepressants, tranquilizers (including buspirone Buspar) or other related substances that act on the central nervous system. One case report in the United States describes kava-kava's adverse interaction with alprazolam (Xanax) in which the herb potentiated the effects of the drug (Almeida and Grimsley, 1996). The potential for this increased effect of alprazolam when taken with kava-kava may be due to the sensitivity of the hepatic P450 system, particularly 3A4, to both herb and synthesized pharmaceutical (Smith 1998). Kava-Kava may also cause excessive drowsiness when used with antipsychotics, sedating antihistamines, muscle relaxants such as carisoprodol (Soma) or cyclobenzaprine (Flexeril), and narcotic pain relievers such as codeine or hydrocodone. Anecdotally, it is believed that it can be used with valerian and is often sold as a combination product. There has been one report that kava interact with levodopa and reduces its activity. Avoid the use of Kava if a person is on such medications.

Herbal Interaction: Avoid the use of Kava with Borage, Comfrey, DHEA, Pennyroyal Oil, Skullcap, Uva Ursi, Valerian, St Johns Wort, and Ginseng.

Food Interaction: Alcohol will increase sedation with Kava. Food that contains tyramine or tryptophan may cause high blood pressure. Foods that may react with MAO-1's agents such as avacados, banana, beans, caffein, caviar, champagne, cheese, chocolate, wine, yeast extracts, and yogurt may be avoided.

Future Prospects: With all the restrictions and toxicity effects on liver, it has very little future unless this issue is cleared. There are contradictory reports on liver damages.

> *"Knowing tree, I understand the meaning of patience. Knowing grass, I can appreciate persistence."*
>
> *—Hal Borland*

Polygonum multiflorum (Fo-TI)

Product(s): *Fo Ti Capsules (Arcadia) Fo Ti extracts (Calibex), Granules (Bio Essence).*
Common Name(s): He-Shou-Wu, Longevity Herb, Chinese Knotweed, Climbing Knotweed.
Latin Name: *Polygonum multiflorum.*
Synonym(s): *Fallopia multiflora.*
Family: Polygonaceae.
Geographical Distribution: China, Japan and Taiwan.
Habitat: It grows in woods, along the bank of the stream and in valley. It prefers light, sandy loamy soil. It requires moist soil. It can go in semi shade or sun.
Part(s) Used: Roots.
Description of the Plant: It is a perennial climber growing 4-5m. It has spear-head shaped leaves. The plant blossoms in September to October. Flowers are white in color and hermaphrodite. The root is tuberous.

Chemical Constituents: The major constituents of fo-ti are polygonimitins, chrysophanol, emodin, anthraquinone, phospholipids, tannins, and tetrahydroxystilbene glucoside. Polygoacetophenoside a new acetophenone glucoside was isolated from *Polygonum multiflorum* together with quercetin 3-O-galactoside and arabinoside. The structure of the new glucoside was deduced to be 2, 3, 4, 6-tetrahydroxyacetophenone 3-O-beta-D-glucoside based on chemical and spectral data by Yoshzaki *et al.* (1987). Antioxidant has been isolated by Yong *et al.* (1999). Recently, nine compounds including emodin, rhein, gallic acid, and an unknown glycoside have been isolated by high-speed counter-current chromatography from the roots of Fo Ti by Yao *et al.* (2006). A new stilbenoid was isolated from the root extract of *Polygonum multiflorum* together with eight known constituents by Kim *et al.* (2008). The chemical structure of unknown was established as the 6"-O-monogalloyl ester of (E)-2,3,4',5-beta-tetrahydroxystilbene-2-beta-D-glucopyranoside based on physicochemical and spectroscopic analyses, particularly by NMR spectroscopic data, i.e., COSY, HMQC and HMBC.

Aphrodisiac Myth: The Chinese common name for Fo-Ti, He-Shou-Wu, was the name of a Tang dynasty man who was cured for infertility by this herb. Since then, it has been used to treat premature aging, weakness, vaginal discharges, numerous infectious diseases, *angina pectoris*, and *erectile dysfunction* in the Traditional Chinese System of Medicine (Foster and Yu 1992).

Scientific Evidence: Survey of literature did not reveal any evidence on its aphrodisiac property except its estrogen-like effects in a recombinant cell bioassay as shown by Oerter *et al.* (2003).

Additional Benefits: The processed root has been used to lower cholesterol levels. Animal studies show that it helps to prevent *atherosclerosis*. Fo-ti stimulates *immune function*, increases red blood cell formation, and exerts an antibacterial action in vitro (Foster 1996). All these findings lack humans' clinical trials. The anthraquinone extract of *Polygonum multiflorum* was studied in vivo for myocardial protective effect and revealed encouraging results (Yim *et al.* 1998).

Dosage: 4-8 grams per day (Bone 1996). A tea can be made from processed roots by boiling 3-5 grams in one cup of water for ten to fifteen minutes. Three cups are suggested each day. Five fo-ti tablets of 500 mg each can be taken three times per day or as prescribed by the physician.

Toxicity: Taking more than 15 grams of processed root powder may cause numbness in the arms or legs. Studies by Cardenas *et al.* (2006) showed product containing *Polygonum multiflorum* caused hepatitis.

Contraindication: Avoid anticoagulant drugs (Appendix O).

Side Effects: The unprocessed roots may cause mild diarrhea (Foster 1996). Some people who are sensitive to fo-ti may develop a skin rash. Photosensitivity in some people has been reported.

Drug Interaction: None known.

Herbal Interaction: Avoid hypoglycemic (Appendix E) and anticoagulant herbs as shown in Appendix D. Avoid valerian, St Johns Wort, damiana and kava-kava and ginseng also.

Food Interaction: Avoid alcohol and fatty food.

Future Prospects: There is no scientific proof for its efficacy as an aphrodisiac or to cure sexual dysfunction. It needs extensive research and scientific data to prove its efficacy as an aphrodisiac. So far, it will stay as myth.

Roupala montana (Bois Bande)

Product(s): Bois Bande Extract, Erectol.
Common Name(s): Bois Bande, Danto Amarillo or Zorillo in Costa Rico, Carne Asada, and Arbol Carne.
Latin Name: *Roupala montana.*
Synoym(s): *Roupala dentata, Hieronyma caribaea.*
Family: Proteaceae.
Geographical Distribution: Trinidad and Dominica, West Indies and Costa Rica. The Country of origin is Grenada. In Brazil it is known as Carne De Vacca, In Costa Ricco, Danto Amarillo, in Dominico as Gimauve.

Habitat: It grows best in soil with good drainage. It likes moist soil and wet forest and low altitude to high with climate temperature of 15^0-25^0 C.

Part(s) Used: Bark of the tree used as an aphrodisiac.

Description of the Plant: It is a slow growing medium size tree that reaches 10-25 meter in height. The gray-brown bark has narrow furrows and leaves are pinnately compound. The number of leaflets is variable even on a singlet branches. Branches and leaves have characteristic odor. Flowers are creamy yellow, blossoms in January to April. Inflorescence is racemes of many flowers. Fruits are flattened capsules that open to release winged seeds from April to July.

Chemical Constituents: Essential oil (Cutler 1998). Very limited information is found about its chemical constituents. Review of the literature does not reveal any information or any clues of its active constituents.

Aphrodisiac Myth: The bark of the tree has been used as an aphrodisiac particularly for erectile dysfunction. Generally, the bark is steeped in water, or a tincture is made in white rum. The extract is also known as macoucherie rum. It is commonly used in a large part of the Caribbean for the tree *Roupala montana* Aubl (Proteaceae). The bark of this tree is famous for its alleged aphrodisiac properties, some say better than Viagra. A strip of the bark is macerated for a week in rum and then filtered. A small glass of the rum extract is recommended every evening. The extract is also known as macoucherie rum. The plant is not only found in Grenada and Trinidad but also in Dominica, where it is known as gimauve or gummier tree, in Costa Rica (danto Amarillo or zorillo) and Brazil (carne-de-vaca). Bois Bande is used to treat erectile dysfunction (ED). It is said that Bois Bande helps people around the world improve their sex lives and appetite.

According to the rumors no other natural herbal remedy has the same amount of success as Bois Bande. It's no wonder over 2 million people around the world use Bois Bande. Bois Bande stimulates sexual appetite in both men and women by increasing blood flow to the sexual organs. Bois Bande can be taken whenever the need arises. *Roupala montana* used in Trinidad and is a documented nervine (Lans 2007).

Scientific Evidence: None reported.

Additional Benefits: Used as tonic, immune system stimulant, in pancreatic and prostate disorder, premenstrual problems and increases blood circulation.

Dosage: One oz water extract. Product is available as liquid, 100g/100ml.

Toxicity: Cases of priapism have been verbally reported.

Contraindication: None reported.

Side Effects: None known.

Drug Interaction: None reported.

Herbal Interaction: None known.

Food Interaction: None reported.

Future Prospects: It is highly recommended for systematic phytochemical investigations since very little information is available on its chemistry. A number of researchers have attempted to isolate the biologically active compounds responsible for inducing erection. It appears that the active substance remains elusive. Survey of the literature could not find any scientific evidence for its use as sex stimulant and for the time being it will remain as myth.

> *"Have a heart that never hardens, and a temper that never tires, and a touch that never hurts."*
> — *Charles Dickens*

Serenoa repens (Saw Palmetto)

Product(s): Cobra, Lust, Chinese Virility Pills, Libidoex, Sextivia A, Viacyn, Vipra.
Common Name(s): Cabbage Palm, Fan Palm, Sabal, Scrub Palm, Serenoa
Latin Name: *Serenoa repens*
Synonym(s): *Sabal serrulata, Sabal fructus.*
Family: Arecaceae, Palmaceae.
Geographical Distribution: South Eastern America. *Saw palmetto* is a native to Florida but also found in the areas of Carolinas to Texas.
Habitat: It germinates from seeds and needs sunlight and well drained soil. It mostly found on sand ridges, flatwood forests, coastal dunes and near marshes.

Part(s) Used: Berries.
Description of the Plant: It is a low palm tree with fanlike, finger like fronds and small berry-shaped fruits. This plant grows about 6-10 feet high and forms *palmetto scrub*. Large leaves form a crown. The flowers are small and white in long panicles. Plant bears deep reddish-brown fruits. The fruit is slightly wrinkled and contains one hard seed. The Seed is pale brown, oval with a hilum near the base (Bennett and Hicklin 1998).

Chemical Constituents: The main constituents include phytosterols such as beta-sitosterol, cycloartenol, stigmasterol, lupeol, lupenone and 24-methylcycloartenol, steroidal saponins, flavonoids, ferulic acid, and volatile oil (Newell *et al.* 1996). The fruits and seeds are rich in triacylglycerol-containing oil. 50% of the fatty acids contain 14 or less carbons such as caproic, caprylic, capric, lauric, palmitic, and oleic acids (Bruneton1995). More recently, Sorenson and Sullivan (2006) have isolated campesterol, stigmasterol, and β-sitosterol from *Saw palmetto.*

Aphrodisiac Myth: Berries from the *saw palmetto* tree were used as early as the 1700s by American Indians to treat enlargement of prostate and erectile dysfunction. Native Americans recognized *saw palmetto* fruits as a food despite its rather unpleasant taste and have been eating them for a long time. According to folk medicine, women have used the herb for breast enlargement and lactation. They also used the berries to treat infertility and painful periods (Duke 1985). Native Americans used it for low libido, impotence, and frigidity (Vogel 1970). Indians of the southern USA were using *saw palmetto* berries for treating testicular atrophy, erectile dysfunction and prostate inflammation in the early 1700s (Hansel *et al.* 1994). It has also been promoted to enhance sexual desire for both men and women; however, no scientific evidence supports any of these uses.

Scientific Evidence: It has a reputation as a sexual stimulant and an aphrodisiac. However, no clinical study seems to be reported on its aphrodisiac properties or use in erectile dysfunction except it has been used in combination with number of herbs. There are plenty of reports available regarding the use of *Saw palmetto* in prostate enlargement. It inhibits androgen and estrogen receptor activity and may be beneficial for both sexes in balancing the hormones.

Additional Benefits: *Saw Palmetto* is an excellent product for treating enlarged prostate, commonly called BPH-benign prostatic hyperplasia (Marks and Tyler 1999; Pytel *et al.* 2002). *Saw Palmetto* inhibits the conversion of testosterone to dihydrotestosteron, the compound that causes prostate cells to multiply excessively (Wilt *et al.* 1998, 2000). Very recently Tacklind *et al.* (2009) reviewed the research data available on *S. palmatto* for the treatment of BPH. He analysed that *S. palmatto* was not more effective than placebo for treatment of urinary symptoms consistent with BPH.

Dosage: As an aphrodisiac 30-60 drops of concentrated extract are recommended in juice. Teas made from *saw palmetto* are of limited value because they do not contain oil-based fatty acids and sterols. Always look for a brand that standardizes the extract in terms of the amount of fatty acid and sterol content. 320 mg daily dose is recommended for prostate problem.

Toxicity: Not known.

Contraindication: Avoid the use of aspirin.

Side Effects: Diarrhea, dizziness, headache, nausea, and upset stomach have been observed.

Drug Interaction: It interacts with antiplatelets such as Plavix, Ticlidand and anticoagulants heparin and warfarin. *Saw palmetto* may interfere with the effectiveness of oral contraceptives and hormone replacement therapy on account of its hormonal activity. The antidiabetic action of the herb may also be enhanced with sulfonamides and antibiotics. Herb is diuretic. It increases the renal excretion of sodium and chloride. The plant may potentiate the hyperglycemic and hyperuricemic effect of glucose-elevating agents. In conjunction with corticosteroids, it can produce hypokalemia.

Herbal Interaction: Avoid the use of anticoagulant herbs as shown in Appendix D.

Food Interaction: None reported.

Future Prospects: *Saw palmetto* has promising future for the treatment or preventing measure of prostate enlargement. However, as an aphrodisiac it needs extensive studies and clinical data to prove its efficacy. For the time being, it will stay as a myth.

Tribulus terrestris (Gokshura)

Product(s): Nirvana, Stamanex, Viacyn, Enzyte, Testrol, Vipra, Erect pills.
Common Name(s): Puncture Vine, Gokshura, Ziozpher and Tribestan.
Latin Name: *Tribulus terrestris*.
Synonym(s): *Tribulus bicornuta, Tribulus orientalis, Tribulus uniflorus*.
Family: Zygophyllaceae.
Geographical Distribution: The herb has worldwide distribution including Asia and the Pacific Domain. *Tribulus terrestris* grows naturally in many parts of the world including the Americas, Australia, Europe, Africa, and the Middle East.

Habitat: It grows well on dry sandy soils. It is found in overgrazed pastures, roadsides, lawn, and neglected areas.
Part(s) Used: Fruits.
Description of the Plant: The genus comprises about 20 species. It is a prostrate vine, generally less than 1ft high, spreading to 5 ft long. It blooms in the fall. Flowers are yellow. It is a branching shrub with distinctive burr fruit with sharp hard spines that grow as weeds in pastures, agricultural land, and road sides. The fruits usually puncture the bicycle tires, thus called "puncture vine". The fruits possess very hard sharp seeds, each one a single-seeded wedge of the intact fruit.

Chemical Constituents: The fruits of *Tribulus terrestris* contain protodioscin, furostanol, glycosides, flavonoids, alkaloids, resins, tannins, sterols, and essential oil (Xu *et al.* 2000; Wu *et al.* 1996; Yan *et al.*1996). Two new furostanol saponins from the fruits of *Tribulus terrestris* L were reported by Xu *et al.* (2009). Three new steroidal saponins were isolated from the fruits of *Tribulus terrestris* by Bedir and Khan (2000). Their structures were assigned by spectroscopic methods (IR, HRESIMS, 1D-and 2D-NMR) Terrestrinins A and B have been isolated by Huang *et al.* (2003) and Sun *et al.* (2003). Protodioscin is considered as one of the active substances within this plant. Analysis of products showed considerable variations of 0.17 to 6.5 % in protodioscin content by Adimoelja and Adaikan (1997). The major constituents include steroidal saponins terrestrosins A, B, C, D, and E, desgalactotigonin, F-gitonin, desglucolanatigonin and gitonin. Hydrolyzed products of saponins include diosgenin, hecogenin, neohecogenin, and neotigogenin (Xu *et al.* 2001). Six Steroidal Glycosides from leaves of *T. alatus* were recently isolated and characterized by Temraz *et al.* (2006). Novel furostanols have been isolated and characterized in *Tribulus terrestris*

(Conrad *et al.* 2004; Kostova *et al.* 2002; De Combarieu *et al.* 2003). The dried foliage contains 44mg/kg of beta carbolines type alkaloids. Among them harmane and norharmane are the major one. One new cinnamic imide derivative, named tribulusimide C , was isolated from the fruits of *Tribulus terrestris*, together with three known compounds, N-p-coumaroyltyramine, terrestriamide, N-trans-caffeoyltyramine by Lv *et al.* (2008). The structure of the noval compound was elucidated based on chemical analysis and spectral methods (IR, 1D and 2D NMR, HR-FAB-MS, EI-MS).

Two new furostanol saponins, tribufurosides B and C were isolated from the fruits of *Tribulus terrestris* by using chemical and spectral analyses, the structures of two new furostanol saponins were established by Xu *et al.* (2008). Five new steroidal saponins were isolated from the fruits of *Tribulus terrestris* by Su *et al.* (2009) and their structures were fully established by spectroscopic and chemical analysis.

Aphrodisiac Myth: For centuries *Tribulus terrestris* has been used in the Indian System of Medicines for genito-urinary problems. It has also been prescribed for impotence in men and infertility in women. Ancient Greeks used it as a tonic. The Chinese System of Medicine also used it for premature ejaculation. In Bulgaria it has been used as a sex booster (Zafar *et al.* 1989).

Scientific Evidence: The active chemical in *T. terrestris* is protodioscin (Adiamoji 1997, 2000), which is considered to increase testosterone level. In a study with mice, *Tribulus terrestris* was shown to enhance mounting activity and erection better than testosterone cypionate (Gauthaman *et al.* 2002). *Tribulus terrestris* is considered to increase testosterone levels indirectly by raising blood levels of the luteinizing hormone (LH). LH is a hormone produced by the pituitary gland and plays a role in regulating natural testosterone production and serum levels. On the other hand, one recent study by Neychev and Mitev (2005) revealed that *T. terrestris* caused no increase in testosterone or LH in young men. The findings in the current study anticipate that *Tribulus terrestris* steroid saponins possess neither direct nor indirect androgen-increasing properties. Further investigations are required to clarify the mode of action of *Tribulus terrestris* steroid saponins. Brown *et al.* (2002) reported that a commercial supplement containing androstenedione and herbal extracts, including *T. terrestris*, was no more effective at raising testosterone levels than androstenedione alone. Whereas Gauthaman *et al.* (2002) concluded that *Tribulus terrestris* extract appears to possess aphrodisiac activity, probably due to its androgen increasing property.

The aphrodisiac and fertility-potentiating action of the plant was studied by Gauthaman *et al* (2003). The furoctanol biglycosides have been active in stimulating spermatogenesis and Sertoli cell activity in rats. Oral application of the saponin terrestrioside-F in male rats increased libido and sexual responses. In female rats, this compound increased fertility. The Study have shown a better than 50% increase in testosterone levels when taking the *Tribulus terrestris* herb. More recently Gauthaman and Ganesan (2008) studied the hormonal effects of *Tribulus terrestris* and its role in the management of male erectile dysfunction using primates, rabbit, and rats. They concluded that the plant increases some of the sex hormones, possibly due to the presence of protodioscin in the extract and may be useful in mild to moderate cases of erectile dysfunction. Adimoelja *et al.* (1997) have proposed that protodioscin extracted from the plant *Tribulus terrestris* improves erectile function via increasing DHEA levels. Clinical studies carried out at The Pharmaceutical Institute in Sofia, Bulgaria on *Tribulus terrestris*, showed improved reproductive functions, including increased sperm production and testosterone levels in men. *Tribulus terrestris* increased the concentration of hormones in women including estradiol, and testosterone; thus improving reproductive functions, libido, and ovulation. A significant benefit of *Tribulus terrestris* is the stimulation of hormone production to a balanced level.

There has been very limited research conducted on the effectiveness of *Tribulus terrestris* in elevating testosterone levels. Some European studies suggest that *Tribulus terrestris* extract can increase testosterone levels 30-50% above baseline levels - but still well within the normal range. These studies also suggest a similar increase in estradiol levels (Bardin 1991). Antonio *et al.* (2000) examined the effects of *Tribulus terrestris* on body composition and exercise performance on 15 resistance-trained males. Subjects received either a placebo or a large dose of *Tribulus terrestris* (1.5mg per pound of body weight per day for 2 months). Results showed no changes in body weight, percentage fat, and total muscle mass or muscle strength related to tribulus supplementation.

Arcasoy *et al.* (1998) reported that *Tribulus terrestris* may work by relaxing smooth muscles and increasing blood flow into the *corpus cavernosa*. The relaxant effect observed is probably due to the increase in the release of nitric oxide from the endothelium and nerve endings. This herb may increase FSH in women, which in turn increases levels of estrogen.

Additional Benefits: The plant carries hypoglycemic activity. Saponins from *Tribulus terrestris* could significantly reduce the level of serum glucose. It could dissolve kidney stones. *Terrestris* saponins could significantly lower the levels of serum total cholesterol and triglycerides. This is also all folkloric information and does not possess any scientific support.

Dosage: *Tribulus terrestris* is most often formulated in combination with other sex-enhancing herbs. *Tribulus terrestris* is also sold by itself often in a dosage ranging from 250 to 750 mg. *Tribulus terrestris* is found in a variety of extract potencies. A typical dosage of 250-1500 mg of tribulus per day is fairly common. Be sure to choose an extract standardized for at least 30-45% steroidal saponins (furostanol).

Toxicity: None well documented.

Contraindication: Avoid in pregnancy and breast feeding.

Side Effects: No side effects in humans so far have been reported from the use of *Tribulus terrestris*. Bourke (1983, 1984) reported that when sheep consumed *Tribulus terrestris* as 80% of their diet, liver damage and other changes occurred. According to

Bourke (1984) neurological disease showing symptoms of weakness of hindlimb was characterized in sheep fed on *Tribulus terrestris* for several months. Sheep studies have suggested the possibility of locomotor (muscle coordination) disturbances or limb paresis following ingestion of *Tribulus* in high quantities.

Drug Interaction: Administration of *Tribulus terrestris* with hormonal agents stimulates ovulation in infertile women.

Herbal Interaction: It should not be taken with herbs that have hypoglycemic effects (Appendix E).

Food Interaction: No record is available.

Future Prospects: The herb is invariably used in combination with other plants. The survey of the literature shows that it has a bright future. Some of the studies are contradictory and need confirmation on its aphrodisiac properties. Toxicity revealed in sheep, can be a guide line to find the appropriate dose.

> *"The mind is everything: what you think, you become."*
> — *Gautama Buddha*

Turnera diffusa (Damiana)

Product(s): IntimX, Potensan.
Common Name(s): Damiana,
Old woman's Broom, Mexican Damiana,
Pastorata, Hierba Del Venado, Oreganello,
Mizibcoc, Bourrique.
Latin Name: *Turnera diffusa and T. aphrodisiaca.*
Synonym(s): *T. microphylla.*
Family: Turneraceae.
Geographical Distribution: Tropical parts of North America, Mexico, and Africa.

Habitat: It prefers hot and humid climates.
Part(s) Used: Aerial parts of the plant are used as aphrodisiacs.
Description of the Plant: Damiana is a small shrub, with pale green leaves and yellow flowers. The small leaf is alternate or in a bunch with serrate margins, and the underside is covered with pale hairs. It has a strong aroma like chamomile, and taste somewhat bitter and resinous. The fruits are small tripartite, slightly curved capsules. The entire bush gives off a characteristic aromatic smell due to the volatile oil present in all parts of the plant.

Chemical Constituents: Volatile oil from Damiana leaves comprises at least 20 constituents. Amongst them are 1, 8-cineole, cymae, p-cymene, alpha-and beta-pinene, thymol, alpha-copaene, caryophylline, caryophylline oxide elemene and calamene are the major components as shown by Alcaraz-Melendez *et al.* (2004). Presence of 5-hydroxy-7', 3', 4' trimethoxy-flavone has been reported in *Turnera diffusa* by Dominguez and Hinojosa (1976). Damiana leaves also contain, flavonoids luteolin, sterols, damianin, the glycosides gonzalitosin, arbutin, and cyanoglycoside tetraphyllin B along with aliphatic hydrocarbons such as hexacosanol, triacontane (Piacente 2002). Tetraphyllin B was isolated and structure was established by Spencer and Seigler (1981) using NMR and MS techniques. More recently phytochemical and pharmacognostical investigation were carried out by Kumar *et al.* (2006). Zhao *et al.* (2007) carried out the detailed phytochemical investigations of *Turnera diffusa* and reported 35 compounds comprised of flavonoids, terpenoids, phenolic, and cynogenic derivatives including five new compounds. The structures were determined by spectroscopic and chemical methods. Zaho *et al.* (2008) isolated two compounds pinocembrin and acacetin from *Turnera diffusa.*

Aphrodisiac Myth: Damiana has been enjoying a great reputation as an aphrodisiac since ancient times. Aphrodisiac effects of Damiana leaves were traced back in scientific literature for over 100 years. Damiana leaf and damiana elixirs were listed in the National Formulary in the United States. Its leaves have been used as an aphrodisiac by the natives of Mexico and Mayan Indians. According to Bradley (1992) the Aztecs used Damiana in their rituals as a sexual stimulant for both men and women.

According to Duke (1985) flower-stage harvested Damiana makes *every nerve tingles with sexual sensation, whetting the appetite of lustful desire.*

In the past, some people have smoked Damiana in a water pipe prior to lovemaking. It was also burned ceremoniously to enable participants to see visions. Damiana has also been called the "lover's herb" in Europe. Damiana is used in men to treat spermatorrhea, premature ejaculation, sexual sluggishness, and prostate complaints. In men, Damiana is believed to cure impotence particularly due to nervous anxiety about sexual performance. It is often used in combination with other herbs to treat impotence (Rowland and Tai 2003; Blumenthal *et al.*1998).

Damiana is mostly used in frigidity-meaning lack of sex interest in female. Damiana is considered to stimulate the release of testosterone. It is not clear how Damiana stimulates testosterone. It is considered to facilitate a feeling of wellness and vitality overall, including sexual energy.

Scientific Evidence: Despite its long history and frequent use in many cultures, scientists have been unable to isolate any active ingredients responsible for domiana's aphrodisiac and hallucinogenic properties. The herb contains a volatile oil that may mildly irritate the genitourinary system could be the cause of aphrodisiac property. Because it carries oxygen and increases circulation to the genital area, it supports both male and female sexuality.

The libido-boosting power of Damiana hasn't been tested in humans, although decoction made from the leaves has long been used as an aphrodisiac in Mexico (Willard 1991). Very few clinical studies are conducted to support the traditional use of the plant for sexual dysfunction and impotence. A study was conducted with the Damiana species *Turnera diffuse* and *Pfaffia paniculata* extracts on the sexual behavior of male rats by Arletti *et al.* (1999). It was reported that sexually potent and sexually sluggish/impotent male rats were treated orally with different amounts of *Turnera diffusa and Pfaffia paniculata* extracts. Extracts had no effect on the copulatory behavior of sexually potent rats. However, both the extracts-singly or in combination-improved the copulatory performance of sexually sluggish/impotent rats. These results seem to support the folk reputation of *Turnera diffusa* and *Pfaffia paniculata* as sexual stimulants. Zaho *et al.* (2008) showed that the extract of *Turnera diffusa* and two isolated compounds pinocembrin and acacetin could significantly suppress aromatase activity. Aromatase is an enzyme which converts androgens to estrogens by a process called aromatizaton.

Most of the Damiana sold in herbal commerce today originates from Mexican and Latin American cultivation project. Damiana is also widely available in most health food and natural product stores in the USA.

Additional Benefits: Used in asthma, cold, and flu situation, and helps digestion and muscular contraction. Damiana works as mood elevator, antidepressant, and nerve tonic (Jiu 1966). Damiana is also believed to balance hormones for menopausal women (Zava *et al.* 1993). The *British Herbal Pharmacopoeia* cites indications for the use of Damiana for "anxiety neurosis" with a predominant sexual factor, depression, nervous dyspepsia, constipation, and coital inadequacy (Newell *et al.*1996).

Dosage: Capsules, tablets and extracts. Damiana is often used in conjunction with other herbs having similar properties, and is often found as an ingredient in herbal mixtures or formulas mostly in combination with Dong Quai, Kava-Kava, Dandelion, Tribulus, Yohimbe, and Wild Yam.

Toxicity: Damiana contains low levels of cyanide-like compounds; exceeding recommended dosage may be dangerous. However, no rigorous scientific studies have examined the effects of long-term use of this herb.

Contraindication: Damiana possesses hypoglycemic effects in animals. Persons suffering from diabetes and hypoglycemia should use this plant with caution under the supervision of their physician. Also avoid the use of Damiana during pregnancy, breast cancer, pshychiatric disorder, Parkinson's disease and Alzheimer' disease.

Side Effects: Damiana appears to be safe when taken occasionally as a sex booster. It has a long history of traditional medicinal use with no harmful consequences reported.

Drug Interaction: It may interfere with insulin and oral drugs for diabetes and synergistic to antidiabetic medications (Appendix U).

Herbal Interaction: Avoid the use of Damiana with herbs such as Eleuthro, Fenugreek, Ginger, Kudzu, and Ginseng. All antidiabetic herbs (Appendix E) need to be avoided. It possesses hypoglycemic activity (Alarcon-Aguilara *et al.* 2002).

Food Interaction: Avoid the use of alcohol.

Future Prospects: It has a great potential provided we find out the reality. There are contradictory reports on Damiana. It will be interesting to investigate the aphrodisiac potential of this plant. More research is required to prove this myth.

> *"Nature is a revelation of God; Art is a revelation of man."*
> — *Henry Wadsworth Longfellow*

Withania somnifera (Ashwagandha)

Product(s): Kama Raja, Kama Rani
Common Name(s): Ashwagandha, Winter Cherry, Calm Button, Indian Ginseng.
Latin Name: *Withania somnifera.*
Synonym(s): *Withania coagulans.*
Family: Solanaceae.
Geographical Distribution: It grows in open and disturbed land in India, Australia, East Asia, and Africa.
Habitat: It prefers sandy loamy soil, rain, sun, and shade.
Part(s) Used: Roots.
Description of the Plant: *Withania somnifera* is an erect, small, evergreen and tomentose shrub. The roots are stout, fleshy, and whitish brown. The leaves are simple ovate and glabrous. The flowers are inconspicuous, greenish, or lurid-yellow, in axillary, umbellate cymes. The fruits are small berries, orange-red when mature,

enclosed in the persistent calyx and have numerous, seeds. Flowering occurs in fall and spring.

Chemical Constituents: Roots of the plant contain steroidal alkaloids with steroidal lactones known as withanolides; the class constitutes withanine, somniferine, cuscohygrine, anahygrine, pseudowithanine and their glycosides sitoindosides Anon (2005). At present 12 alkaloids, 35 withanolides and some sitoindosides have been extracted, isolated, detected, and characterized (Rastogi and Mehrotra 1998). Withanolides is a group of naturally occurring oxygenated ergostane-type steroids generally having a lactone in the side chain and 2-en-1-one system in ring A (Khajuria *et al.* 2004). Alkaloids like tropane, pseudotropine, anaferine, and dl-iso-pelletierine and glycosides sitoindosides are reported by Matsuda *et al.* (2001). A sitoindoside is a withanolide containing a glucose molecule at carbon 27 has been reported by Abou-daou (2002). Pharmacological activity of Ashwagandha has been attributed to two main withanolides, withaferin A and withanolide D. Reverse-phase preparative HPLC analysis of the n-butanol fraction of the methanolic extract of *Withania somnifera* Dunal (leaves) afforded a novel chlorinated withanolide, namely withanolide Z, along with four known withanolides, withanolide B, withanolide A , 27-hydroxywithanolide B and withaferin A. Their structures were elucidated by IR, MS, CD and a combination of 1 D and 2 D NMR spectral analyses by Paramanick *et al.* (2008). Two new and seven known withanolides along with β-sitosterol, stigmasterol, β-sitosterol glucoside, stigmasterol glucoside, and α+ β glucose were isolated from the roots of *Withania somnifera* by Misra *et al.* (2008). Among the known compounds, Viscosa lactone B, stigmasterol, stigmasterol glucoside and β glucose are being reported from the roots of *W. somnifera* for the first time. One of the new compounds contained the rare 16 β-acetoxy-17(20)-ene the other contained unusual 6α-hydroxy-5,7α-epoxy functional groups in the withasteroid skeleton. The structures were elucidated by spectroscopic methods and chemical transformations.

Aphrodisiac Myth: Ashwagandha has been used in the Indian System of Medicine in Ayurveda for the last 5000 years as an aphrodisiac. It is called Indian ginseng (Kulkarni and Dhir 2008). It has been described in the sacred texts of Ayurveda, including the *Charaka Samhita* and *Susruta Samhita.* It is considered to stimulate the reproductive function of both men and women (Brekhman 1980). The root smells like a horse It is from this characteristic odor which its Sanskrit name, "like a horse", derives. *Ashwa* meaning horse in Sanskrit and *Gandha* meaning smell, the distinctive earthy odor of ashwagandha is due to the presence of certain steroidal lactones or withanolides (Nadkarani 1976). Ashwagandha is traditionally used as an aphrodisiac. Its roots, fruits, leaves, and seeds are widely used in the preparation of diuretic, aphrodisiac and adaptogenic agents. Ashwagandha is mentioned in the ancient *Kama Sutra* for stimulating low libido. Ashwagandha is considered to enhance sexual power, prevent impotence, infertility, low sperm count, or seminal debility.

Scientific Evidence: Studies conducted by Iuvone and Esposito (2003) showed that ashwagandha can produce nitric oxide, that is known to help erection of the male organs. *Withania somnifera* purified powder was given 3 g/day for one year to 101 normal healthy male volunteers, age 50-59 years. All subjects showed significantly increased hemoglobin and RBC count, and improvement in hair melanin and seated stature. They also showed decrease in erythrocyte sedimentation rate, and 71.4 percent of the subjects reported improvement in sexual performance. In summary, these studies indicate that *Withania somnifera* may prove useful in younger as well as older populations as a general health tonic (Bone 1996).

In other studies by Ilayperuma (2002) the root extract induced a marked impairment in libido, sexual performance, sexual vigor, and penile erectile dysfunction. These effects were partly reversible on cessation of treatment. These antimasculine effects are not due to changes in testosterone levels or toxicity but may be attributed to hyperprolactinemic, GABAergic, serotonergic or sedative activities of the extract. The investigator concluded that use of *W. somnifera* roots may be detrimental to male sexual competence.

Additional Benefits: Clinical trials and animal research endorse the use of ashwagandha for anxiety, cognitive and neurological disorders, inflammation, and Parkinson's disease. Ashwagandha's chemopreventive properties make it a potentially useful adjunct for patients undergoing radiation and chemotherapy. Ashwagandha is also used therapeutically as an adaptogen for patients with nervous exhaustion, insomnia, and debility due to stress, and as an immune stimulant in patients with low white blood cell counts. Studies indicate ashwagandha possesses anti-inflammatory, antitumor, antistress, antioxidant, immunomodulatory, hemopoietic, and rejuvenating properties. A comprehensive review of the plant *Withania somnifera* has been published by Gupta and Rana (2007). It also appears to exert a positive influence on the endocrine, cardiopulmonary, and central nervous systems. The mechanisms of action for these properties are not fully understood (Mishra *et al.* 2000).

Dosage: Two to three grams of root powder with milk or ghee can be taken as an aphrodisiac to enhance libido.

Toxicity: None known.

Contraindication: Large doses of ashwagandha may possess abortifacient properties. It should not be taken during pregnancy. It is also contra-indicated with sedatives.

Side Effects: Drowsiness.

Drug Interaction: Avoid the use of barbiturates.

Herbal Interaction: Avoid Valeriana, Damiana, St. John's wort, Kava-Kava.

Food Interaction: Avoid the use of alcohol.

Future Prospects: Contradictory studies on this plant regarding its aphrodisiac properties require further confirmation. It has been used in the Ayurvedic System for its aphrodisiac and rejuvenating properties throughout history. Studies conducted by Ilayperuma *et al.* (2002) in the faculty of Medicine at the University of Ruhuna, Sri Lanka, showed that its use may be detrimental to male sexual competence. It needs further confirmation. More clinical trials are required to study its aphrodisiac potential.

"A thing of beauty is a joy for ever: Its loveliness increases; it will never pass into nothingness."
— *John Keats*

Chapter Six B

Truffles as Aphrodisiacs

Truffles are one of the most expensive delicacies in the gastronomic world. Their values are directly proportional to their flavor, taste, and scarcity. The most prized species costing up to $1500-2000 per kg in the US. Their demand is increasing day by day. It has been estimated that the world market could absorb 50 times more truffles than France itself is producing. Truffles are assumed to be aphrodisiacs though it has not been proved scientifically. There are numerous species of the genus *Tuber* and we are describing the most important species that is known for its aphrodisiac use.

Product(s): Summer White Truffles and black-whole brushed in jar by Eugenio Brezzi, Italy. Earthly Delights, Michigan. There are several companies who sell truffles.
Common Name(s): Mushroom Truffles.
Latin Name: *Tuber Alba, Tuber bruchii (White) and Tuber melanosporum (Black).*
Synonym(s): *Tuber magnatum (white truffles); Tuber macrosporum (Black)*
Family: Tuberaceae.
Order: Ascomycetes.
Geographical Distribution: France, Italy, North America, California, British Columbia, Australia, and New Zealand. Every now and then, truffles are being discovered in a new spot on Earth. Recent finds have been in Yugoslavia and China. In North America, many edible species have been documented, but European truffles will remain supreme. In Europe, truffles are a status food, a delicacy and lavishly priced.
Habitat: These are mushrooms that grow in temperate rain forests from November to March while the black species grows from September to January. They grow underground with the symbiotic relationship of oak, chestnut and hazelnut trees.
Plant Description: Truffles are small, like the size of small potatoes and possess nutty or musky odor. These grow entirely underground and draw their nutrition from trees. They are fungus in nature. The genus Tuber comprises many species, differing in color, appearance, and aroma. The black and white truffles are very easy to identify where as other varieties are more difficult to characterize. An experienced mycologist is required to identify them. Moreover, species differentiation has been simplified by the chemists by fingerprinting the volatile compounds and comparing with the known one using cutting edge technology such as capillary gas chromatography coupled with mass spectroscopy.
Aphrodisiac Myth: The Greeks and the Romans considered the rare truffle to be an aphrodisiac. Napolean, the Marquis De Sade and Madame Pompadour all ate them for their amatory power. The white variety is from Italy and very expensive ($2000 per pound) whereas the black one is $500 per pound. Black truffle is the jewel of the French cooking. The Black Diamond of the PerioGard has been at the origin of the passions of the gastronomes since the antiquity, 2500 years BC in Egypt during

the reign of the Pharaoh Cheops. The Greeks and the Romans prepared the truffles during their official banquets. In Italy, Spain and France, hunters traditionally used female pigs to search forests for truffles on account of its constituent androstenol which is also the sexual attractant emitted by the boar. Since the pigs love the mushrooms and eat them, modern hunters have switched to dogs since they do not like to eat them. Truffles are very hard to find, and have a very short season; growers are trying to cultivate them on farms. However, wild truffles taste better than the cultivated in controlled condition.

Scientific Evidence: Until recently, there was no scientific evidence for the aphrodisiac use of truffles, Murat and Martine (2008) findings kindled some hope for turning myth into reality.They reported first evidence for the use of truffles as aphrodisiacs.

Chemical Constituents: Truffles contain androstenol, an essential hormone required for the body. Aroma is due to volatile compounds. Volatile organic compounds (VOCs) of nine Tuber species and two corresponding forms are identified via solid-phase micro-extraction-gas chromatography-mass spectrometry analysis. Seventy-five compounds are identified. The most abundant are dimethyl sulphide, 2-and 3-methylbutanal, 2-methylpropanol, and butanone by Mauriello *et al.* (2004). The sterol composition of *Tuber malanosporum* was examined by Harki *et al.* (1996) using high performance liquid chromatography, nuclear magnetic resonance, and mass spectroscopy. Ergosterol and brassicasterol were identified as the major components. Buzzini *et al.* (2005) isolated twenty-nine yeast strains from the ascocarps of black and white truffles (*Tuber melanosporum* Vitt. and *Tuber magnatum* Pico, respectively), and identified using a polyphasic approach. A novel sterol polyhydroxy ergosterol (tuberoside) from Chinese truffles *Tuber indicum* was isolated along four known ergosterol derivatives and characterized on the basis of chemical and spectroscopic means by Jinming *et al.* (2001). Gioacchini *et al.* (2005) identified six different species of truffles by analyzing their volatile compounds using static headspace solid-phase micro extraction coupled with GC/MS and characterized 29-66 compounds depending upon the species.

Toxicity: Truffles can be toxic unless these are picked up by experts. Some of the mushrooms grow wild that can be poisonous. These can be neurotoxic, hepatotoxic and nephrotoxic and also cause GI and respiratory problems (McPartland *et al.*1997; Pinillos *et al.* 2003).

Drugs Interaction: Truffles can interact with many prescription drugs.

Herbal Interaction: Truffles can have synergistic effect with many herbal products.

Food Interaction: Unknown.

Future Prospects: Unless the aphrodisiac properties of Truffles are proven clinically, they will remain as myth. Very recently, Murat and Martin (2008) have come with first evidenc of black truffle as an aphrodisiac; however, the studies are in progress to confirm their findings.

Chapter Six C

Aphrodisiacs from the Animal Kingdom

Deer Antler

Products: VP-RX, Inferno, Boom, Deer antler plus, Virility Ex, Velvita.™
Family: Cerevisae.
Description: The product is derived from the deer antler. Antler velvet is the only deer organ that completely regenerates every year. This includes bone, tissues and nerves. Deer antlers grow rapidly. When the antlers mature, the cartilage within them get converted into bone. In the final stage, the antler's blood supply and nerves are lost. When the antlers have fully hardened, it is harvested. Deer velvet is composed of the bone cartilage and skin of the rapidly regenerating antler. While the antler is growing, the developing antlers are covered in a soft skin with short furry hair, that resembles velvet in texture and thus it is called velvet antler.
Geographical Distribution: Alaska, USA, Canada, New Zealand and Asia.
Habitat: Cold climate.

Chemical Constituents: Valet antler contains about 40 compounds including valvatins, pantocrin, proteins, 15 Free amino acids, Free fatty acids, gangliosides, lecithin, phospholipids, steroids, prostaglandins, glycosamino-glycans, chondroitin, monoamine oxidase inhibitors, and plus some minerals like magnesium, potassium, selenium, calcium and phosphorus.

Aphrodisiac Myth: Deer antler has been used as an aphrodisiac for about 2000 year in Asia. Russian acclaimed its healing powers, which triggered deer farming in the 1840's. Velvet deer antler product called **Pantocrin,** is manufactured by a Russian state pharmaceutical company. Velvet deer antler is extensively used in Asia, New Zealand, and Korea. In the Chinese System of Medicine, it is used in balancing the endocrine system and in the treatment of penile erection dysfunction in men. It is also recommended in anti-ageing treatment.

Scientific Evidence: Research has shown that deer velvet antler demonstrates androgenic and gonadotrophic effects, meaning that it helps to regulate the activity of the sex organs. A series of investigations have shown that Pantocrin (velvet antler extract) contains biologically active substances of both the male and female sex hormone types. Over 250 articles, summaries, and reviews have been published on the studies and research on velvet antler. Research has been conducted in Russia, China, Japan, Korea, New Zealand, and Canada (Ewashkiw 2001). Zheng (1997) used pilose antler to treat sexual dysfunction in 297 cases of male sterility. Sperm count and sperm motility was increased in most cases.

The sex hormones estrone, testosterone, and a substance similar to progesterone have been identified at low levels in velvet, and the estrogen hormone most affected by velvet is estradiol, which is a precursor to testosterone. Deer velvet is used as a sex stimulant along with ginseng and Ganoderma mushroom. It is also known as antiaging supplement.

Additional Benefits: Immunomodulating, anti-inflammatory, osteoarthritis, and rheumatoid arthritis, restores iron levels, reduces cholesterol and blood pressure, and enhances muscular strength and endurance. It stimulates memory. It is traditionally used for yang deficiency which indicates depression, cold, lower back pain, exhaustion, weak pulse and leukopenia. It is used for the immune system, whole body rejuvenation and bringing the body to its peak physical performance and has been used by many Olympian athletes over history.

Dosage: Preparations that are sold over the counter or on the Internet contain 100 to 250 mg powder of deer velvet antler.

Toxicity: Toxicity studies in rats demonstrated no adverse effects on a short term usage.

Contraindication: According to study, people with circulation problems, congenital heart disease, or angina should not take velvet deer antler. It should be avoided during pregnancy and breast feeding. People suffering from blood pressure should also avoid its use. People who have an enlarged prostate should not use velvet antler, as it may promote continued enlargement of the gland.

Side Effect: The product can cause allergic reaction. (Bensky *et al.,* 1993; Dalefield and Oehme 1999) Too much velvet deer antler can cause mild upset stomach which disappears if use of the product is stopped.

Herbal Interaction: None well documented, however anticoagulant and hypertensive herbs may interact with velvet deer antler.

Food Interaction: Unknown.

Future Prospects: Limited studies are available for its use in sexual dysfunction. More clinical data is required to prove its efficacy in sex dysfunction. For the time being its use as an aphrodisiac will remain as myth. However, it has a wide range of benefits and could be used in other ailments such as arthritis and ulcers.

Spanish Fly

Product(s): Liquid drops, Solutabs, Kriptonite drops, Creams.
Common Name(s): Spanish Fly
Latin Name: *Lytta vesicatoria*
Synonym(s): *Cantharis vesicatoria*
Family: Meloidae.
Morphological Description: The Spanish fly is the dried, crushed body of the green blister beetle known as *Cantharis Vesicatoria*, or the Spanish fly. The beetle is 15-22 mm long and 5-8 mm wide and is found on plants belonging to the Caprifoliaceae and Oleaceae plant families.

Geographical Distribution: It is found in Spain, the Mediterranean region and in Russia.
Part(s) Used: Crushed beetle powder.

Chemical Constituents: Beetles contain cantharidin (5%).

Aphrodisiac Myth: Spanish fly has long been used for enhancing the libido of males and females. It arouses the female libido and improves the orgasm as advertised by the sellers. It has been wrongly reputed to stimulate sexual desire over the centuries. The irritation of the urethra will increase the blood flow to this region and might result in priapism, a persistent abnormal erection of the penis. It is likely that the priapism is the origin of the use of Spanish fly as an aphrodisiac. It can cause a prolonged painful erection in men and in women an engorged clitoris. However, an elevated level of sexual arousal does not accompany this erection.

Scientific Evidence: It is all a myth. Survey of literature does not reveal any scientific evidence to support this issue. It is believed to stimulate the penis and vagina by irritating the mucous membranes of genitourinary tract. Cantharidin the active constituent in Spanish fly is very toxic. Even the small quantity can cause toxicity. Its use should be avoided. The notorious Marquis De Sade found the poisonous nature of the cantharidin in June 1772. For his evening entertainment, he fed his prostitutes aniseed sweets amply laced with Spanish fly believing that this would "set them on fire." Two of them died on the spot and three became seriously ill.

Additional Benefits: Very dilute preparation of Cantharidin in oil is used as hair tonic to treat baldness and to remove warts (Rowland 2005).

Dosage: Very poisonous, even 10 mg can cause death.

Toxicity: A consumption of 1.6 grams of pulverized beetles containing the toxic chemical cantharidin led to death after 26 hours. Cantharidin is excreted by the kidney and during excretion irritates the entire urinary tract. Spanish fly is widely available. It is a notorious aphrodisiac, and considering the fact that ingestion is frequently inadvertent, cantharidin poisoning may be fatal. Cantharidin poisoning should be suspected in any patient presenting with unexplained hematuria or with GI hemorrhage associated with diffuse injury of the upper GI tract. In the event of a coma the outcome will usually be death within 24 hours. It can produce blister when applied topically (*Karras et al.* 1996; Tagwireyi *et al.* 2000). A case of a man who died after ingestion of cantharidin used as an aphrodisiac was reported by Marcovigi *et al.* (1995).

Future Prospects: It will remain as a myth. Its use as an aphrodisiac is very dangerous. It is illegal to sell this product in many countries. Various formulations containg Spanish fly (canthradin) are sold in the form of lubricating creams, spray, and solutions. One should be careful not to be carried away with illusive advertisements.

"The trees and plants show respect for each other by the way they live in harmony. This also applies to the animal kingdom."

—Masuru Emoto

Miscellaneous Animal Products Sold as Aphrodisiacs

Rhinoceros Horn: It has become the most valuable animal product in the world on account of its reputation as a sex stimulant. According to the myth, by ingesting the powder of rhinoceros horn a person is supposed to mount his woman for hours like a bull rhinoceros. It is used in the Traditional Chinese System of Medicine. A few tribes in Indian subcontinent use rhino horn as an aphrodisiac. Unfortunately such folkloric reputation has greatly endangered the existence of this species. Its cost is very high about $12,643 per kilo. It's true that rhino horns and tiger parts have been used for thousands of years in the Traditional Chinese Medicine (TCM). During this process the species are being wiped off (Gay 2002).

However, there is no scientific data available in support of its aphrodisiac property. The Swiss pharmaceutical firms Hoffmann-La Roche researchers after working on this project for a long time found that rhino horn had no effect whatsoever, good or bad on the human body (McCarthy 1999).

Tiger Penis: The tiger penis is valued as an aphrodisiac in China, Taiwan, and South Korea. The soup made from tiger penises is sold $300-400 a bowl. Again very little information is available in the literature to support its use as an aphrodisiac.

Turtle Eggs: Turtle eggs were eaten raw with salt and lime juice. Caviar is another one of the delicacies that supposedly nourish the nerve cells that stimulates romantic feelings. Similarly fish milt is the seminal fluid of fish, mollusks, and certain other water-dwelling animals that reproduce by spraying milt onto roe (fish eggs).

Sea Cucumber: The sea cucumber is an echinoderm of the class Holothuroidea, with an elongated body and leathery skin, which is found on the sea floor worldwide. Sea cucumbers are an important food source in many parts of the world such as China, Indonesia and have renowned aphrodisiac properties. Fresh raw flesh of sea urchin is not only a great culinary delicacy but also inflames erotic passion (Ratsch 1999).

Mollusks: The class constitutes oyster, octopus, snails, cuttlefish, and scallops that are commonly used by the people in many countries as aphrodisiacs. Similarly crabs, crayfish, lobster, prawn, and shrimps are known as aphrodisiacs. Likewise fishes such as carp, fugu, and sea horse are consumed as aphrodisiacs. In Chinese System of Medicine dried seahorses are powdered and ingested as aphrodisiacs. They are considered to strengthen the yang aspect of a person.

Reptiles: Snake, turtles, eggs of crocodiles in wine are consumed in Taiwan and China as aphrodisiacs. Men have eaten animal products and hundreds of other home remedies since the time immemorial in a futile attempt to regain their lost potency. *However, there is always lack of scientific evidences in support of these products and will remain a myth. Most of the products contain toxic chemicals and can be injurious to health* (Kweli 2007).

Chapter six D

Minerals as Natural Aphrodisiacs

Myth

Minerals such as rocks, ores, gemstones, fossils, and organic stones are also used as aphrodisiace. History reveals that gemstones, gold, silver, rubies, sapphires, diamonds, emeralds, topaz, pearls and corals were consumed by kings and rulers in the form of powder to increase low libido. Many fossils like Shame Stones, Mother Stones, Shark's Teeth, Red Coral (*Corallium rubrum*) and the thick spines of sea urchin (*Heterocentrotus*) worn as amulets. All these were believed to affect love and viewed as love charm. Quartz crystals have been worn as amulets stones and used for meditation since the time immemorial. It is believed that the rays reflected from the magical color stones can have a positive effect on the human body.

The biological importance of minerals has come to light during the last decades. Nutritionists have found that human body needs certain minerals and shortage of these can cause general debility, and physical decline. The body needs minerals to build and maintain healthy teeth and bones, carry nerve signals to and from the brain, carry oxygen to the cells, regulate blood sugar levels, and maintain a healthy immune system. Minerals are cosidered the key ingredients of life. Activity of enzymes in the body depends upon the presence of minerals, and this is a known fact. A total of 60 minerals have been discovered in the body, and 22 are essential to health. Some of the minerals such as boron, zinc, magnesium, and selinium are very essential for stimulating low libido and play a significant role in the quality and motility of semen. Nielsen *et al.* (1987) studied the effect of dietary boron on mineral, estrogen, and testosterone metabolism in postmenopausal women. According to their findings boron also boosts serum levels of estradiol and testosterone yielding enhanced sex drive. Zinc is a critical trace mineral for male sexual function. It is involved in virtually every aspect of male reproduction, including hormone metabolism, sperm formation, and sperm motility. Zinc found in the seminal fluid, increases sperm count and mobility, and blood testosterone levels (Scot *et al.* 1993; Prasad 1983). A deficiency in Zinc can result in impotence, decreased sperm count, or decreased motility possibly resulting in infertility. There are over 200 enzyme systems in the body that require zinc as on of their constituents. Zinc plays a major role in the manufacturing of protein. Similarly phophorus is an essential trace minerals of seminal fluid and its deficiency can lead to neurosthenia and low libido.

Homeopathy, Ayurvedic and the Chinese System of Medicine also believed in the importance of minerals for human body. Ayurvedic System of Indian Medicine has tremendous amount of information on the use of minerals to stimuate low libido. The formulations are combinations of minerals and herbs. However, there is no significant clinical data to support their use as aphrodisiacs. One very important and famous product of Ayurveda is 'Shilajit' that comprises about 85 minerals in ionic form; humic acid and fulvic acid are the main ingredients (Winston and Maimes 2007). Shilajit is a *rasayana* herb and is an *adaptogen*. Clinical studies are being conducted in India to confirm the ancient claims. 'Shilajit' is a *Sanskrit* word meaning *conqueror of mountains* and *destroyer of weakness*. Use of Shilajit to stimulate low libido along with some herbs like Ashwagandha and Safed Musli is very common in India.

Reality

Wearing of stones, bones and teeth of animals as amulets is nothing but a myth. Consumption of the pulverized precious stones does not make any sense besides psychoogical effect, elevating the mood of the individual that he is taking precious stones. It is just like a placebo. Corals and related fossils mostlty are rich in calcium and phosphorus along with other trace minerals. It is understandable that these can help the body like other trace minerals. Most of the unknown trace minerals present in these fossils and stones can be harmful on ingestion.

Do we really need mega doses of trace minerals for enhancing low libido?

There is no doubt that trace minerals play a significant role in the human body and are considered key ingredients of life. However, there presence in the body should be balanced. These minerals such as boron, calcium, phosphorus, potassium, mangnesium, zinc and many more are naturally available in vegetables and food we consume. Eating a variety of foods is the healthiest way for a body to receive the entire mineral it needs. Magnesium plays a key role in regulating the heart beat and high blood pressure. It also boosts fertility by making sperm more vigorous. One can get about two thirds of his daily requirement of magnesium from a breakfast of two cups of shredded wheat, skim milk and a banana. Taking extra quantity of these minerals leads to health problems. For example zinc formulations are being advertized to stimulate low libido. Excessive quantities of zinc can decrease copper, iron, calcium, and cadmium absorption because they all compete for the same binding sites. Zinc is believed to decrease selenium absorption too. So instead of stimulating low libido by consuming extra amount of zinc one can find himself in the hospital bed. Likewise, intake of high doses of chromium, magnesium and other trace minerals can cause health problems. Healthy food and regular exercise can be better approdisiacs than having additional doses of minerals.

> *"Sex is first of all a matter of energy. The more energy you have, the more blisful you can be, and the better sex becomes."*
> — Margo Anand

Chapter Seven

"Gravitation is not the cause of falling in Love; hormones can be."

—*Ravi K. Puri*

Hormones as Natural Aphrodisiacs

Chemistry of Love

Every culture has ample literature on the philosophy of love. Artists, poets and novelists have defined the philosophy of love and romance in their own styles since antiquity. Artists exhibited their expressions through their brushes and colors on the canvas whereas poets and novelist with their pens on the paper. However, they did not explain the basis of falling in love. Why is there love at first sight? Why do we like or fall in love with a person at first sight? Is there any chemistry existing between the lovers? It is still unexplored. Scientists are trying to explain the chemistry of love. Their findings indicate that love is a biochemical reaction between the two people that releases many neurotransmitters in the body at the various stages of love interaction which are responsible for falling in love.

There is not any magic love potion that one can use to make someone fall in love, but chemistry does play a significant role in the development of a relationship. First, there's attraction, nonverbal communication, and the eye contact during the initial attraction. This may involve pheromones and the later stages of development of love are attributed to neurotransmitters. It has been found that lust is due to high level of testosterone in the body and infatuation is caused by higher than normal levels of norepinephrine. The *high* of being in love is contributed to phenylethylamine and dopamine. Oxytocin level is increased in men during kissing. Both men and women had a decline in cortisol after smooching, an indication their stress levels declined (Schmid 2009). Stabilized production of serotonin and oxytocin in the body prolong love even after the honeymoon is over. Even infidelity is blamed on chemistry. Researchers have found that suppression of vasopressin can cause males to abandon their beloved and seek new mates. The chemistry of love, infatuation and infidelity is summarised in Figure 10.

It is quite a popular saying that *the chemistry of the couples should match*. Chemistry of love is directly proportional to our hormones. During attraction hormones monoamines are released into our system. Phenylethylamine (PEA) is the chemical of romance; it kindles love at first sight. It is also known as the molecule of love. PEA is a natural amphetamine, a mood elevator for men and women.

Dopamine is another biochemical associated with falling in love and reinforcing sexual desire (Giuliano and Allard 2001). Similarly, oxytocin is released during sex and the amount released is directly proportional to the quality of the sex, duration of the orgasm and sense of closeness. All these biochemicals are important parts of *love making*. The male and female sex hormones are the force in controlling sex drive. The slight fluctuation in hormone level can affect the whole system of the body. Even the swinging of the moods and behavior of an individual are hormone dependent. Hormones control sex life. In a majority of the cases, we should blame the hormone imbalance instead of blaming an individual partner. It is very pertinent to understand the chemistry of hormones before the use of any aphrodisiac or substance affecting directly or indirectly the reproductive organs.

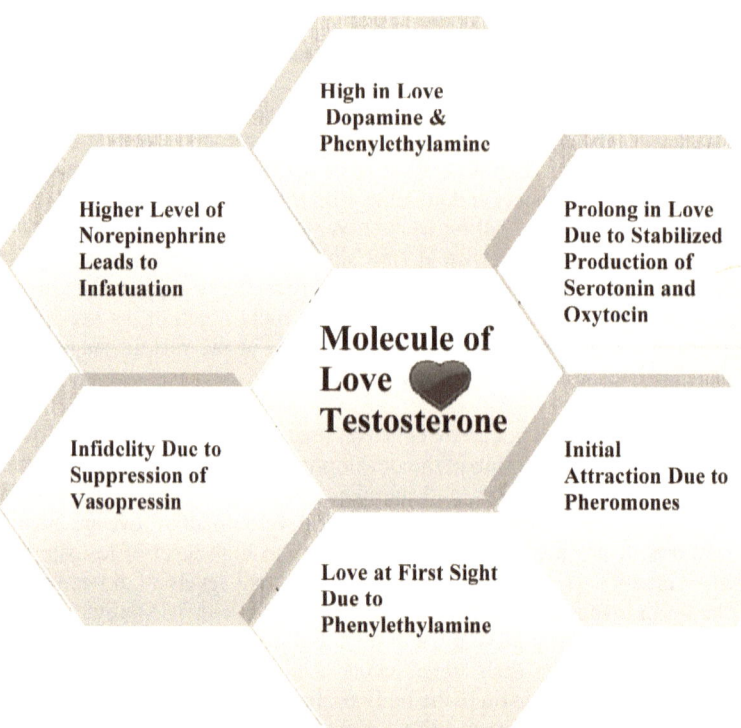

Figure 10: Chemistry of Love, Infatuation & Infidelity

Definition

The word *hormone* comes from the Greek word *hormao* meaning to drive, send, set in motion, or excite. It refers to the fact that each hormone excites or stimulates a specific part of the body, known as the target gland. Their job is to transmit information between cells and organs. They are known as "messengers." Hormones control metabolic processes in their target cells. A gland secretes a hormone into the blood stream. It is programmed for a specific target cell or organ that the gland regulates. Once released, the hormone circulates in the blood until a receptor cell on its target picks it up. It binds the receptor and delivers the message that triggers a reaction in the target tissues which is aroused to activity. Hormones are in a concentrated form and are released in a very small quantity parts per trillion. They possess very powerful effcts. Hormones are secreted by endocrine glands and passed from the cells of the gland directly into the blood.

Production of Hormones

The two sections of the pituitary gland, the anterior pituitary and the posterior pituitary produce a number of different hormones that act differently with target glands or cell.

1. Anterior Pituitary

Adrenocorticotropic Hormones (ACTH)
Thyroid stimulating Hormones (TSH)
Luteinizing Hormones (LH)
Follicle Stimulating Hormones (FSH)
Prolactin (PRL)
Growth Hormones (GH)
Melanocyte-Stimulating Hormones (MSH)

2. Posterior Pituitary

Anti-Diuretic Hormones (ADH)
Oxytocin and Vasopressin

Types of Hormones

There are four types of hormones categorized on the basis of their structural groups.

a. Steroidal Hormone
b. Peptide Hormone
c. Amino Acid Derivatives
d. Fatty Acid Derivatives-Eicosanoids

a.   **Steroidal Hormones:** These hormones regulate reproduction and control the maturation and development of the male and female sexual characteristics. They are called sex hormones. Sex hormones are steroidal in nature and consist of cyclopentano per hydrophenanthrene ring known as steroidal ring. Discussion will be focused on only sex hormones in this chapter.

b. **Peptide Hormones**: Any natural or synthetic compound containing two or more amino acids linked by the carboxyl group of one amino acid to the amino group of another is known as peptide. Gastrointestinal hormones insulin and the pituitary hormones oxytocin and vasopressin are called peptide hormones.

c. **Amino Acid Derivatives**: Thyroid hormones thyroxine and catecholamines are the examples

d. **Fatty Acid Derivatives-Eicosanoids:** These are large group of molecules derived from polyunsaturated fatty acids such as prostaglandins, prostacyclins, leukotrienes, and thromboxanes. Arachadonic acid is the precursor of thse hormones.

Sex Hormones (Steroidal Hormones)

The sex hormones are produced mainly in the sex glands such as gonads, the ovaries in women, the testes in men. They are carried to their target organs by the bloodstream. Estrogens and progesterone are **female sex hormones**. Androgen represents **male sex hormone** such as testosterones. DHEA (dehydroepiandrosterone) is also part of the androgen family. It is an important precursor that can be converted to testosterone. Testosterone is considered to cause sexual arousal. Women also produce testosterone in small quantities which is very essential to maintain normal sexual desire.

Male Sex Hormones
When the Luteinizing Hormone (LH) secreted by the anterior pituitary in male reaches the testes via the bloodstream, the cells there begin to produce a hormone called testosterone. Testosterone molecule plays a significant role in the development of manliness. Testosterone reaches the targeted tissues of the male genital organs and enters the cells. Inside the cell, it unites with an enzyme specially created for it whereby its effects are greatly stimulated. This newly formed hormone unites with a special receptor designed especially for it. The resulting molecular combination unites with the cell's DNA and uses the information received from the DNA to bring about a new protein synthesis. This operation ensures that the distinction between male and female bodies, and their different sexual functions, will continue.

This system is so flawlessly created that the mechanism formed from the three-fold union of testosterone-enzyme-receptor finds the place allotted to it from among the countless data codes in the DNA and on the basis of this information, ensures production. For example, for the growth of the beard, they act on the relevant regions in the DNA's of the hair root cells. To lower the voice, they act on the appropriate region in the DNA's of the vocal cord cells.

Testosterone is a molecule produced from carbon, hydrogen and oxygen atoms. The unconscious, lifeless molecule knows that the information it needs to perform its function is located in the DNA. Nevertheless, it can also find so quickly and accurately the few letters it is looking for from among three billion letters in the DNA. Scientists, who have been working for many years on the Human Genome Project, are only today able to map the DNA. Inspite of all, they still do not know which section of the DNA is related to which organs, proteins or hormones of the human body. However, estrogen and testosterone molecules know this information quite well and have applied it without error since antiquity. It is a miracle of nature.

Functions of Male Sex Hormone
Testosterone molecules help the multiplication of muscle cells. The increase in muscle mass gives the male body its characteristic appearance.

Testosterone

Testosterone controls male sexual development, sex drive, and physical stamina. It is required for sexual arousal (Mikhail 2006). It controls sperm production, erection, elevates mood, confidence and sexual fantasies. Testosterone molecules influence the cells in the roots of the hair, causing the beard and mustache to occur. Testosterone affects the vocal cords causing a male voice to be of lower pitch than that of a woman. In addition, the testosterone molecule gives the male body its ability to fertilize the female egg.

The hormone testosterone is present in females in smaller quantity. Women produce about 0.2 mg of testosterone per day where men produce about 20 mg per day. It is responsible for the sex drive and desire in both the sexes. Testosterone stimulates sex in women. Testosterone receptors in the nipples, clitoris, and vagina make these areas sensitive to sexual stimulation. Excessive concentration of testosterone in the body leads to aggressive, abusive, violent, irritable, intolerant, and disrespectful behavior. Low levels of testosterone in the body lead to a lack of sexual desire and derives poor erection, accelerates aging process, decrease in bone mass, depression, and increase in fat and flabbiness (Buvat *et al.* 2006). By age 65 more than 60 % of the men have low levels of testosterone. It is not out of the way to mention here that sex drive is determined by a number of factors. It does not depend upon testosterone level only. Hormone levels are at peak between the age of twenty and thirty in men and women. After the age of 30, hormone levels starts declining, resulting in poor stamina and low libido. Metabolism becomes slower and an individual starts gaining weight. Role of testosterone is summarized in Figure 11.

During menopause, the hormonal level changes dramatically in women. This causes many problems in females including a diminish sex drive. The decrease of female hormones causes vaginal dryness and thinning of the vaginal lining that makes intercourse painful. It results in a lack of libido, stress, depression, and weight gain. These symptoms lead to many problems in relationships and most of the time to divorce. However, if people understand the chemistry of hormones it can mitigate their problems to a great extent. Testosterone levels in females determine their sexual activity. The cause of low levels of testosterone in women is mostly during and after menopause. The ovaries produce androstenedione, testosterone, and dehydroepiandrosterone (DHEA). The adrenal glands produce androstenedione and dehydroepiandrosterone sulfate (DHEA-S). The DHEA-S can be further metabolized to testosterone or estrogens. In addition the testosterone through the enzyme of 5-alpha reductants converts the serum testosterone to dihydrotestosterone (DHT) or estradiol (E2); these are the active hormones that work within the cells.

Causes of Low Androgen Level in Female

Androgen is also present in low concentration in female and is responsible for their sexual desire. Most of the problems in females occur during or after menopause. Age leads to a drop in androgen level in females and is due to the age-related drop in adrenal production of androgen and the loss of the mid-cycle surge in ovarian testosterone.

Hysterectomy or removal of the ovaries results in a reduction of 50 percent in testosterone and androstenedione in females. Chemical oophorectomy, chemotherapy, GnRH (Gonadotropin-releasing hormone) inhibitors, radiation therapy, glucocorticoids and the administration of exogenous estrogens are other causes for decrease in androgens level.

Oral postmenopausal estrogen therapy and oral contraceptives will suppress free testosterone by increasing sex hormone-binding globulins (SHBG) and suppressing pituitary luteinizing hormone (LH).

Consumption of steroids by mouth suppresses pituitary secretions of adrenal corticotropic hormones and therefore adrenal androgen production as well. This probably explains the bone loss frequency in patients who are taking long-term steroids. Lastly, hypothalamic amenorrhea and hypoproaccelerinemia are usually associated with low testosterone and many women with premature ovarian failure have low testosterone levels. Therefore, the use of oral contraceptives in older women or women with amenorrhea or premature ovarian failure may actually worsen their androgen deficiency.

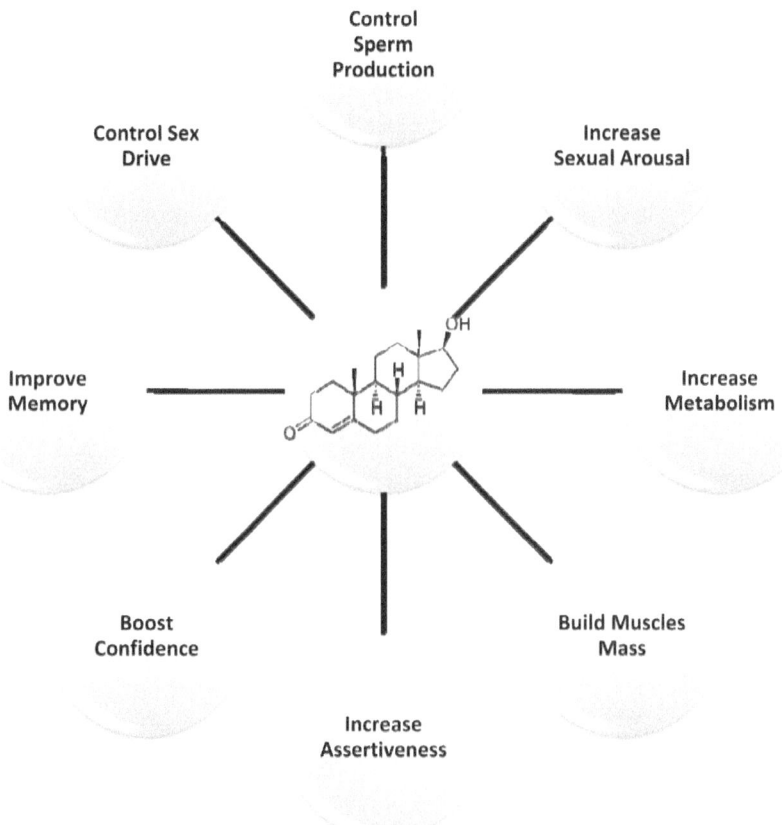

Figure 11: Role of Testosterone in the Body

Although, it is a clinically known factor that testosterone affects female sex life, how testosterone therapy affects female sexuality is not well understood. The male hormones may work directly on androgen receptors or may be a precursor for additional estrogen production in tissue such as fat, bone, brain, blood vessels or possibly by lowering sex hormone binding globulins (SHBG) and therefore causing an increase in the levels of bioactive steroids such as androgen.

There is no doubt that the administration of testosterone to women with sexual desire problems improves the intensity of sexual desire, arousal, and frequency of sexual fantasies, satisfaction, pleasure and relevancy and importance of sex to daily life. Postmenopausal women who are probably treated with estrogen therapy should be offered androgen replacement to improve this symptom complex.

A more difficult question deals with the pre-menopausal women who complain of decreased sexual drive and libido and who have low bioavailable testosterone. Studies have not been done; each case should be individualized especially in those individuals in whom other factors do not appear to play a role in desire and where the psychosocial and sexual history indicates hormonal problem as being the basic ideology of their libido decrease.

Oral testosterone undecenoate has not been studied in women and doses as low as 20 milligrams appear to cause undesirable side effects and therefore is not recommended.

Subcutaneous implants of testosterone are not available in the United States at this time, but have been in Australia and the United Kingdom for many years and have found to be quite effective for up to six months. Doses of 50 to 100 milligrams appeared to effectively raise the levels of testosterone for up to six months to adequate levels to treat sexual desire problems. In conclusion, androgen deficiency in women causes various symptoms including poor sexual desire is in the menopausal and pre-menopausal female.

Recently transdermotestosterone patches have been manufactured and approved for use by men and newer technology is developing androgen replacement patches for women. Patches that increase testosterone levels greater than 25 nanograms per DL appear to produce significant masculinization and side affects. Patches should not be used.

Transdermotesosterone cream using a transvaginal testosterone impregnated cream is available in the United States by specific prescriptions.

The aphrodisiac herbal preparation for female sexual deficiency should be taken with lots of care. The chemical constituents present in the products can interfere with female hormone equilibrium.

Contraindications to Testosterone Treatment Include

Acne, hirsutism, alopecia, and circumstances in which enhancing libido would be undesirable. Absolute contraindications include pregnancy and lactation as well as known or suspected androgen dependent neoplasia. Side effects from excessive testosterone include virilization, fluid retention and an adverse lipoprotein profile which more likely occur with the oral administration of the drug.

Female Sex Hormones

Estrogens and progesterone are female hormones. Estrogens control the development of feminine body characteristics. Progesterone regulates the menstrual cycle and pregnancy

Estradiol

Estrone

Estriol

Estrogens

The three major naturally occurring estrogens in women are estrone, estradiol, and estriol. Estradiol is the dominate form in nonpregnant females. Estriol is the primary estrogen of pregnancy whereas estrone is produced during menopause. Estrogens maintain sexual drive and promotes frequency but fluctuate throughout a women's life.

Formatiom of Estrogens

Estrogens are produced in the ovaries, adrenal glands and in the placenta of a developing fetus. They are also produced by fat cells even after menopause. The production is controlled by the hypothalamus. Special estrogen receptors are located in the breast, lining of the uterus, cervix, brain, bones and in the vagina.

The pituitary gland secretes the LH hormone. It reaches ovaries via the blood stream. Inside the ovaries are thousands of immature egg cells. Under the influence of the LH hormone from the pituitary gland, part of these egg cells begins to mature. Normally, only one of these developing cells will fully mature and be released from the ovary as an egg cell. If two cells are released and each one is fertilized, twins will be born. The egg cell in process of development with its surrounding nutritive layers is called a "follicle." The follicle stimulating hormone (FSH) sent by the pituitary gland has a major effect on the follicle, suddenly causing it to produce a special molecule. This molecule is the estrogen hormone.

Function of Estrogens

Estrogen is responsible for development of female reproductive organ during puberty, regulating menstrual cycle and preparing the uterus for pregnancy. Moreover, in women, development and shape of the breast, hips and thighs during puberty is due to the influence of estrogen. Estrogen also produces the high pitched woman's voice.

Estrogen facilitates fertilization. The amount of estrogen in the blood rises significantly during the time when the egg is ready for fertilization. This results in the secretion of a special fluid from the uterus to the vagina. This fluid captures the male sperm, increases its mobility and carries them upward towards the egg cell.

Estrogen also plays a significant role in the development of uterus where the fertilized egg is implanted and grows. Estrogen helps the uterus to prepare for pregnancy. The uterus wall increases in thickness to protect the fertilized egg and is surrounded by capillary vessels that supply the nutrients to the zygote. A Female's womb is a highly septic environment. Estrogen protects the unborn baby and the mother too. When estrogen molecules reach the epithelial cells in the mother's womb, these cells begin to secrete an acid. This acidic environment is a suitable environment for the multiplication of beneficial bacteria *Döderlein's bacillus* and, at the same time, it protects the vagina from infections. The chemical molecule estrogen produced by a tiny follicle not only gives shape to the fetus from head to toe, but also makes the necessary arrangements for a new baby to be born.

Estrogen can be used to reduce lactation after birth. It may be used in men for the treatment of prostate cancer (Oh 2002). Estrogen prevents depression and reduces stress. Estrogen protects against schizophrenia and Alzheimer's disease.

Progesterone

Progesterone is produced by the ovaries, the adrenal glands and during pregnancy, by the *corpus luteum.* .

Progesterone

The progesterone hormone has a very special purpose. It stays in the ovaries but influences cells located far from it. When progesterone molecules reach the uterus they help maintain the lining of the uterus. Progesterone also prevents the release of a new egg. Otherwise, as an embryo develops in the mother's womb, a second egg would be fertilized, producing a great danger for both the developing embryo and the mother.

Another special function of progesterone is to diminish the influence of the oxytocin hormone secreted by the pituitary gland. Oxytocin is a hormone that contracts the muscle and help in the expulsion of the fetus at the time of birth. As a result of these contractions, the baby comes out of the mother's womb more easily. If oxytocin affected the uterus during the first days of fertilization, these muscles would eject the fertilized egg as it attached itself to the uterine wall, and abort the pregnancy. Progesterone goes into effect at this stage and inhibits the influence of oxytocin, preventing the fertilized egg from being ejected. Progesterone negates the effect of oxytocin.

The moment a fertilized egg reaches the wall of the uterus and begins to grow there, it becomes something foreign in the mother's body. The immune cells in the mother's body would inevitably attack this group of cells as they multiply. This attack would end the life of the baby even before it began and pregnancy would never occur. But progesterone prevents the cells of the immune system from attacking the zygote in the wall of the uterus. Besides its other functions, progesterone also protects the developing group of cells from being attacked.

Secretion of progesterone occurs in the second half of a four-week period. If fertilization does not occur in this period of time, the amount of progesterone and estrogen in the blood quickly decreases. These preparations, the multiplying capillary vessels to feed the fertilized egg in the wall of the uterus are expelled from the body, a process called menstruation. The secretion of the FSH hormone four weeks later in the pituitary gland corresponds to a new egg cell that begins to mature in the ovary, and another new four-week preparation period starts.

Oxytocin
Oxytocin is a peptide of nine amino acids. The sequence is *cysteine-tyrosine-isoleucine-glutamine-asparagine-cysteine-proline-leucine-glycine* (CYIQNCPLG). Oxytocin and vasopressin are almost similar in structures with a little variation. Oxytocin also plays a significant role in sex life. Oxytocin is released from the posterior lobe of the pituitary gland that stimulates the contraction of smooth muscle of the uterus during labor. The secretion is also stimulated by touch. Oxytocin promotes a desire to touch and be touched. Oxytocin makes an individual feel good about the person who causes the oxytocin to be released, and it causes a bonding between the two persons. Even thinking of someone you love can stimulate this hormone. So it is connected to sex, love, and romance. Higher levels of oxytocin result in greater sexual receptivity. It increases testosterone production which ultimately stimulates sex in both men and women. When coupled with estrogen it increases libido. Nevertheless, oxytocin increases the sensitivity of the penis and the nipples, improves erections, and results in stronger orgasm and ejaculation.

Oxytocin was classified as a pregnancy hormone, promoting contractions during labor and aiding breastfeeding. In the 1970s, oxytocin was more than just a hormone-it was also a neurotransmitter, released from the hypothalamus during social interactions and sex. Oxytocin is detected by receptors throughout the brain's emotional center, the limbic system. This discovery prompted scientific interest, as oxytocin is now one of the hottest topics in neuroscience

Kosfeld *et al.* (2005) showed in their study that oxytocin increased trust in humans. Oxytocin's ability to connect social contact with feelings of pleasure and well-being has got researchers excited about potential therapeutic uses, since so many mental illnesses involve disorders of sociability or empathy. An obvious starting point is autism, which is marked by difficulty understanding the minds of others, aversion to human contact, and repetitive behaviors such as rocking. Hollander *et al.* (2007) studied the effect of oxytocin to autistic adults. He found that it improved their ability to recognize emotions like happiness and anger in people's tone of voice, something autistic people struggle with. A single intravenous infusion produced improvements that lasted two weeks.

Oxytocin has been called the love hormone, the cuddle chemical and liquid trust. It peaks with orgasm, makes a loving touch magically melt away stress and increases generosity when given as a drug. Oxytocin is the essence of affection itself, the brain chemical that warmly bonds parent to child, lover to lover, friend to friend, and it could soon be unleashing its loved-up powers far and wide.

Conclusion

Sexual desire is directly proportional to the rise and fall of the hormone levels of an individual in both the sexes. This directly affects the relationship of the couples. Instead of solving the problem, we start blaming the partner. We should try to understand the chemistry of our body before jumping to conclusion. When a woman is not ovulating her sex drive is still there, but it varies depending upon the level of hormones. Similarly in a man, the androgen's level is responsible to their swinging mood. Hormone levels are constantly changing daily in our bodies and this change continues over the years.

"It is not just giving that brings pleasure; it is also just receiving. In concern, two essential human acts join in a circle of interaction that expands with use. When the circle is complete, the more you give, the more you get and vice-versa."

— George Leonard

Chapter Eight

Aromas as Aphrodisiacs

The use of aroma or perfume is as old as the civilization. Undoubtedly, aroma plays a significant role in our day to day life for improving physical and mental awareness. Our moods can swing on the nature of aroma. However, there is not enough scientific explanation to this fact. Scientists are curious to investigate the effect and mechanism of action of aromas on the human body. This branch of science is known as aromachology and is gaining importance these days.

Today science has begun investigating fragrances and pheromones with respect to how they affect our attraction to the opposite sex and our sexual chemistry. This chapter will focus on the various fragrances that are used in daily life as aphrodisiacs.

Ancient Origins

Throughout history, aromatic oils have been used for healing and are still the main constituents of many of the world's healing systems. Some aromatic oils are considered aphrodisiacs since they have the potential to affect the circulatory, hormonal, and the nervous systems. A blend of the exotic, stimulating, calming, and euphoric essential oils would make excellent aphrodisiac formulation. Jasmine is the most popular fragrance in the perfume industry on account of its aphrodisiac effect on humans. Likewise, amber, angelica, basil, bergamot, cinnamon, frankincense, hyssop, lavender, myrrh, patchouli, rose, rosemary, saffron, sandalwood, and ylang-ylang oil are also known to possess aphrodisiac properties. Distribution of the aromatic oils in different organs of the plants is shown in Figure 12.

The Queen of Sheba used fragrance to seduce Solomon. Nevertheless, when the famous Egyptian queen Cleopatra visited Mark Anthony, the sails on her ship were scented to indicate her arrival. Cleopatra used opiates and perfumes to seduce her lovers. Legend says the palace where she entertained Mark Anthony was often saturated with precious essential oils.

A first record of fragrance use was traced back to ancient Schumer civilization, which occurred in Mesopotamia (modern Iraq) some 15,000 years ago. Clay tablets from that age delineate ancient priests burning scented plants and oils in religious ceremonies and celebrations. In Greek mythology, Aphrodite, the goddess of beauty was supposed to be the first perfume user. Elysian Fields displaying millions of flowers of various types of pleasing fragrances are famous in Greece. All Greek cities-states had hundreds of perfume shops. Aromatic herbs such as marjoram, lily, thyme, sage, anise, rose and iris, incorporated into olive, almond, castor and linseed oils were in great demand.

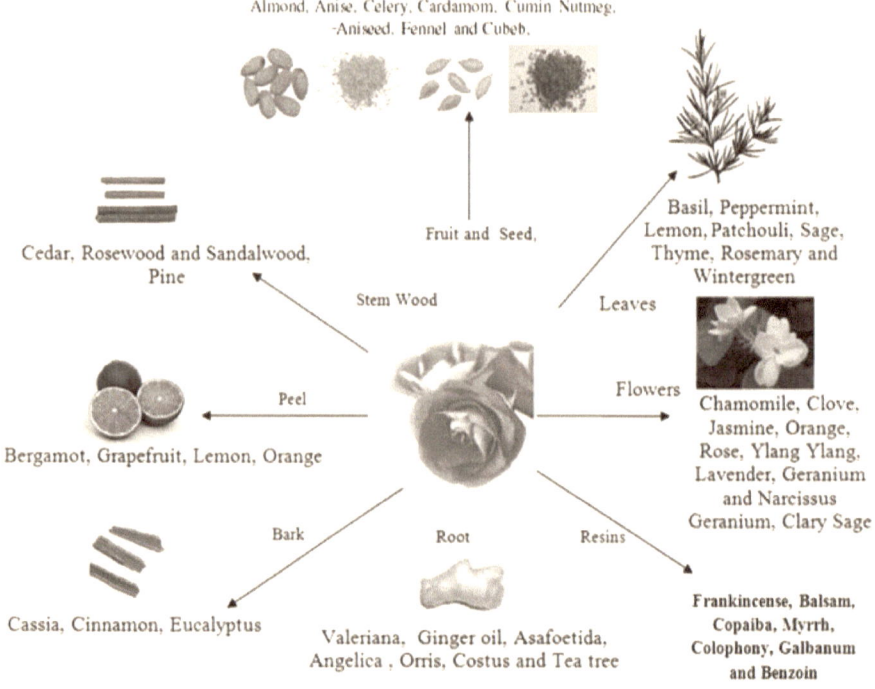

Almond, Anise, Celery, Cardamom, Cumin Nutmeg, Aniseed, Fennel and Cubeb.

Fruit and Seed,

Basil, Peppermint, Lemon, Patchouli, Sage, Thyme, Rosemary and Wintergreen

Cedar, Rosewood and Sandalwood, Pine

Stem Wood

Leaves

Peel

Flowers

Bergamot, Grapefruit, Lemon, Orange

Chamomile, Clove, Jasmine, Orange, Rose, Ylang Ylang, Lavender, Geranium and Narcissus Geranium, Clary Sage

Bark

Root

Resins

Cassia, Cinnamon, Eucalyptus

Valeriana, Ginger oil, Asafoetida, Angelica, Orris, Costus and Tea tree

Frankincense, Balsam, Copaiba, Myrrh, Colophony, Galbanum and Benzoin

Figure 12: Aromatic Oils in Various Parts of the Plants

After his conquest of Asia, Alexander the Great learned to love perfumes and organized frequent expeditions in search of the new and exotic scents. He brought perfume to Greece from India.

Perfumes were found in tombs of Egyptian Pharaohs. The hieroglyphs found inside pyramids explained that fragrance played an important role in the daily lives of ancient Egyptians. Studies conducted on bottles left near female mummies indicated that women used perfumes made from flower essences. Archaeologists have also found sachets around mummies' necks containing aromatic grains.

Perfumes were an integral part of worship as described in ancient papyruses. Perfume from irises, heliotrope, saffron, cinnamon, cedar oil, myrrh and many more helped create delicate scents for aristocrats in the Egyptian court.

Fragrances were also contemplated as aphrodisiacs in the Roman Empire. Julius Caesar (100-44 BC) was very fond of perfumes and promoted their use in Rome. During that time the Romans indulged in the practice of applying perfumes three times a day.

In Western Europe, Napoleon Bonaparte was very fond of wearing perfumes of jasmine and violet cologne. The ancient city of Cologne where Gian Pasto Feminis created first Eu De Cologne is famous.

Even the *Bible* had several references cited on essential oils such as frankincense, myrrh, cassia, rosemary, and hyssop. These oils were used for anointing and healing the sick. Indus Valley, one of the greatest civilizations of the world rendered strong evidence about the use of perfumes by its people. Vedic texts *Rig-Veda*, *Mahabharata* and *Ramayana,* originated around 10, 000 BC also delineate the use of perfumes in the form of incense and body massage. The *Kama Sutra*, the sex manual of Indian origin, suggests a blue or pink lotus and jasmine oil massage for sensual effects. A tantric sexual ritual from India called *The Rite of the Five Essentials* uses scent to arouse the woman. She is anointed with five different oils on various areas of her body to lift her spirits. Thus she may manifest as a goddess and be worshipped as the embodiment of the creative force, Shakti. On her hands, jasmine is applied; to her neck and cheeks, Patchouli (*Pogostemon patchouli*); to her breasts, amber, or musk; spikenard (*Nardostachys jatamansi)* is rubbed in her hair; and sandalwood (*Santalum album*) caresses her thighs.

The Ayurvedic System of Medicine in India traditionally uses essential oil fragrances to obtain the right doshik balance needed for good health. Indian sages believed fragrances affected man's consciousness, and encouraged rituals of worship that incorporated flowers and burning of incenses.

To this day, flowers and incenses are an integral part of daily worship throughout India. It is believed that the constant exposure to these highly evolved fragrances refines and elevates consciousnes to superconsciousness.

The use of essential oils had been cited in Chinese culture too. The first Great emperor of China, Shi Huang (246-210BC) was devotedly attached to perfumes from natural sources. He ordered his court official to wear perfume on daily basis while on duty.

Role of Aromas as Aphrodisiacs

Dr. Hirsch, Neurologist and Director at the Smell & Taste Treatment and Research Foundation in Chicago, reported at the March 1998 American Psychosomatic Society annual meeting the results of his studies relating certain scents to changes in blood flow to the genital areas of men and women. According to Hirsh (2004), combination of lavender and pumpkin pie increased the blood flow to the male organ by 40 percent. A combination of black licorice and donut scents increased penile blood flow by 32 percent. Surprisingly, these scents did not show any effect on women. This suggests that males and females need different scents to stimulate their libido. The smell of pumpkin pie includes the spices ginger, cinnamon, and clove. However, ginger (*Zingiber officinale*) and cinnamon (*Cinnamomum zeylanicum*) essential oils are mentioned as aphrodisiacs in the myth world.

Aphrodisiacs can be of great value when dealing with sexual disharmony between partners. The root cause of impotence and frigidity during youth is mostly emotional and psychological and in this case aromatherapy can be beneficial. The role of aromas as aphrodisiacs is that they help to set mood by rendering pleasant feeling. Smell, hormones and sexuality are interlinked. Every individual reacts differently to the same perfume. One person likes it whereas another dislikes it. It can be allergic to someone and pleasant to another one.

It has been believed that exotic essential oils optimize the circulation of the blood in the organs of the pelvis minor, normalizing the work of excretory and exocrine glands, increasing cellular cycle, removing stagnant processes and slowing reaction of decomposition. Though, it is hard to believe without any scientific evidence.

It is also believed that aromas possess spasmolytic and antiphlogogenic actions, thus removing depression, feeling of aversion and uncertainty. Some essential oils can evoke strong emotional and psychological responses when inhaled. These have been shown to act through the olfactory tract directly on the brain's limbic system where sexual feelings and behavior arise. Aromas actually can trigger the release of hormones and neuro-chemicals in the brain that create a feeling of well-being. Their action is quick. It takes only a few seconds. However, according to Damian and Damian (1995) "Olfactory research is still in its infancy stage, we are now gaining rudimentary knowledge of how and why essential oil's fragrances affect human psychology and physiology."

Essential oils help to create harmony in body, mind and spirit, and promote relaxation, confidence and receptivity. Sandalwood and ylang-ylang oils are considered to possess powers of arousal. Their fragrances supposedly produce stimulating effects on moods to create erotic sensations and awaken desire. Some essential oils can have a direct stimulating effect on the body whereas some have hormone-like influences that may trigger sexual desire and performance. However, these speculations need further investigation.

Mode of Action of Aromas

Essential oils are absorbed into the body either through the skin by massage or by inhalation through nose. The tiny chemical molecules within the oils enter the nervous system through the blood stream influencing emotional and physical well being. Volatile oils can affect the body through the highly sensitive olfactory system. When cells located in the upper part of the nose capture odor molecules, signals go to the brain's limbic region, a primitive portion of the brain. This region controls the body's basic survival functions, in part, by influencing key hormone-secreting glands affecting the entire body. Hence, a smell can quickly influence our entire body.

Oils absorbed through skin pores and hair follicles enter bloodstream capillaries and circulate throughout the body. We smell the fragrances as the oil is rubbed on the skin. It is difficult to separate the synergistic effects from inhalation due to topical administration.

Although no androgen or progesterone-like activity has been found so far in essential oils, however, there have been reports of estrogenic activity. The chemical oestrone has been found in beetroot, yeast, potatoes, and palm kernel oil. Anethole is methyl ether of oestrone and has been found in fennel, aniseed, coriander, and many other volatile oils. It has been suggested that it is the plant polymers of anethole, such as photoanethole and dianethole that are responsible for the action. Polymerized anethole is similar to the stimulant diethylstilboestrol (McCormack 2000).

Red roses, jasmine, lavender and French perfumes are all symbols of passion and love. The essential oil from red roses is a strong aphrodisiac. It is said that Cleopatra used to take bath with milk and rose petals. Maury *et al.* (2004) attributed aphrodisiac properties to rose oil and prescribed it for frigidity. She believed rose oil was a great tonic for women who were suffering from depression.

According to myth, clary sage and geranium are supposed to regulate the female hormones and can be used in a bath or in a room fragrance during premenstrual tension (PMT). Lemon grass is considered to strengthen the emotions and is used as inhalant or room fragrance during the periods. Tension, irritability and emotional disturbance are very common during PMT.

Chemistry of Aromatic Oils

Mostly essential oils containing about 80 to 350 different chemical constituents, aromatic in nature when analyzed by capillary gas chromatography coupled with mass spectroscopy. These constituents are long chains of hydrocarbon, forming a ring-like chemical structure known as hydrocarbon terpenes. The molecule is known as Isoprene unit. Isoprene unit combine to make various compounds. These are monoterpenes, sesquiterpenes or diterpenes and triterpenes in nature. These compounds are oxygenated such as phenol, aldehydes, ketones, lactones, ether, esters, and coumarins. These constituents have different fragrances and pharmacological activities depending upon their chemical structures. Volatile oil constituents differ in the same species depending upon their ecological conditions, harvesting and distillation techniques.

Adulteration in Aromatic Oils

The fragrance alone, when it comes from good quality pure oils, is exquisite. Altered or artificial versions do not have the same aroma and effect. So these products should be pure and of therapeutic grade. Adulteration is very common in essential oils and should be avoided. Since the volatile oils are costly, greedy people and cut-throat competition in this trade tempt people to adulterate these oils. Adulteration product can be toxic due to the type of adulterated components.

Types of Adulteration

1. **Addition of diluents or solvents material**
   Vegetables or mineral oils are used. Sometimes solvents and perfumery materials are added. Various diluents or solvents used as adulterants are shown in Appendix L.

2. **Addition of cheaper essential oils and adjuncts**
   To meet the customer target price or to make additional profit, cheaper oils or substituents are added. Second and third distillation products are cheaper since they loose 50% of the original constituents, are added. This is known as *sophistication in adulteration.* Examples are shown in Appendix M.

3. **Addition of cheaper synthetics**
   Sometimes synthetic chemical substances are added as adulterants so that the chemistry of oils is not altered. These substances do not impart the same effect as that of the natural oils. These are devoid of natural isomer molecules present in the oil that is responsible for their pharmacological properties. Examples are given in Appendix M-1.

4. **The addition of isolates or natural components to essential oils.**
   Natural adulterants are also added to increase the profit. An example is (-) linalool added to lavandula and bergamot.

5. **Addition of reconstituted oils to genuine oils**
   Oils are made from natural identical synthetics and added to the natural essential oils to make profit. It is done in the case of rose, jasmine oils, lavender and more valuable oils.

6. **Addition of racemic synthetics**

These substances will increase the concentration of the oils. However, addition of racemic material alters the pharmacological action of the v. oils have been demonstrated by many researchers. Huenberger *et al.* (2001) have found that inhalation of (+)-limonene caused increases in systolic blood pressure and changed alertness and restlessness in subjects, whereas (-)-limonene only affected blood pressure. Similarly (-)-carvone was reported to increase pulse rate, diastolic blood pressure and restlessness whereas (+)-carvone increased systolic and diastolic blood pressure. Traynor (2001) reports that (+)-rose oxide confers relaxing physiological effects, while (-)-rose oxide from Bulgarian Rose oil and geranium oils possesses a significantly higher stimulatory effect.

Sugawara *et al.* (2000) investigated the effects of inhalation of the different linalool isomers [(-)-linalool purified from lavender, (+)-linalool from coriander, and synthetic (+/-)-linalool] inhaled before and after work. They found variation with synthetic and natural isomers.

Consequences of Adulteration in Aromatherapy

Adulteration is very common in essential oils on the basis of demand and supply. There is more demand and less supply. Adulterated oils are of less therapeutic values and can be toxic depending upon the nature of the adulterants.

1. Phthalates such as Diethyl phthalates are still occasionally found as adulterants in essential oils. Phthalate esters have been found very toxic. Traces of residual organic solvents (such as hexane and cyclohexane) in oils and absolutes are found as a result of extraction and co-distillation practices.

2. The presence of pesticides in essential oils is also a serious health & safety issue.

3. Addition of cheaper oils or synthetic will not give the benefits of aromatherapy.

4. Addition of racemic synthetic would change the pharmacological actions of the oils. Buchbauer (1998) maintained that each constituent of an essential oil contributed to the beneficial or adverse effects of the oil. He believed that changing the distribution of chiral components of oils by adulteration with racemic synthetic odorants might in fact change the beneficial properties of the oil.

5. Some adulterants are photosensitive, irritants or contain allergens that cause allergy reaction as well. Photo-toxicity is light-related irritation, and involves percutaneous penetration and bio-distribution of a light-activated substance in the dermis, followed by skin exposure to light after the application of the essential oil. If photo-toxic oils are applied to the skin, and exposed to bright light/UV lamps/ sunshine over the next 12-24 hours, it cause skin reaction known as phototoxicity. Photo-toxic affects are due to the oil contents such as furanocoumarins or related constituents present in volatile oils. Precautions should be taken when using photosensitive oils. Photosensitive aromatic oils are listed in Appendix M-2.

Analysis of Aromatic Oils

Now a day, more emphasis is on the chemistry of the volatile oils rather than their physical characteristics such as specific gravity, optical rotation, refractive index, molecular refraction, solubility etc. Scientific research has isolated hundreds of chemical constituents in essential oils. Some of them contain more than 200 identified chemical substances. Essential oils can be analyzed qualitatively and quantitatively.

A. **Qualitative Analysis**

   Qualitative analysis can be done by thin layer chromatography observing the Rf values of the fingerprint region of the main constituents. Most of the time standard oil is chromatograhed simultaneously with the oil to be tested. It is known as cochromatography.

B. **Quantitative Analysis**

   Capillary Gas chromatography hooked with mass spectroscopy or ion trap spectroscopy coupled with infrared spectroscopy is a very useful tool for the identification of the constituents of essential oils. A skilful chemist can decide the type and length of the column he needs for the analysis of v. oils. GC/MS is the most frequently used technique for analyzing essential oil. Gas chromatograph is coupled with the detector mass spectrometer or ion trap spectrometer. A small sample of an essential oil is introduced into the GC, where it is heated and carried along the column by inert gas helium. As it vaporizes through the column, it separates into individual molecular constituents that are adsorbed on the stationary phase of the column. The separated constituents are ionized and carried through the module. The ionized constituents are amplified and detected as current by the MS. Each constituent is represented by a peak in a chromatograph and identified by superimposable mass spectra of standard component. Some lab use infrared spectrometer and NMR along with the mass spectrometer. The peak is cryogenically trapped and subjected to IR and NMR spectroscopy. This technique ensures a complete characterization of that peak.

   High-pressure liquid chromatography coupled with mass spectrometer is also being used for the analyses of volatile oils. Marriot *et al.* (2001) described a comprehensive review on gas chromatograhic technologies available for the analysis of essential oils. More recently the combination of gas chromatography-olfactometery and multidimensional gas chromatography for the characterization of essential oils is very beneficial technology as shown by Eyres *et al.* (2007).

Toxicity of Aromatic Oils

Concentrated essential oils should not be used internally or externally. They should be diluted before use. Essential oils are blended with carrier oil before applying to the skin for massage. Avoid getting essential oil into your eyes. Some essential oils are photosensitizes. They increase the skin's reaction to the sun and cause sunburn. One should avoid the use of essential oils such as lemon, bergamot, angelica, orange, tagetes etc. before going into the sun. Certain essential oils may also irritate sensitive skins. If irritation occurs, stop using the oil immediately. Some can cause allergic reaction.

Pregnant women should avoid the use of essential oils. Aromatherapy massage should not be performed on people who are ill or who have torn muscles or broken bones. People with high blood pressure, cardiac problems, cancer, and epilepsy, kidney or bladder disease should consult their physician before aromatic massage.

There are mixed opinions on the toxicity of essential oils. Those who have limited knowledge of the chemistry of essential oils think these oils are safe. However, the oils consist of various organic compounds that can be toxic if ingested in large quantity. Essential oils used in massage, in baths or in the form of inhalation are not toxic owing to their low dosages. However, survey of the literature revealed very little information of any fatalities via topical application of essential oils or due to excessive ingestion of specific essential oils.

Certain essential oils can cause problems to people taking medication, including those taking anti-coagulant and anti-depressive drugs. If you are on medication, until you have consulted your physician, or have otherwise sought expert advice, avoid undue exposure to essential oils.

Essential oils should be stored in amber colored or blue bottles at a cool temperature. Essential oils are highly flammable and should therefore not be used near naked flame.

Classification of Aromatic Oils

A. **Chemical Classification**
B. **Pharmacological Classification**

A. **Chemical Classification**: Essential oils are classified on the basis of the functional group of their active constituents as shown in Figure 13.

B. **Pharmacological Classification:** Aphrodisiac essential oils can be classified on the basis of their pharmacological action as shown in Figure 14.

B.1 Direct stimulants: V. Oils that invigorate the body and rejuvenate the spirit. These are known as depression eliminators such as bergamot, clary, sage, frankincense, geranium, lavender, lemon, sandalwood, tangerine, and vetiver.

B.2 V. Oils promoting hormonal stimulation or hormonal regulation in the body: Essential oils used to balance or promote hormonal activity include clary sage, anise, fennel, angelica, geranium, cypress, pine, and spruce.

B.3 V. Oils that tone balance and regulate our bodily functions: Bergamot, orange, rose; affect love-juniper, jasmine, sandalwood, lavender, ylang-ylang, Affect confidence-jasmine, sandalwood, patchouli, peppermint, rosemary, and ylang-ylang.

Such as bergamot, ginger, lavender, rose, valerian and Roman chamomile.

Application of Aromatic oils

Aphrodisiac oils can be used in many ways. They can be applied to the body by massage, added to the bath tub, spread in the environment with candles, and be diffused with room diffusers or on a humidifier. Essential oils are highly concentrated and are usually diluted before being applied to the skin through oil-based mixtures, such as salves, creams or lotions; alcohol or water-based tinctures; or with a water-soaked cloth. In therapeutic aromatherapy, essential oils treat medical conditions. For example, they are claimed to fight infections, promote wound healing, reduce inflammation, affect hormonal levels, stimulate the immune system, heat the skin in a liniment, promote blood circulation and digestion, and lessen sinus or lung congestion (Ryman 2004).

Metabolism of Aromatic Oils in the Body

Unlike many chemicals or drugs, essential oils do not accumulate and are quickly excreted from the body. Furthermore, unlike medications that must be swallowed and systemically absorbed, locally applied essential oils bypass the stomach and liver and, therefore, do not undergo any metabolic alteration. They go directly to the site of the sore and aching muscle, where they are needed the most. Their actions are below the threshold of consciousness.

Sensual massage allows aphrodisiac essential oils to be absorbed as well as inhaled and, of course, includes the erotic element of touch. Massage oil is a simpler blend, for example combining four drops each of sandalwood, lemon, and jasmine absolute essential oils with one ounce of sweet almond oil. Likewise, many formulations are prepared and sold as massage oils.

Aromatic oils as Aphrodisiacs

Appendix N contains important essential oils used as aphrodisiacs. Their name source, chemical constituents and compatibility with other essential oils are given in details. However, there is not sufficient scientific data available to support their aphrodisiac activities.

The essential oils of coriander (*Coriandrum sativum*) and cardamom (*Elettaria cardamomum*) are derived from the seeds of the plants and have long-standing reputations as aphrodisiacs. It is interesting to note that the seeds of the two plants share properties helpful in digestion and sweetening the breath, being especially effective in mitigating garlic odors. Both essential oils have spicy *aromas* and blend well with a wide range of other oils. Coriander has a light sweet edge to it, while cardamom has a warm floral undertone. Ylang-Ylang (*Cananga odorata*) and jasmine (*Jasminum grandiflorum*) essential oils are sweet floral scents with warm, sensual, euphoric properties. These are considered to promote a sense of calm and relaxation, and the potential to stimulate low libido.

Clary sage (*Salvia sclarea*), patchouli (*Pogostemon cablin*) and neroli (*Citrus aurantium*) essential oils are warm, rich, sweet, and considered to be exotic. Musk, clary and sage are known for their euphoric properties. Patchouli has a balancing, grounding component in its earthy fragrance that helps to ease anxiety. Naroli scent and effect run deep, lending support to love and sexual-emotional connection. Rose maroc (*Rosa centifolia*) and rose otto (*Rosa damascena*) are harmonizing, comforting, and romantic. They are reputed to be appealing to women. Though *Rose otto* has a more profound and rich floral fragrance, it is *Rose maroc* that is associated with passion. Both rose essential oils are supposed to blend well with clary sage, lavender, and ylang-ylang oil. One of the famous perfume formulations is Aura Cacia's recipe for Aphrodite's Amorous Blend, comprising two drops neroli, three drops each of rose absolute (*Rosa centifolia*), jasmine absolute, and sandalwood (*Santalum album*) with four drops ylang-ylang.

Amber is used in low libido. It is believed to reduce stress and provide harmony and balance Ancient Greeks and Egyptians considered amber to be an important healing remedy against mental illness, fever, throat infections, and rheumatism. The Greeks called amber the "sun stone" and believed it connected them to the sun god, creating an atmosphere of renewal and alertness. In ancient Greece, amber was called "electron", the root of the word "electricity." When rubbed this resin becomes electrostatic

Basil (*Ocimum basilicum*) as an aphrodisiac invigorates body and spirit. It helps refresh the mind promoting concentration. It has a sweet liquorice-like fragrance. Italian women displayed basil to alert possible suitors and the men would present the women with basil buds. In India the plant is held sacred to Lord Krishna and Vishnu. Vishnu's wife Lakshmi, the goddess of wealth was transformed into Tulsi or holy basil.

Bergamot oil (*Citrus bergamia*) rejuvenates the spirit and emotions with its fresh and invigorating citrus fragrance particularly in frigidity. The name is derived from the city Bergamo in Lombardy, Italy, where the oil was first sold. The oil is one of the most widely used in the perfumery and toiletry industry and forms, together with neroli and lavender, the main ingredient for the classical eau-de-cologne fragrance. It is used to flavor 'Earl Grey tea.'

Lavandula oil is used as aphrodisiac. Lavandula oil is obtained from the flowers of *Lavandula angustifolia.* The name is derived from the Latin word 'lavera' which means 'to wash' and the Romans used it frequently in their daily bath. It is said to have been introduced by them into England, where it soon became favorite. It has been associated with spiritual love. Aphrodisiac properties are by reducing stress. It is supposed to calm the shattered nerve and restores balance. It is used for women with low libido.

Matricaria chamomilla is commonly known as Chamomile-Blue (German). The blue color is from azulene, which is formed during the distillation of the oil. The odor is sweet and adds a warm, long-lasting undertone in perfumes. Chamomilla is derived from Greek word *Kamai melon*, meaning *ground apple* with its fruity smell resembling that of a fallen apple. The plant chamomilla is dedicated to St. Anne, mother of the Virgin Mary. It stands for patience in adversity due to its delicate but hardy nature. It is used as massage oil blend with lavender and neroli oil as an aphrodisiac.

Clary Sage (*Saliva sclarea*) as an aphrodisiac regenerates energy and inspires both mind and spirit. It can induce euphoria or feeling of elation. Clary sage is calming to the nervous system, particularly in cases of depression, stress, insomnia, and deep-seated tension. It is a good tonic for the females in general and helps in painful and scanty menstruation and relaxation during labor, thus encouraging a less painful birth. The Romans called it *herba sacra* or sacred herb because it was believed to banish all grief from mind and body. Its infusion in water was one of the most popular drinks in U.K. before the Indian and Chinese tea came to existence.

Cinnamon (*Cinnamon zeylanicum*) increases blood circulation towards the pelvic and is used as sexual stimulant. The Greek word *kinnamomon* or cinnamon means 'tube' or 'pipe' refering to the curved bark. Cinnamon oil was used as temple incense by the Greeks while the Egyptians used it for foot massage. It was also used as an ingredient for wines, love potions and as a sedative during birth. It was an important trade commodity between India, China, and Egypt.

Rose oil is obtained from *Rosa centifolia.* It stimulates and elevates the mood creating sense of well being. Fascinating smell of rose is intoxicating and acts as an aphrodisiac. The rose has been recognized by all cultures. The Turks believe it came from the blood of Venus. All the rituals from birth to death embrace the rose. Known around the world as the *Queen of Flowers* indeed, images of Cupid and Venus are crowned with roses. The 17th-century English physician Culpeper wrote that red roses strengthen the heart. This flower has come to symbolize innocence, love, passion, and desire. Rose petals were scattered at weddings to ensure a happy marriage and are still a symbol of love and purity and are also used to aid meditation and prayer.

Oil from jasmine flowers is another potential aphrodisiac. It is usually blended with lavender, rose, and bergamot. The name *Jasmine* is derived from the Persian word *Yasmin*. The Chinese, Arabians, and Indians used it medicinally, as well as for an aphrodisiac and for other ceremonial purposes. Jasmine grows along the Nile in Egypt and it is associated with the Egyptian mother goddess who held the secret of fertility, magic and healing. The plant is known as the queen of the night, meaning *Rat Ki Rani* in India. This wonderful fragrance has proven to have sensual effects in humans. The Fragrance and Flavor Association (New Jersey, USA) has found that it stimulates our senses without any negative side effects.

This very expensive and rare essential oil has been used as an aphrodisiac for hundreds of years in India. Women in India use a small garland of jasmine flowers in their hair at night to lure their husbands. *Moonlight of the Grove* is an Indian name given to this tropical flower. The rich, narcotic fragrance is most potent at night and the scent becomes more seductive when applied to the skin

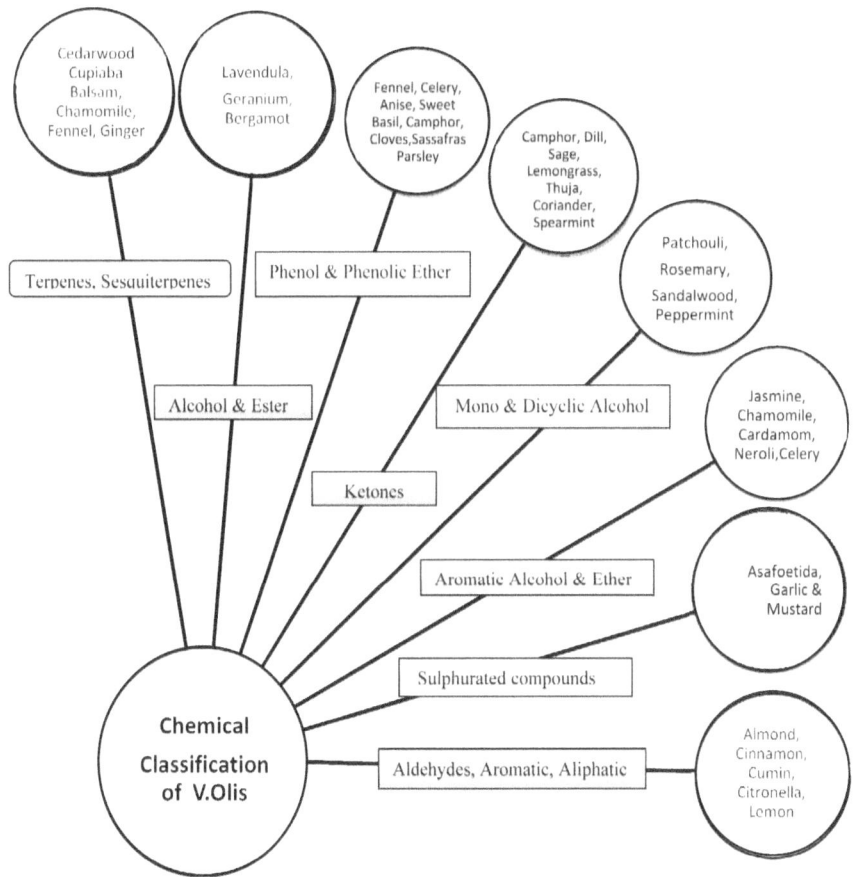

Figure 13: Chemical Classification of Aromatic Oils

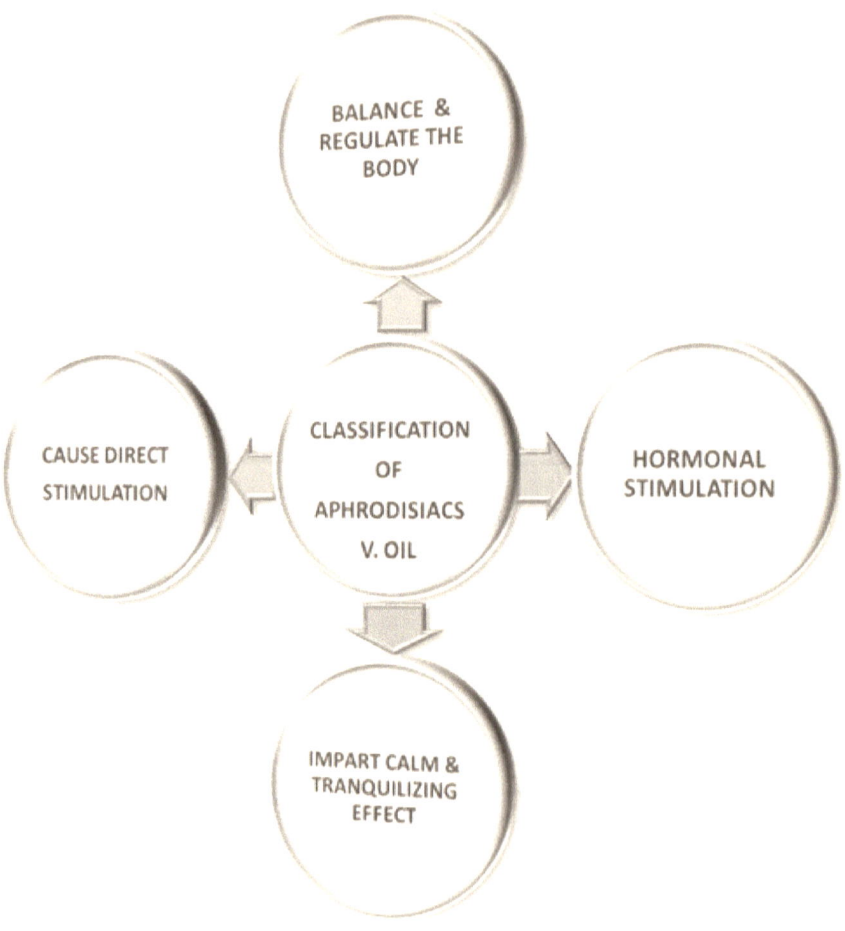

Figure 14: Pharmacological Classification of Aromatic Oils

The sweet aroma of vanilla obtained from the vanilla pods of *Vanilla planifolia*, also has aphrodisiac effects according to myth. The green fruits are picked and cured. The immature Vanilla pods or beans are fermented and dried to turn them into fragrant brown vanilla beans. It was so precious to the pre-Colombian Indians that they considered it suitable for their Gods.

Ylang-ylang oil also known as Cananga oil. It is obtained from the flowers of *Cananga odorata*. This extremely fragrant essential oil has a calming effect on the mind and body and is used in cases of frigidity and impotence. It has an excellent balancing and stimulating effect on the skin, and is also used to stimulate hair growth. It increases sexual energy and enhances relationship. It is used for evoking feelings of deep, calm that melt away anxiety. This oil has an intense and sharp fragrance that is much less expensive than either rose or jasmine. In Indonesia, the flower petals are scattered upon the bed on wedding nights. The oil was once a popular ingredient of hair preparations in Europe and was known as macassar oil.

Patchouli oil is derived from the *Pogostemon patchouli*. In the East, it is used in potpourris and sachets. It is used as an aphrodisiac because of its seductive smell.

Since antiquity, the saffron has been used as an aphrodisiac, and even being cited in the *Kama Sutra*. Cleopatra used saffron in her baths. The Persians used saffron as an aphrodisiac (Willard 2001). To the ancient Greeks and Persians, saffron was used to treat melancholia. Recent studies have shown that saffron is comparable to prozac and impiramine for treating depression. There is no research available on the aphrodisiac properties of saffron, but given its ability to influence neurotransmitters, there may be something to it. They significantly express a tangible tranquility, hard to understand in its totality. The stamens and stigma of the flowers of the plant are edible part of the saffron. A balanced consumption of saffron could create a state of physical and psychical comfort. Saffron has an aphrodisiac effect, being cited by Plinius the Old. The food with saffron can create a general state of wellness. Saffron is recommended to the women, especially the ones with genital illnesses. Saffron in milk is good for both partners before the sex; it creates an atmosphere of serenity in intimate relationships.

Rosemary oil obtained from *Rosmarinus coronarium* has been used for low libido. Rosemary means *dew of the sea* since it grows in proximity to the sea. Its leaves were traditionally burned in hospitals to purify the air. The ancient Romans used rosemary as incense and according to legend, Mary with Christ child stopped to rest on her journey, laid her blue cloak upon a rosemary bush full of white flowers and they all turned blue in their compassion. Although rosemary essential oil has many benefiting qualities, there are some contra-indications. Rosemary oil is also helpful in treating cellulite. This essential oil balances intense emotions and controls mood swings. It lifts 'Spirit' and reduces depression. It also assists in managing stress and overcoming stress-related disorders, nervous exhaustion, and low libido.

Sandalwood obtained from heartwood of *Santalum album* is one of the most recognized scents on earth. It has been used as an essential oil for 4,000 years in temple decorations, incense, perfumes, and cosmetics in India. Egyptians used it in embalming. This ancient perfume was used in Buddhist and Hindu rituals. Indian holy manuscripts *Vedas* cite sandalwood to be sacred. It has been used as ritual oil to achieve higher consciousness in meditation. Mysore, India is known worldwide as 'Sandalwood City.'

Aphrodisiac properties are due to its seductive smell. Sandalwood in Indian is believed to create an exotic, sensual mood with a reputation of an aphrodisiac. Major perfume companies use it as important constituents in fragrance formulation such as the fragrance Joy by Palou, one of the most expensive women's fragrances. Dior's fragrance Diorissimo, Chloe by Lagerfield, and Samsara by Guerlain comprise sandalwood as a base to add texture and depth to the perfumes. Sandalwood is much in demand as incense and has a calming effect during meditation. Yogis recommend it for the union of the senses and Tantric yogis for the awakening of sexual energy. This relaxing oil has a harmonizing and calming effect which reduces tension and confusion and is ideal for use in depression, hectic daily lifestyles and states of fear, stress, nervous exhaustion, chronic illness and anxiety. It enhances meditation by opening third eye meaning the super consciousness. Its blend with ylang-ylang, patchouli, spruce, lavender, frankincense oil makes it more seductive.

Frankincense oil obtained from the steam distillation of oleoresin of *Boswellia carteri* soothes and calms the mind, slowing down and deepening breathing and is excellent for use when meditating. It also helps to calm anxiety and obsessive states linked to the past. The deep benefits of the oil to focus consciousness and gain transcendent awareness have been recognized by spiritual traditions all over the world. The word *Frankincense* is derived from the French word *Franc* meaning 'luxuriant' or 'real incense' and together with Myrrh, it was the first gum to be used as incense. Also known as *Olibanum*, frankincense was used by the ancient Egyptians as an offering to the gods and as part of a rejuvenating face mask. It was also used to fumigate to banish evil spirits. It is one of the most important incenses. It was an offering to the baby Jesus by the wise men and it has been cited many times in the *Bible*. It was also dedicated through the ages to varieties of Sun Deities-Ra of Egypt, Apollo of Greece, and the Babylonian god Bail.

Musk oil formulations (Traynor 2001) with other herbs are used to stimulate low libido in many cultures. Musk has a great impact on Islamic culture. Muslims regarded it above all other scents and Muslim legend maintains that beautiful maidens who are the living embodiment of musk inhabit the Garden of Paradise. The scent's mystical properties were so highly valued in Persia that musk was mixed into the mortared walls of mosques at Tabriz and Kara Amed. In the Ayurvedic System of Indian Medicine, it has been used to stimulate low libido. In Europe, it has different reputation. The fragrance was discredited as unclean, immoral, habit-forming, and even poisonous. They accused musk of causing stomach problems in men, and nervous disorders from loss of consciousness to hysteria and insanity in women.

Conclusion

Perfume industries run in billions of dollars and very well indicate their influence on our lives. Choosing a perfume as an aphrodisiac is a very complicated task. It is a matter of personal choice. A perfume pleasant to one can be unpleasant to other. This programming starts at birth and continues until death. It may be a scent that brings back memory of a time of dejected love, or the cologne worn by your first love. Sometimes a particular woody or earthly smell can bring back the memory of the past and the environment prevailing at that time. Everyone has had the experience of smelling something that brings you back not only some embedded memory but also the emotions existing at that time.

It doesn't matter how many books recommend a certain fragrance as a sexual stimulant if you don't like its smell. The perfume reminds you of something sad from your past; you are not going to like it. Someone wearing a perfume that brings back the memories of a past love could stimulate or depress him.

Perfumes are the mood elevators and impart good feelings. Apart from elevating the emotions they do not have any logic to stimulate low libido. During ED or any other case of impotence they won't work. There is no single or cocktail of essential oils available that can stimulate the sex organs and make them ready for action. These are not a quick fix like Viagra. There is no scientific evidence to support the myths. At the same time, it is true that essential oils have been used since antiquity. They represent the life force, spirit, or soul of the plant (Berwick 1994). Pleasant odors can be soothing and may enhance relaxation. It could stimulate the mood of a person. However, except for the myths and anecdote evidences there is no scientific evidence that aromatherapy products act as aphrodisiacs.

Perfumes comprising natural essential oils are sold for several millions of dollars in the USA and even more all over the world. Most people wear these costly perfumes. Companies are making dollars from the scents. How beneficial these products are as aphrodisiacs is not known, however, they certainly stimulate the moods, if somebody is wearing a seducing scent. Again, how good are these mood elevators? It depends upon the liking of an individual.

> *"Odors have a power of persuasion stronger than that of words, appearances, emotions, or will. The persuasive power of an odor cannot be fended off, it enters into us like breath into our lungs, it fills us up, imbues us totally."*
>
> — *Patrick suskind*

Chapter Nine

Human Pheromones as Natural Aphrodisiacs

Pheromones in animals and insects have been known for a long time. These are the primary communication system for animals used to sense danger, food and mating. In fact, animals rely on pheromones for their survival.

Human pheromones may be defined as natural chemicals produced by an individual and transferred by air that affects the sexual physiology of another individual. They are believed to send out subconscious scent signals to the opposite sex that trigger very powerful responses.

The word pheromone is derived from the Greek word *pherein*-to carry, and *hormon*-to excite. Karlson and Luscher (1959) coined the term *pheromone* for the first time. Pheromones are called ectohormones, meaning chemical messengers that are transported outside the body and have the capability to trigger responses like physiological and behavior changes. Human pheromone detection has also been proposed to be the reason of instant attraction or dislike when first meeting someone.

Currently, human pheromones remained ambiguous bioactive compounds, as only a few have been identified. Standard bioassays have suggested that they are nonvolatile, activate vomeronasal sensory neurons, and regulate innate social behavior and neuroendocrine release. Recent discoveries of potential pheromones reveal that they may be more structurally and functionally diverse than previously defined (Stowers and Morton 2005).

There are overwhelming advertisements on the Internet, newspapers, magazines and TV shows that claim the effectiveness of human pheromones as sexual attractants and aphrodisiacs. According to these advertisers, the concept of seduction in a bottle is no longer a myth, rather becoming a reality these days. Love at first sight is well-known. However, love at first smell is also being recognized. Seduction, sexual attraction, lust, sexual drive, lovemaking, and falling in love are being attributed to human pheromones.

Advertisers claim that human pheromones make a person irresistible to the opposite sex. When added to a favorite perfume they are believed to enhance pheromone power into one's romantic life. Do they really work? Can these costly sex drops in a tiny bottle find the person of our choice and stimulate the sex desire or it is all in the head?

There are many pheromone websites on the Internet that make wild claims about their products. Exotic advertisements make emphatic claims that their products can lure any person with low libido to buy these products. Some commercially available human pheromones are being advertised using claims that these products contain sexual pheromones and can act as aphrodisiacs. Nevertheless, newspapers and magazine add fuel to the fire with emphatic news and articles about human pheromones. From time to time, there have been stories about human pheromones on 20/20, Dateline ABC, and various other television programs. Newspapers, medical journals, and many popular magazines have featured stories about the astonishing discovery of human pheromones as follows-

> *"Pheromones can improve one's sex life, pheromones send out subconscious signals to the opposite sex that naturally trigger romantic feelings."-(Willis 2002)*

> *"Scientific studies have actually shown that subjects who used synthesized pheromones had sex more often."-(Knowlton 1994)*

"The power of smell is undeniable humans are influenced by airborne chemicals undetectable as odors, called pheromones. Researchers at the University of Chicago say they have the first proof that humans produce and react to pheromones."   *-(Rowland, 1998).*

> *"Secret perfume ingredients that may boost your love life."*
> *-(Utton 2002)*

In spite of all these claims and propaganda, human pheromones often lack faith due to marketing by unsolicited e-mail and lack of consistent scientific data. Contradictory scientific reports cause doubts about their efficacy.

There are some people out there who want to know if human pheromones really work, and if so, which pheromone products are the best on the market. Do they really act as aphrodisiacs? To know the reality about human pheromones, it is very pertinent to understand the definition, occurrence, function, chemistry, and mechanism of their action (detection) as sex attractants or stimulants.

Occurrence

The apocrine (sweat) glands of humans secrete pheromones. They carry characteristic odor. Freshly produced apocrine secretion has no odor. The microbial conversion causes the odor (Zeng *et al.* 1992).

They are chemically similar to hormone dehydroepiandrosterone and are secreted by endocrine glands. Apocrine glands occur in the armpits, face, nipples, and anal and genital regions of both sexes. The apocrine glands become functional after reaching puberty which, some believe, could contribute to people developing a sexual attraction for others at that time (Cohn 1994).

Cutler *et al.* (1986) conducted the first controlled scientific studies to reveal the existence of pheromones in humans. Prior to their landmark research there were no conclusive indications that humans excreted pheromones. A few well-controlled scientific studies have been published demonstrating the possibility of pheromones in humans. The best-studied case involves the synchronization of menstrual cycles among women based on unconscious odor cues, the so called *McClintock effect*, named after the first investigator (McClintock 1971). In the later stages, a similar synchrony was observed between mothers and daughters living in the same house. They also have the menses at the same time. Study by Stern and Mclintock (1998) states that human female axillae release two types of pheromones. One, produced prior to ovulation, shortens the ovarian cycle, and the second, produced just at ovulation, and lengthens the ovarian cycle.

Pheromone production starts to decline with the level of declining of sex hormones. That means the production is inversely proportional to age. This links between apocrine gland function and puberty reflects that function is closely related to level of sex steroid hormones in humans.

Function
It is assumed that human pheromones inhaled may cause sexual attraction, relaxation, excitement, euphoria, and regulation of menses. Pheromones may affect women and men differently. Every man and woman produces pheromones. Every person has an odor print as unique as his or her fingerprint. Odorprints depend upon diet, gender, heredity, health, medication, occupation, and mood. Human pheromones can be used in menstrual regulation and can be used in marriage therapy to improve interpersonal relationships. Its use can protect people from sex offenders and rapist using a characteristic odor that may dispel the offenders. The significance of pheromones is summarized in Figure 15.

Detection
Human pheromones are detected by vomeronasal organ (VNO) unlike fragrances, which are detected by the olfactory glands in the nose and are a special part of the olfactory system. The VNO is located in the nasal pit directly under the nose. When the VNO receives a pheromone signal it sends a message to the brain, going to the hypothalamus, the center in the brain that control one's basic drives and emotions. The signal stimulates the body and creates a subconscious increase in sexual desire in opposite sex; women get attracted to men and men to women (Monti-Bloch *et al.,* 1991; 1994). However, so far there is no concrete evidence that the human VNO is connected to a functional accessory olfactory system. This has caused significant controversy among the scientists even about the presence of pheromones in human (Martinez-Macros 2001; Regelson 2002).

Figure 15: Significance of Pheromones

Berliner *et al.* (1996) tried to prove that humans have a functional VNO. They measured electrical activity in VNO tissue in response to chemical stimulation, which may indicate that the cells are transmitting signals. However, cells that generate electrical activity are not necessarily transmitting signals to the brain, says Wysocki, neuroscientist at the Monell Chemical Senses Center, Philadelphia. There must be chemical communication among humans. Infants respond to breast pads worn by their mothers but not those worn by other women and a mother can identify the T-shirt worn by her infant from a pile of T-shirts. "We know its chemical, but is the information being transferred pheromonal?" asks Wysocki. "No one knows." Other researchers found Berliner *et al.* results intriguing, but not conclusive. "Clearly there's something present in humans but it is quite different than in animals," says Meredith, Professor, of the Department of Neurobiology at Florida State University, Tallahassee, Florida (Ettienne 2002; Wyoscki and Preti 2000).

Monti-Bloch *et al.* (1994) purified several potential pheromones found in sweat and other human secretions. One secretion, purified from the skin of men seems to affect mood in women. In one study, Monti-Bloch applied either the secretion or a placebo directly to the VNO of 40 women. The women exposed to the secretion displayed a statistically significant decrease in negative affect. Although VNO is often regarded as only pheromones detector, evidence is emerging that suggests it might respond to a much broader variety of chemosignals (Brennan & Keverne 2004).

Chemistry

As discussed above, human pheromones are largely produced by the skin's apocrine sebaceous glands, which develop during puberty. They are usually associated with sweat glands, other glandular secretion and to skin flora present in moist area of the body. The substances produced by these glands are imperceptible by the human nose; because these are not the fresh glandular secretions of the skin but rather the bacterial breakdown products of these glandular secretions. The sebaceous secretions themselves comprise mostly of lipids such as squalene and other esters. After degradation by bacteria present on human skin it give rise to free fatty acids including hircine which is very unpleasant. The most prominent examples of these hircine fatty acids are butyric acid, caproic acid, and caprylic acid.

Chemically, the active constituents of human pheromones are similar to DHEA (dehydroepiandrosterone). The most commonly known pheromones (Figure 16) are:

Androstenone	(5 alpha-androst-16-en-3-one)
Androstenol	(5 alpha1androst-16-en-3-ol)
Androsterone	(3 alpha hydroxy-5 alpha-androstan-17-one)
Androstadienone	(delta 4,16-androstadien-3-one)
Androstadienol	(delta 4,16-androstadien-3-ol)
Copulins	(C2-C5 aliphatic acids)
Estratetraenol	(estra-1,3,5(10) 16-tetraen-3-ol)

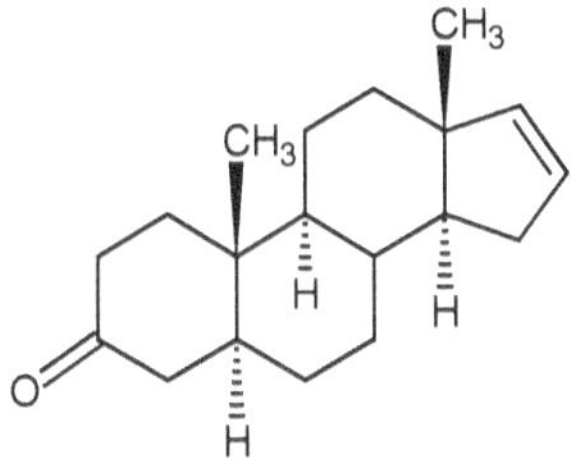

Androstenone

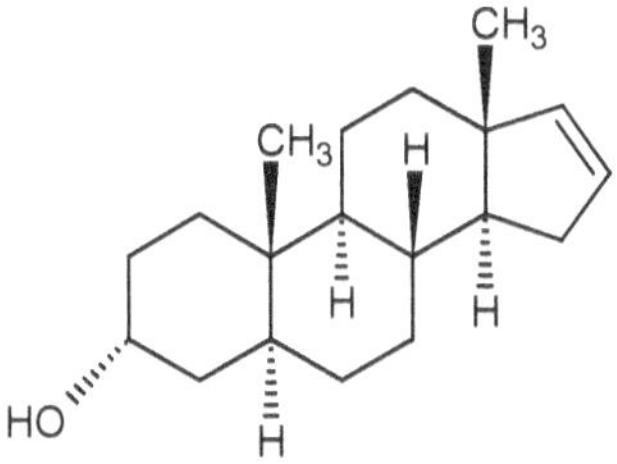

Androstenol

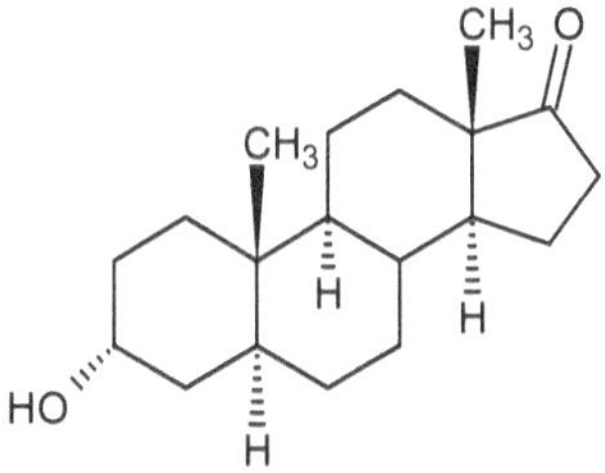

Androsterone

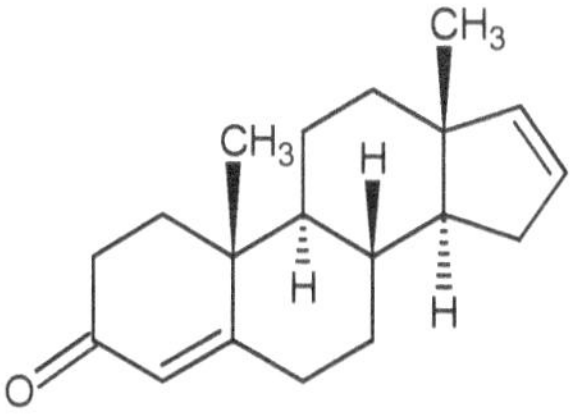

Androstadienone

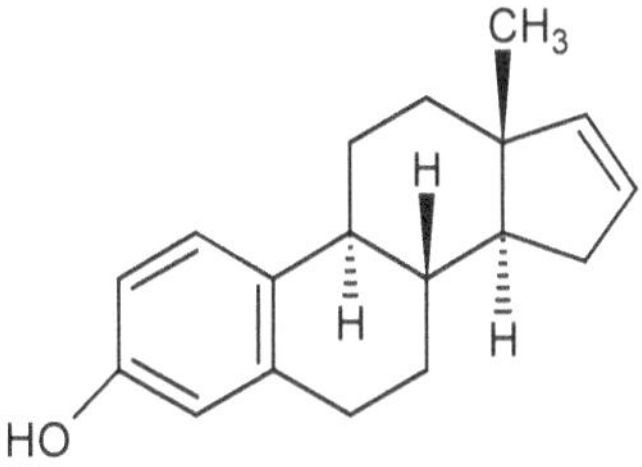

Estratetraenol

Figure 16: Commonly Used Human Pheromones

Androstenone, androstenol, androsterone, androdienone, and androdienol have been identified as the male pheromone that increases the luteinizing hormone (LH) in women. Presence of testosterone and estrogens has also been shown in the pheromone secretions.

Estratetraenol is supposedly the female equivalent to androstadienone. It stimulates the VNO and acts as a mood elevator.

Female pheromones are known as copulins. Copulins do behave like pheromones but chemically dissimilar to any other compound that is considered pheromone. These are found in human vaginal secretion and are secreted into the vagina during ovulation. Copulins believed to increase male testosterone levels that are directly linked to increased sexual drive. Chemically copulins are volatile C2-C5 aliphatic acids (Michael *et al.,* 1974). Huggins and Preti (1976) studied the chemical Composition of copulins in 12 patients for 44 ovulatory cycles by means of gas chromatography, tandem with mass spectroscopy to identify organic volatile components. These vaginal secretions contain a mixture of aliphatic acids, alcohol, hydroxy ketones, and aromatic compounds.

Some manufacturers use oil base for human pheromones and claim that a drop of alcohol can completely destroy pheromones (Pheromones-pheromones.com) where as the other recommends the use of alcohol as base. They believe that alcohol keep the solution sterile and increases shelf life. It kills bacteria on the skin where the pheromones are applied. Skin bacteria are capable of breaking down pheromones fairly quickly, which destroys their effectiveness (pherone.com). The company further states that alcohol does not change the chemical nature of synthetic pheromones. However, there is no evidence or supportive studies in both the cases. Mostly human pheromones are steroids in chemical nature and are soluble in organic solvents. These compounds are lipophilic and like nonpolar solvents. Some of the manufacturers are selling pheromones in the form of shave lotion or cologne and as additives to perfumes. Some believed that deodorants and antiperspirants often destroy the natural human pheromones that humans do produce.

Classification of Human Pheromones
During the past twenty-five years, there has been a significant increase in the number of studies conducted on human pheromones.

The scientific literature recognizes four classes of human pheromones:
a. Olfactory recognition (Mother-infant recognizer)
b. Primer (Menstrual synchrony)
c. Modulators (Human sex-attractant pheromones).
d. Releaser (Behavioral effect)

Recognizer pheromones provide information such as mother-infant recognizer. An infant can smell her own mother's pheromones through the air and mother can also recognize the shirt of her infant from a pile of shirts.

Pheromones that affect endocrine glands are primers. These can influence long term changes in hormones levels such as menstrual cycles, puberty, and pregnancy.

Pheromones that modify emotions or moods are known as modulators such as sex attractants. These pheromones elevate the mood and alleviate stress.

Finally, those that affect behavioral releaser responses are known as releaser. However, according to Wysocki and Preti (2004) there is no good evidence for releaser effects in adult humans. They further emphasized that no bioassay-guided study has lead to the isolation of true human pheromones, a step that will elucidate specific functions to chemical signals.

Role of Human Pheromones

a. Regulating Menstrual Cycle
Grammar and Jutte (1997) suggested that male pheromones androstenol/ androstenonone from male sweat have a direct impact on the female menstrual cycle and ovulation. Effects of male pheromones on females are summarised in the Figure 17.

Cutler (1999) showed that pheromones in men's bodies can cause their female sex partners to be more fertile, have more regular menstrual cycles and milder menopause. Women who have sex with men at least once a week are benefited the most from these chemicals. They have regular menstrual cycles and fewer fertility and menopause problems, apparently because of exposure to pheromones. Cutler's studies show that women are affected by pheromones from men. Women with unusually long or short menstrual cycles get closer-to-average cycles after regularly inhaling male essence, described as a compound of male sweat, hormones, and natural body odors.

b. Synchronizing Menstrual Cycle
Cutler et al. (1986) confirmed a long-observed phenomenon (Mclintock 1971) that women exposed to another woman's *female essence* menstruated at the same time. After a few months, women who live together menstruate at the same time. However, according to Trevathan et al. (1993) this synchronization was not observed among lesbian couples. Morofushi et al. (2000) found that women whose menstrual cycles synchronized with roommates had higher olfactory activity for androstenol.

c. Increase Sex Attraction
Scientific studies (Gorner 2000) have actually shown that people who used synthesized human pheromone had sex more often. Researchers at the University of Chicago said, "The power of scent is undeniable, humans are influenced by airborne chemicals undetectable as odors, called pheromones."

Human pheromones also act as sexual attractants. The secretion of pheromones by humans is believed to significantly increase the desirability and sexual attractiveness in both males and females suggested by Thorne et al. (2002) and Grammer et al. (2005).

Willis reported (2002) that scientists at San Francisco State University found that women who had pheromone added to their perfume reported a more than 50% increase in sexual attention from men. The study, which was conducted by McCoy and Pitino (2002), found that 74% of the women saw an overall increase in 3 or more of the following socio-sexual behaviors: frequency of dates, kissing, heavy petting and affection, sexual intercourse, and sleeping closer to their partner.

Copeland and Link (2001) published the findings of an Australian organization, Bennett Research, which conducted a survey of 306 men using human pheromones. Their findings showed that use of pheromones had increased attractiveness to women. They reported an increase in various socio-sexual behaviors, as follows: making conversation-61%; starting up a conversation-52%; expressing an interest in the man-43%; being responsive to him-40%; paying unsolicited compliments-36%; and overt flirting-34%.

Sobel and Brown (2001) reported that men and women's brains respond to two putative pheromones-related to testosterone and estrogen. When men smelled a compound similar to estrogen, increased blood flow to the hypothalamus was documented. In turn, female participants experienced increased blood flow to the hypothalamus when exposed to the testosterone-like compound. These researchers found that the testosterone-like compound existed in men's sweat at levels 20 times that of women's. They monitored the results using sophisticated brain imaging techniques.

d. Human Pheromones and other Related Attractions
Men are strongly attracted to women with large breasts, thinner waists, and broader hips being attributed to female pheromones. Similiarly blondes, brunettes and red head smell differently (Kohl 2002). Likewise, women prefer tall and handsome men. Their selection is linked to an increase in male pheromone production (Kohl 2002).

Countless similar studies have since been conducted and the results have always been stunning. Pheromones appear to have the ability to dramatically increase sexual attraction between men and women. No doubt, these studies have ignited hopes in people who are assiduously buying them.

e. Send Specific Signals to Potential Mates:
Human pheromones may play a big role in selecting a mate according to a new study conducted by researchers at the *Monell Chemical Senses Center*, Philadelphia. They found that your preference for another person's body odor is influenced by the gender and sexual orientation of that person as well as your own gender and sexual orientation (Martins *et al.* 2005).

"Our findings support the contention that gender preference has a biological component that is reflected in both the production of different body odors and in the perception of and response to body odors," says Charles Wysocki, neuroscientist at the Monell Chemical Senses Center.

f. Regulate Ovulation Time

Stern and McClintock (1998) found that odorless compounds from the armpits of women in the late follicular phase of their menstrual cycles accelerated the preovulatory surge of luteinizing hormone of recipient women and shortened their menstrual cycles. Axillary (underarm) compounds from the same donors that were collected later in the menstrual cycle (at ovulation) had the opposite effect: they delayed the luteinizing-hormone surge of the recipients and lengthened their menstrual cycles. By showing in a fully controlled experiment that the timing of ovulation can be manipulated, this study provides definitive evidence of human pheromones.

Preti *et al.* (2003) studied the effect of male axillary extracts containing pheromones. Human under arm secretion when applied to a women recipient, altered the length and timing of the menstrual cycle, reduced tension and increased relaxation. These results demonstrate that male axillary secretions contain one or more pheromones that can act as primer (endocrine affect) and modulator pheromones.

It is now evident that human pheromones, a type of social chemosignal, modulate endocrine function by regulating the timing of ovulation. In animals, pheromones not only regulate ovulation but also female reproductive motivation and behavior. There is no extant evidence that humans produce social chemosignals that affect human sexual motivation or reproductive behavior as occurs in other mammals.

Spencer *et al.* (2004) demonstrated that natural compounds collected from lactating women and their breastfeeding infants increased the sexual motivation of other women, measured as sexual desire and fantasies. Moreover, the manifestation of increased sexual motivation was different in women with a regular sexual partner. Those with a partner experienced enhanced sexual desire, whereas those without one had more sexual fantasies. These results are consistent with previous pheromonal effects on endocrine function, and require further study of these social chemosignals as candidates for pheromonal processes.

g. In Postmenopausal Women

In the past twenty years hundreds of peer-reviewed studies have provided significant information to guide women during and after their menopause phase. One of the guidelines is hormonal replacement therapy. It is a successful treatment in a majority of the cases; however, hormonal replacement therapy does not work for everybody.

Some other regimens for menopause have the potential to produce disease, especially over-the-counter remedies like dehydroepiandrosterone and the formulas that contain estrogen. All sex hormones influence physiologic systems, including the cardiovascular system, bones metabolism, cognitive function, sexual response, and sexual attractiveness. Menopause is an extremely complex phase. During this period, some women increase their estrogen levels to new lifetime highs; others start an unequivocal decline, and still others vary from month to month. In addition to estrogen, changes in progesterone and androgen secretion by the ovary also occur. Many women show increases in circulating androgens while many others show deficiencies. Both the adrenal and the ovarian sources of these hormones show age-related changes that alter a woman's capacity to attract sexual attention through both her physical appearance and her pheromonal excretions. The menopause phase differs from woman to woman. The phenomena is complex and very complicated and most of the time difficult to understand. In most of the cases, a hysterectomy is suggested but there is no guarantee that hysterectomy works. Estrogen, progesterone, and androgens all tend to be compromised by hysterectomy; all should be considered for replacement. Hormonal regimens can be prescribed to enhance the quality of life; the review of the available research on pheromones can allow the medical art to greatly benefit mature women. Avoidance of hysterectomy helps prevent its side effects such as sexual deficiencies, acceleration of cardiovascular disease and more rapid aging. Cutler and Genovese-Stone (1998) concluded that human hormones if prescribed appropriately, best serve the menopause patient. They suggested that regular exposure to human pheromones from both sexes can provide health benefits that include better endocrine secretion, healthy estrogens level and a reduced risk of developing osteoporosis and heart diseases.

Cutler and Genovese-Stone (2002) further reviewed pheromones and their effects with a special emphasis on their potential contribution to sexual attractiveness in the menopause. Key topics included were biological functions of pheromones in humans and the source of pheromones in humans. The axillary extract studies that led to the independent synthesis of pheromones, olfactory mechanisms for mediating pheromones, aging, attractiveness, and sexual dysfunction. Physical attractiveness is important for a better quality of life. Three separate, double-blind, placebo-controlled investigations, using the same protocol, demonstrated that a synthesized human pheromone, topically applied, increased sexual attractiveness. If partners are available, sexual attractiveness can increase affectionate intimate behavior, which, in turn, increases well-being and quality of life. However, more research is needed to address ways in which postmenopausal women can benefit from these pheromones

h. In Socio-Sexual Behavior
Friebely and Rako (2004) studied the pheromonal influences on socio behavior in postmenopausal women to determine whether a putative human sex-attractant pheromone increases specific sociosexual behaviors of postmenopausal women. They tested a chemically synthesized formula derived from research with underarm secretions from heterosexually active, fertile women that was recently tested on young women. Participants(n=44, mean age 57) were postmenopausal women who volunteered for a double-blind placebo-controlled study designed, to test an odorless pheromone, added to their preferred fragrance, to learn if it might increase the romance in their life. During the experimental 6-week period, a significantly greater proportion of participants using the pheromone formula (40.9%) than placebo (13.6%) recorded an increase over their own weekly average baseline frequency of petting, kissing, and affection. More pheromone (68.26%) than placebo (40.9%) users experienced an increase in at least one of the four intimate sociosexual behaviors. These results suggest that the human pheromone formulation worn with

perfume for a period of 6 weeks have sex-attractant effects for postmenopausal women.

To find the influences of human pheromones on sociosexual behavior in young women a double blind, placebo-controlled study of a synthesized putative female pheromone was conducted by McCoy and Pitino (2002) with regularly menstruating, university women (N=36, mean age=27.8). The pheromone formula was derived from their earlier work investigating the underarm secretions of fertile, sexually active, heterosexual women. A vial of either synthesized human pheromone or placebo was selected blindly and added to a subject's perfume. Subjects recorded seven sociosexual behaviors and reported them weekly across three menstrual cycles. Beginning with Day 8 of each cycle, the first cycle contained a 2-week baseline period followed by an experimental period of as many as 3 weeks each from the next two cycles for a maximum of 6 weeks. The 19 pheromone and 17 placebo subjects did not differ significantly in age, weight, body mass index, dating status or ethnicity or in reported accuracy, back-filling data, and perception of a positive effect or perfume use. Placebo subjects were significantly taller than pheromone subjects. Except for male approaches, subjects did not differ significantly at baseline in average weekly sociosexual behaviors.

A significantly greater proportion of pheromone users compared with placebo users increased over baseline in frequency of sexual intercourse, sleeping next to a partner, formal dates and petting/affection/kissing but not in frequency of male approaches, informal dates or masturbation. Three or more sociosexual behaviors increased over baseline for 74% of pheromone users compared with 23% of placebo users. They conclude that this synthesized human pheromone formula acted as a sex attractant pheromone and increased the sexual attractiveness of women to men.

Cutler *et al.* (1998) also studied the influence of human pheromones on sociosexual behavior in men. This study tested whether synthesized human male pheromones increase the sociosexual behavior of men. Thirty-eight heterosexual men, ages 26-42, completed a 2-week baseline period and 6-week placebo-controlled, double-blind trial testing a pheromone "designed to improve the romance in their lives." Each subject kept daily behavioral records for 6 sociosexual behaviors: petting/affection/kissing, formal dates, informal dates, sleeping next to a romantic partner, sexual intercourse, and self-stimulation to ejaculation (masturbation). Significantly more pheromone than placebo users increased above baseline in sexual intercourse and sleeping with a romantic partner. There was a tendency for more pheromone than placebo users to increase above baseline in petting/affection/ kissing, and informal dates, but not in self-stimulation to ejaculation or in formal dates. A significantly larger proportion of pheromone than placebo users increased in sociosexual behaviors involving a female partner. Thus, there was a significant increase in male sociosexual behaviors in which a woman's sexual interest and cooperation plays a role. These initial data need replication but suggest that human male pheromones affected the sexual attractiveness of men to women.

i. In Homo and Heterosexuality

Using a brain imaging technique, Swedish researchers at the Karolinska Institute in Stockholm, Sweden, have shown that homosexual and heterosexual men respond differently to two odors that may be involved in sexual arousal, and that the gay men respond in the same way as women. Lesbian women and heterosexual women respond differently to the scent of human pheromones. This research suggests a possible role for human pheromones in the biological basis of sexual orientation. More specifically, heterosexuality and homosexuality is determined by one's preference for a specific odor or pheromones that of a man or woman (Berglund *et al.* 2006; Hitti 2006; Kohl 2006; Martins *et al.* 2005).

Most Commonly Known Human Pheromones

A number of human pheromones have been isolated and studied. Some of them have been found effective for attracting members of the opposite sex. These are androstenone, androstenol, androsterone, androstadienone, and copulins.

Androstenone appears to be the essence of male aggression and dominance. Women are attracted to men who naturally secret a large quantity of these pheromones. However, it has been shown that females find the odor of androstenol to be attractive or the perception of this odor of androstenol results in stimulating female sexual arousal while androstenone induce negative emotions particularly when they are menstruating or close to that period (McCollough *et al.* 1981). Commonly available preparations containing androstenone are Scent of Eros (Men's Version), Perception, Alter Ego (Men's and women's versions), and Perfect 10.

Androstenol is produced from the start of puberty until the early twenties, after that it starts declining. Androstenol has been considered to alter people's impressions of a man attractiveness, intelligence, and confidence. Cowley and Brooksbank (1991) reported that females exposed to androstenol engage in sex more, and had a longer duration and deeper interaction with men. Androstenol is known as the friendly pheromone. It tends to make the person nice and friendly. On the other Shinohora *et al.* (2000) reported that androstenol retarded the growth and maturation of ovarian follicles and consequently delayed the timing of ovulation.

Preparations containing androstenol are Scent of Eros (Men's version), Alter Ego (Men's and women's versions), Perception, Perfect 10 (Men and Women's versions), Chikara, Passion Pheromone Attractant (women's version).

Androsterone is known as a mood elevator. It is assumed that men with a larger ratio of androsterone are more masculine and dominant. When women wear this pheromone, it elevates their mood. Preparations containing androsterone are Scent of Eros, Alter Ego, Perfect 10, and Perception.

Androstadienone is a very popular pheromone that has a very specific effect on the brain activity of women. It is considered to affect attention and social cognitive areas of the brain. It can elevate a woman's mood and even alleviate PMS (Premenstrual

Symptoms) stress. It has been known to increase intimacy and caring feelings of women and is thus known as *Love Pheromone*. Preparations comprising androstadienone are Pheromax (Men's version), and Realm (women's version). Grosser *et al.* (2000) observed that administration of androstadienone results in a significant reduction of nervousness, tension and other negative feelings in women. Bensafi *et al.* (2004) reported that sniffing androstadienone elevates the mood and autonomic arousal in women. Wyart *et al.* (2007) revealed that smelling androstadienone found in the sweat of man increases the cortisol level in women. Villemure and Bushnell (2007) studied the effects of the steroid androstadienone on the mood and pain perception of men and women. They found that it diminished pain and elevated the mood of men and women. Today there are many companies that are selling different kind of pheromone products. These products contain one or all of the human pheromones such as androsterone, androstenol, androstenone, and copulins. Copulins are found in the products designed to lure men.

Copulins

Copulins are found exclusively in human vaginal secretions as discussed earlier (Waltman *et al.* 197; Michael *et al.* 1974, 1975). It is assumed that once a man smells copulins on a woman she is deemed to be more attractive. Secretions as well as quality and quantity of copulins vary with menstrual cycle phase of an individual. The odor of copulins and its behavioral effect also vary with the menstrual cycle. Sokolov *et al.* (1976) isolated short chain fatty acids from the fractionation of human vaginal secretion by chromatographic techniques and confirmed that there is a possible correlation of the rise and fall of hormone level during the female menstrual cycle. These acids were found predominant at midcycle and during the luteal phase (Huggins and Preti 1976). Female pheromones copulins, influence male perception of females and induce hormonal changes in males (Grammer and Jutte 1997). The preparations comprising copulins are being sold as female pheromones to lure males. Most common pheromone products that contain copulins are as follow:

Essence of Woman (EW)
Primal Instinct (PI)
Passion Copulins Concentrate (PCC)
Passion Pheromone Attractant (PPA)
Perfect 10 (P10-Women's version)
Scent of Eros (SOE-Women's version)
Pheromax (PMAX-women's version)
Realm (contains Estratetraenol-Men's version)

Toxicity of Pheromones

Adverse reaction: Higher concentration may cause aversive effect.
Contraindication: None so far.
Side effects: None known.

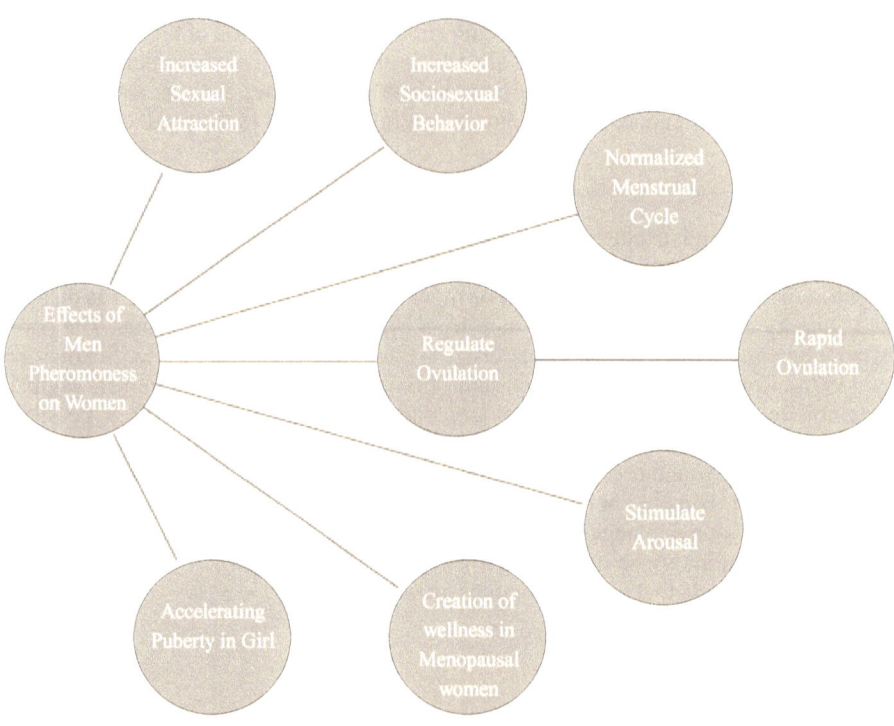

Figure 17: Effects of Men Pheromones on Women

Toxic Effects

These could cause skin reaction or can be allergic to hypersensitive person. An overdose may cause headaches, impression of weakness.

Shelf Life

Shelf life of pheromones is not very well studied. However, shelf life depends upon the purity, concentration, and conditions of storage of the finished products. If the product is devoid of microflora, it may last longer.

Products on the Market

There are three kinds of pheromone products in the Market:

A. Pheromone colognes
B. Pheromone sprays
C. Pheromone releaser

Products are sold in various forms as concentrated pheromones or diluted in some base or sometimes these products have perfume additives. However, these perfume additives do not enhance their action as revealed by Winman (2004). These products are called pheromones perfumes.

There are a number of manufacturers who produced pheromones products and pheromones perfumes. Search of any engine on the Internet will give you lots of information about these preparations. Each manufacturer claims the inclusion of human pheromones in his products. The first is Realm Corporation, formerly known as Erox Corporation claiming that their products are designed to enhance sexual attraction, confidence and feeling of well being in both the sexes. They promote pheromones perfumes such as **Realm Perfume, Pherose and Realm Cologne** for male and females. Similarly Estee Lauder (NY) a famous perfume manufacturer is also launching pheromone perfumes. Likewise, Athena Institute of Chester Springs, Pennsylvania founded by Dr Winnifred Cutler, co discoverer of pheromones in humans has also produced products like **Athena pheromone 10X** for men and **Athena pheromone 10-13** for women. These products are in the form of shave lotion or cologne or cosmetics and supposedly enhance sex appeal and romance in life. Another company, Health Freedom Nutrition, Santa Rosa, California has produced a pheromone containing product for men and women known as Eroscent. Erosscent is placed on the upper lip so that it directly affects the VNO (Vomeronasal organ). The same product works for both the sexes. It contains the pheromone derived from dehydroepiandrosterone (DHEA). They claim that just one drop of Eroscent applied to the upper lip of either sex imparts mood improvement, social relaxation, improves ones self-image, self-esteem, and sexual confidence, and attracts the opposite sex. Another company known as Pherone, Wilson, North Carolina sells pheromone products such as **Pherone Formula M-11, Pherone formula M-15, and pherone formula W-1** for men and women. The company claims that their preparations are very effective to enhance sexual interactions, attractions and create a light euphoria.

Pheromone preparations are very expensive and range from $50 to $110 for a small quantity (10ml). It is very difficult to find the right product that suits to an individual chemistry. There are thousands of brands and formulations of pheromones available on the market in the USA with different concentrations and ingredients. Only a few meet high standards in quality, research, and product knowledge (Regelson 2002).

Conclusion

During recent years, there has been much speculation on human pheromones. They are considered sex attractants, mood elevators, and aphrodisiacs. The fundamental question arises in mind: are pheromones in humans a myth or reality? At this point, it is very difficult to answer this question. It can be both. The field is in its rudimentary stage. Contradictory findings make the question difficult to answer. Some researchers demonstrate the existence of human pheromones, other refute their existence. Moreover, a majority of the work in this field is qualitative and without adequately controlled studies and statistical data. Majorities of the earlier studies have been criticized on the methodology, insufficient sample size, inadequate control groups, and lack of ecological validity and double blind studies. Moreover, human response is variable from person to person. The results vary from one lab to the other, meaning there are too many uncontrolled variables.

It is well recognized that pheromones play a significant role in the life span of animals. Ever since scientists discovered pheromones 30 years ago, they have found pheromone communication in hundred of species. They depend upon them for their food, mating, and signals of danger and survival. There is no doubt; this invisible and undetectable force runs their life. Animal action after sensing the pheromones is without reasoning. They can be lured by them. How about human beings? Are we lured by such unambiguous signals? Humans behaving in a similar manner do not make any sense, having intellect, emotions that govern by physical attraction. Man makes his move on the basis of physical, intellectual, and spiritual data instead of *sniffing*. Sexual identity, preference, performance depends upon the stimulation of sexual awareness through body appearance, body odor, vision, hearing and emotions. It is a complicated phenomenon and does not depend on a single factor (sniffing).

For years, scientists have been researching the same latent force in humans that can draw to one being to another. They have been successful to some extent in isolating and characterizing human pheromones that can attract the opposite sex. However, some of the scientific data is controversial (Wysocki and Preti 2004). After two decades of research, the field of pheromone is still in the infancy stage. More scientific evidences are required to support this hypothesis. The idea of applying a few drops of human pheromone formulation by an individual to lure the opposite sex sounds bizarre. It is difficult to accept the tall claims of the advertisers; however, people are buying them.

The chemistry of love is a complicated phenomenon and depends upon many factors in addition to these sex attractants. Love is natural not synthetic. One cannot be in love just sniffing the other person. It is a blend of physical attraction, emotions, and the right chemistry between the two persons. The pheromones industry does not care for this notion. They have sufficient raw data to their advantage to support the production and marketing of the synthetic version of these human pheromones as perfumes or colognes. In spite of the lack of consistent clinical data, these pheromone products are being sold like hot cakes. Love at first sight is known forever. But love at first sniff sounds weird.

It is hard to admit that pheromones act as aphrodisiacs. Do they stimulate the sexual desire or low libido? However, manufacturers claim that pheromones increase sexual desire and name them as olfactory aphrodisiacs. How much these products help to stimulate low libido, nobody knows. They may be acting as a sex attractant or mood elevators when added in a perfume. Furthermore, it is not clear whether the perfume is luring the opposite sex or the pheromones. Testimonials given by the people vary from person to person. Men and women react to aromas like perfume, cologne and pungent footwear and armpit sweat very differently. The same goes with the pheromone preparations.

Human pheromone products may alleviate the mood or sense of well being in some person depending upon their liking and prejudices. The multi-billion dollar perfume industry indicates that the power of smell is undeniable. That is the reason people are buying them assiduously. However, buyers should be aware of the fact that FDA does not control these pheromone products either. Moreover, there is no way to determine the validity of claims of the advertisers or toxic side effects of these products.

The future of human pheromones is bright; many commercial enterprises are in hot pursuit of sexual fragrances. In old age, physical and sexual attractions diminish with the decline of hormones. Quality of the odor of human pheromones and copulins are also changed at this age and rejuvenation of the entire body is required. Either the pheromones or perfume industry should take advantage of this fact and imbibe youthful spirit by the help of suitable pheromone preparations in aging men and women. Women particularly, during and after menopause, have low libido and sexual interest. Instead of HTR (Hormone Treatment Therapy) pheromone preparations should take their place. If they motivate the emotions, hormones can be motivated since hormones and emotions go hand in hand. Preliminary studies with human pheromones have shown some encouraging results in postmenopausal hormonal therapy as discussed earlier. The pheromones can be used as an alternative or complementary treatment to HRT.

At present, the number of good studies available on human pheromones and their role in human reproduction is very limited. Active research can be conducted to isolate specific pheromones from men and women that could be used as fertility agents for couples who wish to conceive, and as contraceptive who do not want to conceive. Human pheromones can be used as mood regulators and could help to control depression. Human pheromone treatment could control prostate activity in men to reduce the risk of cancer since the prostate gland is highly hormone dependent, and sexual activity plays a significant role in determining the prostate cancer risk.

The field of human pheromones needs more research and consistent clinical data. The subject is still in an elementary stage and studies are contradictory. It needs to be completely explored. There is a possibility of finding an intriguing sexual mystery that the researchers are just beginning to unravel. History reveals that scientific data can turn myth into reality. Without the concrete scientific evidences, pheromone preparations will stay as myth.

"There is nothing like an odor to stir memories."
— William McFee

Chapter Ten

Natural Aphrodisiacs: Interaction with Drugs, Herbs and Food

There is no doubt that the demand of aphrodisiacs is increasing day by day after the evolution of Viagra. It is also a naked truth that Viagra is one of the quick fixes but has certain limitations regarding its side effects on the heart and eyes. Thus people are assiduously looking for safe Viagra from the natural sources. However, we are ignoring most important aspects regarding the safety of these natural aphrodisiacs. There are many questions that arise in the mind. How safe are these products? Are these products compatible with other drugs patients are taking? What are their side effects? What are their contraindications? How about their interaction with other herbs, drugs, or food we are consuming at the same time? What is the right dose for the appropriate remedy?

Aphrodisiac formulations are comprised of natural products, mostly herbs. Majoritiy of the consumers and suppliers of herbal remedies believe that these natural products are safe. This, however, is a misconception. Herbal drugs can produce adverse drug reactions as explained in details by de Smet Pagm (1997) and Philp (2004). A further complication is that, these herbal products are self-prescribed and most of the time, patients do not inform the general practitioners about their intake. If they do tell their physician then most of the general practitioners who are not interested in the herbal remedies discourage their patients about the use these products.

Nevertheless, these herbs may not be safe (Friedman 1996). The rule of the thumb is to consider all substance-based alternative therapies as drugs. These are in the form of crude extracts or plants' parts in the form of powder as pills or capsules or tablets. These substances interact with drugs, herbs, and food.

It is well established that patients taking more than one pharmaceutical medicine can experience drug interaction, which may be detrimental to their health. Though limited information is available in the literature concerning the interaction of herbs with conventional medicine, yet it is pertinent to know that certain interactions can be fatal. In recent years there have been an increasing number of reports of poisoning related to medicine. This may be due to the expanding use of these products or increased awareness among physicians and allied professionals of the toxic potential of some of these substances. The WHO database currently lists 8985 adverse drug reaction reports in which herbal preparations are the prime suspects.

Types of Interactions

    a. Herbal/Dietary Supplement and Herbal Aphrodisiacs Interaction
    b. Prescription Drugs and Herbal Aphrodisiacs Interaction
    c. Non-prescription Drugs and Herbal Aphrodisiacs Interaction
    d. Food and Herbal Aphrodisiacs Interaction
    e. Unknown Interaction due to Contamination in Herbal Aphrodisiacs.

a. Herbal/Dietary Supplement and Herbal Aphrodisiacs Interaction

The term herb used in the text refers to herbaceous and woody plants. These aphrodisiac preparations are commonly available over the counter without prescription. Aphrodisiacs are sold as health nutrients or supplements. The available forms of aphrodisiac products include dried plant parts, liquid extracts, tincture extracts or concentrated powdered extracts and capsules comprising powdered herbs. Aphrodisiac formulations administer orally, applied externally or consumed by inhalation are focused in the text. One single herb contains 5-20 active constituents. The aphrodisiac preparation that is sold over the counter is comprised of at least 3-10 herbs. That means there is a possibility of 50-200 active components in each dose. The chances of interaction increase with the number of active ingredients. If a person is having more than one herb, it is very pertinent to know their incompatibility. There can be herbal interaction. For example, ginseng is widely used in the preparation of aphrodisiac or sex boosters' formulations. Ginseng lowers blood sugar level. So all the herbs that have antidiabetic activity (Appendix E) needs to be avoided such as aloe, bitter melon, fenugreek, horse chestnut, mushrooms, milk thistle, and rosemary. These will have a synergistic effect.

Ginseng can also affect blood pressure so the herbs that influence the blood pressure used along ginseng can be dangerous. Aconite, arnica, betal nut, bilberry, black cohosh, cayenne, cola, ephedra, flax seeds, garlic, ginger, ginkgo, goldenseal, hawthorn, Indian tobacco, jaborandi, licorice, polypodium vulgare, oleander, periwinkle, turmeric, and wild cherry influence blood pressure (Appendix G,H).

Ginseng may increase the risk of bleeding when taken with herbs and supplements that are considered to increase the risk of bleeding. Some cases of bleeding have been reported with the use of aphrodisiacs such as ginkgo, garlic, and saw palmetto. There are some herbs and supplements that may increase the possibility of bleeding. Some examples are alfalfa, asafoetida, capsicum, celery, chamomile, clove, dong quai, evening primrose oil, fenugreek, fish oil, flax seeds, horseradish, licorice root, turmeric, vitamin E, wild carrot, wild lettuce, wintergreen, and yucca (Appendix D).

If ginseng is combined with supplements that have monoamine oxidase inhibitor (MAO) activity, it may cause some headaches, mania, and insomnia. Some examples are 5-hdroxytryptamine, California poppy, chromium, DHEA, diphenylalanine, ephedra, St. John's Wort, tyrosine, valerian, vitamine B6, and yohimbine bark extract that can create these symptoms. Some of the interaction of aphrodisiacs with herbs and drugs are given in Appendix K.

b. Prescription Drugs and Herbal Aphrodisiacs Interaction

The constituents contained in the aphrodisiac herbal preparations can combine with the normal prescribed drugs and form complex molecules that can be toxic. On the other hand, some constituents can add to the effect, giving a synergistic effect and thus create problems. Sometimes these active principles could be antagonistic to the prescribed drugs and cause problems. In some cases existing toxicity of conventional pharmaceutical drugs may be enhanced.

In some cases, herbal preparations may increase the anticoagulant effect of warfarin. These may increase bleeding if taken along with aspirin, heparin, clopidogrel and antiinflammatory drugs ibuprofen, and naproxen. It should be avoided also with anti-depressants drugs like monoamine oxidase inhibitors such as isocarboxazid, phenelzine, and tranylcypromine. Some steroidal preparations and oral contraceptives are also known to cause interaction with herbal remedies. Do not combine ginseng with blood pressure and heart medications. The analgesic effect of opioids may be inhibited by ginseng. Ginseng can induce mental disorders when used with phenelzine.

Many herbs can induce or inhibit liver and intestinal drug metabolizing enzyme cytochrome P450s, leading to alter drug concentration and clearance. Clinical studies have documented that St John's Wort reduced the plasma concentrations of cyclosporine, amitriptyline, digoxin, indinavir, nevirapine, oral contraceptives, theophylline, and simvastatin. Garlic supplement decreased the plasma concentration of the protease inhibitor saquinavir used in suppressing HIV, while licorice increases the plasma concentration of prednisolone, a synthetic steroid used in various compounds as an anti-inflammatory, immunosuppressive, and anti-allergic drug. Induction or inhibition of cytochrome P450s is considered the major reason for these observed herb drug interactions. St. John's Wort, Licorice, and Garlic are used in many aphrodisiac formulations.

The clinical importance of herb-drug interactions depends on factors that are related to co-administered drugs such as dose, dosing regimen, and administration route, herbs (species, dose, dosing regimen, and administration route), and patients (genetic variations, age, gender, and pathological conditions). Generally, a doubling or more in drug plasma concentration has the potential for heightened adverse effects. Less marked changes may still be clinically important for drugs with a steep concentration-response relationship or a narrow therapeutic index. In most cases, the extent of herb drug interaction varies among individuals, depending on differences in drug metabolizing enzymes, existing medical condition, age, and other factors.

Ginkgo, another commonly used herb in aphrodisiac preparations, raised blood pressure when combined with a thiazide diuretic used in controlling high blood pressure, and coma when combined with trazodone used as an anti-depressant. In addition, garlic produced low blood sugar level when taken with chlorpropamide, which is used to lower blood sugar in diabetes. Kava caused a semicomatose state when taken concurrently with alprazolam, a drug for treating anxiety.

c. Nonprescription Drugs and Herbal Aphrodisiacs Interaction

Some drugs like aspirin, paracetamol, ibuprofen, cough syrups, expectorant, and antihistamine can increase or decrease the affect of herbs when taken with aphrodisiac preparations. Aphrodisiacs such as angelica, horny goat weed, garlic, ginkgo, potency wood and zallouh root should be avoided. Some of them contain coumarins and increased potential for bleeding by reducing activity (Appendix D).

d. Food and Herbal Aphrodisiacs Interaction

Drug interactions occur not only between drugs but with foods and beverages too. For example alcohol enhances the anticoagulant of warfarin and sedative effect of antidepressants. Verapamil used for the treatment of arrhythmia and hypertension may increase blood alcohol levels. Similarly, grapefruit juice increases the plasma concentration of some dihydropyridine calcium channel blockers such as nifedipine, amlodipine, and nocardin used for the treatment of hypertension (Baily *et al.* 1994).

Among vegetables, broccoli interferes with warfarin (Kempin1983) and can result in excessive anti-coagulation. On the other hand, cheese, beans and yeast contain tyramine which may exert presser effects on monoamine oxidase inhibitors drugs (MAOIs) such as phenelzine used as an antidepressant resulting in high blood pressure and causing throbbing head-aches (Anderson and Philippson 1985). Hypericum, withania, capsicum, licorice, ginseng when used in aphrodisiac preparations will have a synergistic effect with MAOIs drugs.

Herbs interaction can occur with coffee, tea, cheese, cocoa, chocolate, cayenne, cola-nut, and wine. The interaction can be synergistic or antagonistic depending upon the constituents of the herbs, and it is very pertinent to avoid this interaction during the intake of herbal aphrodisiacs.

e. Unknown Interaction due to Contamination in Herbal Aphrodisiacs

Unknown contaminants in the herbal aphrodisiacs formulations can cause serious effect that can sometimes lead to death. There is always a possibility of contamination or adulteration while collecting the herbal drugs from the wild sources. Sometimes it is done intentionally to increase the profit. There have been some cases where allergic reaction or liver damage occurs due to the intake of herbal preparations. Physicians blamed such occurrences to the toxicity of the drugs while the herbalist claims that the symptoms are due to some contamination or adulteration in the collection of herbs and processing of the herbal formulations. Aphrodisiac formulations are comprised of many herbs and there is a possibility of adulteration or contamination. Even trace metals contamination like lead, iron and arsenic as reported by Saper *et al.* (2004) in herbal formulations can cause toxic effects. Ang *et al.* (2003) reported the presence of lead and mercury in herbal aphrodisiacs particularly in *Eurycoma longifolium* (Tongkat Ali).

Mechanism of Drug Interaction

Herbal aphrodisiac formulations interact with drugs in two ways,

1. **Pharmacokinetically**
2. **Pharmacodynamically.**

1. Pharmacokinetic interaction results in alteration of drugs or natural medicine by **absorption, distribution, metabolism, or elimination**. These interactions affect drug action by quantitative alterations, either increasing or decreasing the amount of drug available to have an affect.

1a. Absorption

Absorption is the physical passage of herbs or drugs from the outside to the inside of the body. A majority of the absorption occurs in the intestine, where herbs and drugs must pass through the intestinal wall to enter the blood. Several mechanisms may interfere with the absorption of drugs through the intestine. The absorption of herbs may be adversely affected when the herbs are given together with some drugs, due to the binding in the GI tract. Drugs such as Cholestyramine, Colestipol and Sucralfate may bind to certain herbs, forming an insoluble complex and decreases absorption of both the substances. Absorption of herbs may be further affected when the herbs are given with some drugs, which change the pH of the stomach. Drugs such as antacid, Tagamet (cimetidine), Pepcid (famotidine), Axid (nizatidine), Zantac (ranitidine) and Prilosec (omeprazole) may neutralize, decrease or inhibit the secretion of the stomach acid. This will increase the pH of the stomach and herbs may not be broken down properly leading to poor absorption in the intestines. Thus, aphrodisiac formulations comprising horny goat weed, licorice, capsicum, ginseng, yohimbine, and withania should be avoided when antacids are prescribed.

Absorption of the herbs is also affected by GI motility. GI motility is the rate at which the intestines contract to push the content from the stomach to the rectum. Slower GI motility means the herbs stay in the intestines for a longer period and there will be an increase in absorption. Faster GI motility means that the herbs stay in the intestine for a shorter period of time and there may be a decrease in absorption. Drugs such as Reglan (metoclopramide) and Propulsid (cisapride) increase GI motility and possibly decrease absorption of herbs. An antipsychotic drug such as Haldol (haloperidol) decreases GI motility and may increase absorption of herbs. So when a patient is taking a drug, which affects the GI motility of herbs, he or she should increase or decrease the dosage of herbs accordingly. Aphrodisiac formulations should be adjusted as advised by the physician.

1b. Distribution

This is another factor, which plays a significant role in the interaction of herbs with drugs. Distribution refers to the process in which herbs or drugs are carried and released to different parts of the body. Interaction occurs during the distribution phase if the drug has a narrow range of safety index and is highly protein bound. For example, warfarin is an anticoagulant medication which is very highly bound to protein and has a very narrow range of safety index. Warfarin interacts with various drugs, vitamins, herbs, and foods via different mechanisms. Some known examples that interact with warfarin include aspirin, ibuprofen, vitamin K, and green tea. The patients who are taking herbs or natural aphrodisiacs concurrently with warfarin have to be very careful (Fetrow and Avila 1999).

1c. Metabolism

Metabolism also affects the interaction. The liver inactivates derivatives and metabolizes herbs and drugs. The rate at which the liver metabolizes these herbs and drugs determines the length of time these herbs and drugs stay active in the body. If the liver was induced to speed up the metabolism, the herbs and drugs would be inactivated at a faster rate and overall effectiveness of the ingested substances would be lower. On the other hand, if liver was induced to slow the metabolism, herbs and drugs would be activated at a slow pace and overall effectiveness of the substances would be higher. Drugs such as phenytoin (Dilantin), carbamazepine (Tegretol), phenobarbital and rifampin speed up liver metabolism. Therefore, the herbs may be inactivated faster and their overall effectiveness will be low. Under such scenario, dose of the herb should be higher to get the desired effect. Since the aphrodisiac formulation comprises number of herbs, one has to be very careful selecting its dose particularly when taking the above drugs.

On the other hand, drugs that inhibit liver metabolism have an immediate onset of action. The rate of liver metabolism may be greatly impaired within a few days. Therefore, there will be a higher risk of herbs accumulating inside the body. Drugs such as Tagamet (cimetidine), erythromycin, Diflucan (fluconazole), Sporanox (itraconazole) and Nizoral (ketoconazole) should be avoided. In this case one may need to lower the dosage of herbs to avoid side effects. In an aphrodisiac formulation, it is not easy to adjust the dose.

1d. Elimination

Elimination is another process, which is quite potential in eliminating herbs and drugs from the body. If the kidneys are damaged, then the rate of elimination of herbs or drugs by the body would be slowed down leading to an accumulation of herbs or drugs in the body. In this case, the dose of the herbal remedy may be lower to avoid side effects. Such drugs that affect the kidneys are amphotericin B, methotrexate, tobramycin and gentamycin. The elderly population has to be more careful since their elimination process is decreased with age. Aphrodisiac formulation is usually taken by people over 50 when they are experiencing low libido. This population should be more careful.

2. Pharmacodynamic Interactions

Pharmacodynamic interactions cause alterations in the way a drug or natural medicine affect a tissue or organ system. These interactions affect drug actions in a qualitative, either through enhancing effects or additive action or antagonizing effects.

The highest risks of clinically significant interactions occur between herbs and drugs that have sympathomimetic, cardiovascular, diuretic, anticoagulant and antihypertensive and antiseizure drugs. The classic example of sympathomimetic herb is Ephedra (Ma Huang) which contain ephedrine and related alkaloids. This herb should be used with great precaution in patients who have hypertension, seizure, diabetes and thyroid conditions (Krauss 1996; Cowley 1996). Similarly herbs with diuretic effects may have additive or synergistic effects making hypertension more difficult to control. Herbs with anticoagulant effects include herbs that have blood-activating and blood-stasis-removing functions. Such herbs may interfere with anticoagulant drugs such as coumadin (warfarin) include *Salvia miltiorrhiza*, *Angelica sinensis*, *Carthamus tinctoria*, *Chamomile species, Tanacetum parthenium,* and *Hydrastis canadensis* (Fugh-Berman 2000). Vitamin E, fish oil can also enhance the effect due to antithrombin activities (Schmidt 1997). Various examples are discussed (Almeida 96; Cupp 1999; D'Arcy 1991, 93; de Smet Pagm 1997; Lambreecht *et al.* 2000). Antidiabetic herbs (Appendix E) may interfere with antidiabetic drugs (Appendix U) by enhancing hypoglycemic effects. The dose of herbs and drugs should be balanced very carefully to control blood glucose level (Miller 1998). Under the scenario aphrodisiacs such as Ginseng, Safed Musli and Damiana need to be used very carefully.

Herbal Aphrodisiacs in Pregnancy

Women have turned to alternate medicine due to premenstrual Syndrome (PMS) menopause and depression. The psychological conception is there that herbs are safe containing natural supplements. Black Cohosh (*Cimicifuga racemosa*) is commonly used for women for menstrual and menopause problems. Adverse effects include nausea, vomiting, dizziness, nervous system and vision disturbance and reduced heart rate. It contains salicylic acid that interferes with anticoagulants. Evening Primrose oil (*Oenothera biennis*) is used in PMS. The oil may cause gastro-intestinal disturbance. Kava-kava and St. John's Wort (*Hypericum perforatum*) are used in aphrodisiac preparations and can cause nausea and vomiting.

Aphrodisiac formulations are a combination of many herbs, which contain different chemical constituents and some of these can be detrimental to pregnancy. There are some herbs, which are teratogenic in nature. Teratogenic herbs are known to cause harm to fetus during pregnancy and thus leading to birth defects or abortion. Herbalist should be consulted before taking herbs. The use of some herbs is prohibited during pregnancy. Prohibited herbs include *Abrus precatorius, Achyranthes aspera, Adhatoda vasaka, Hibiscus* species, *Ricinus communis, and Gossypium* species (Chaudhri 1999). It is beyond the scope this review to list the name of many more. The bottom line is herbal preparations should be used during pregnancy under the supervision of a herbist. Nevertheless, consumers often self-medicate without consulting an appropriate healthcare professional. This can be an unwise decision with serious consequences, especially in pregnancy and later on in breastfeeding mothers. All herbal aphrodisiacs mentioned in this text can be harmful in pregnancy.

Herbal Aphrodisiacs and Breastfeeding

The benefits of breastfeeding greatly surpass that achieved by formula feeding. These benefits are two folds, improving the health of both nursling and mother. The nursling will tend to have a lower incidence of ear, respiratory and gastrointestinal infection due to antibodies that are transferred from mother to the child through breast milk (Wright *et al.* 1998). Furthermore, breast feeding also decreases the risk of other diseases such as childhood diabetes, Crohn's disease, and lymphoma (Koletzo *et al.* 1989).

According to the Ross Laboratories Mothers' survey, there has been a significant increase in the initiation of breast-feeding and in maintaining nursling at six month of age (Ryan 1997). The following Herbs should be used very cautiously during breastfeeding. *Symphytum officinale* (Comfrey root) is a tropical herb that contains pyrrolizidine alkaloids with the potential for hepatotoxicity and antimitotic activity. Pyrrolizidine has been found in the breast milk of lactating rats (Kopec 1999; Ridkar1989). *Tanacetum parthenium* (feverfew) is used for severe migraine. The main constituent is parthenolide, a sesquiterpene lactone. Thujone a neurotoxic is another compound found in Tansy (Feverfew) and Absinthe preparations. Use of this herb in breast-feeding cannot be assured. Purple Coneflower (Echinacea) is a very popular herb commonly used to boost immune system. It contains nontoxic form of pyrrolizidine. However, it is also not considered safe yet. Ephedra, a potent herb indicated for the alleviation of Asthma-associated bronchoconstriction due the presence of ephedrine. Given the potential cardiovascular risks and addiction complications, the use of this herb is not recommended in breast-feeding. Aloe, Cascara sagrada, Rhubarb, and senna are used as laxatives. Their use is prohibited during breast-feeding due to anthraquinone constituents that may cause potassium deficiency (Blumenthal *et al.*1998). Licorice root is used in the treatment of peptic ulcer/duodenal ulcers and also in the aphrodisiac formulations. The active components glycyrrhizic and glyrrhetinic acids account for anti-inflammatory and anti-allergic effects. However, the components of licorice root possess mineralocorticoid properties leading to toxic effect such as sodium and water retention, hypokalemia and hypertension which can lead to cardiac arrest (Conn *et al.*1968). *Rauwolfia serpentina* (Serpgandha), contain reserpine used in hypertension. It is also used as CNS depressant. It is, therefore, contraindicated in breast-feeding. *Ginkgo biloba* used as an aphrodisiac may affect coagulation process in the mother and nursling (Shoup and Larson 1999). Ginseng another aphrodisiac may cause estrogenic side effects as well as platelets changes. Until more information is known, ginseng root should be avoided in lactating mother (Howard and Lawrence 1999). In short, breast-feeding mothers who use herbs should consult a physician or pharmacist who has information concerning the safety of herbal medication (Hardy 2000). Herbal aphrodisiacs are not recommended for women during breast feeding.

Herbal Aphrodisiacs and Surgery

People undergoing any kind of surgery including plastic surgery who take aphrodisiacs such as garlic, ginkgo or ginseng, St. John's Wort and kava-kava are susceptible to risk of excessive bleeding and other complications. Many formulations may interact with drugs prescribed for surgery and prevent clotting or increase the affect of anesthetics. Ginseng may aggravate low blood sugar, kava and valerian can stimulate the impact of anesthetics, St. John's Wort can accelerate the metabolism, and echinacea poses a risk of poor wound healing and infection. Patients and physicians are often unaware of the risks, and some patients are reluctant to admit to their doctors they used natural aphrodisiacs, increasing the likelihood of serious complications. The preparations can speed up or slow down the heart rate, inhibit blood clotting, alter the immune system and change the effects and duration of anaesthesia. The researchers, from the University of Chicago, have published guidelines on when patients who are undergoing surgery, should stop taking herbal medicines.

The American Society of Anesthesiologists has recognized the potential for adverse reactions and suggests that patients stop taking all herbal medications including natural aphrodisiacs two weeks before surgery.

Quality Control and Standardization of Natural Aphrodisiacs

FDA does not regulate QA/QC and standardization of aphrodisiac products since these are sold over the counter as food supplements. Aphrodisiacs are not regulated as drugs, so no legal standards exist for their processing, harvesting or packaging. Consumers are depending upon the integrity of the manufacturers. In spite of the good intentions of majority of the manufacturers, it is very difficult to control the quality of these products. Natural aphrodisiacs are herbal formulations and their quality control depends upon many factors. Proper identification of the raw material at the time of collection is very important. Wrong identification of the herb is detrimental to the quality of the products and most of the time manufacturers are not even aware of this problem. The herb suppliers, who supply the raw material, are not devoid of the identification's problems especially in the case of expensive herbs, less expensive substitutes are used. Thus, proper authentication of the raw material is very pertinent to avoid adulteration (But 1993). Herbs are generally collected from the wild source by unskilled native labor. The chances of substituting related species are more. Secondly, the season at which the herbs are collected is usually very pertinent as the nature and amount of the active constituents varies from time to time. The age of the plant is also another important factor and governs not only the total quantity of active constituents produced but also the relative proportion of the components of the active mixtures. A well-known aphrodisiac herb ginseng is the right example. The six-year-old root of ginseng is recommended for the pharmacological activity due to the age and concentration of the active mixtures. Nevertheless, the concentration of the active constituents also varies throughout the day in these species. Appropriate time of harvesting is required in most of the herbs. Those companies who are cultivating their own species are better off.

Underground organs such as roots and rhizomes of the medicinal plants must be freed from all soils and contaminants befor drying them and subjecting to the formulation process.

Proper drying and storage of the collected plant parts is another important aspect. Improper drying and storage will definitely impact the active principles. If enzymatic action is not desired then the drying should be done immediately after collection. Volatile oils are liable to loose their aroma if not dried properly or if the oil is not distilled from them immediately after collection. Moist plants are liable to develop mould. As a general rule leaves, herbs and flowers may be dried between 20 and 40 °C and bark and roots between 30 and 65 °C respectively.

During growth and storage, pesticide residues, microorganisms, aflatoxins, radioactive substances and heavy metals can contaminate crude plant material that can be very toxic to human health (Chan 2003). Proper storage conditions for collected plant material under appropriate temperature and containers are required. These conditions vary from one plant to another.

Clinicians should adopt proper strategies to minimize toxic herb drug interactions. Early identification of drugs that interact with herbs and the mechanisms involved is important. Identification of drugs and herbs that interact with each other can be incorporated into the early stages of development of synthetic drugs and herbal medicines. If a drug has to be used in combination with an herb, the physician must ensure a safe drug combination regimen, dose adjustment, and discontinuation of therapy should toxic herb-drug interaction occur. Aphrodisiac formulations should be properly studied and patients be advised accordingly.

Another approach for circumventing possible toxicity is proper design of drugs and/or modification of the aphrodisiac herbal formulation (e.g. removal of specific constituents) to minimize the potential for undesired interaction.

Conclusion

Wheaton *et al.* (2005) reported in their survey that 5-20% of the general public use herbal medicines, resulting in global annual sales of about $60 billion. Out of that 43% of individuals taking herbs were also taking a prescription drug or over-the-counter drug (Baily and Dresser 2005). Herbal products can be purchased without any restriction. Most of the people think that natural products are safe and they do not consult their primary physician while taking their regular prescriptions. They are also quite ignorant about the herbal-drug interactions. Conversely, physicians are also not accustomed to asking their patients about their consumption of herbal medicines (Clement *et al.* 2005). Lack of communication between the patients and physicians can result drastic consequences and undesirable clinical outcome. Health care professional take extra care and educate their patient about the limitations of herbal products. The use of herbal products those lack scientific, clinical and safety data should be discouraged. Patients should be advised for the adverse effects, side effects, contraindication, drugs herbal and food-herbal interactions.

> *"Standardization of our educational systems is apt to stamp out individualism and defeat the very ends of education by leveling the product down rather than up."*
>
> — *Harvey Cushing*

Chapter Eleven

Natural Aphrodisiacs: Myth or Reality

Youth is full of energy and fun. In a healthy young person, enthusiasm is at its zenith, and one can take any risk to fulfill his or her mission. This phase of life is in fact full of dreams, love, and fantasies. Prime of youth is full of pleasure. Sex is at its peak. Hormones are kicking in both the sexes. Fun and frolics are among the priorities. However, the fountain of youth passes by with fun and frolics. Middle age rolls over in struggle to strive for a good career. Later on, the middle age sets in with some physiological changes. Most of the changes in middle age go unnoticed except the sex life. The slightest decline of sexual power becomes a serious problem. There is no doubt that sexuality is an integral part of mankind that gives cardinal pleasure. Moreover, virility is an essential part of the male ego and the human mind would not leave any stone unturned to regain its declining pleasure or ecstasy and take the help of sex stimulants or sex boosters or aphrodisiacs. This clearly explains that these products have not lost popularity even today and selling like wildfire throughout the world.

Sex is one aspect of human life that has always gained prominence within every culture. It is a biological instinct and basic need of all the universal species. For humans it becomes more than physical need and leads to mental and spiritual satisfaction. To keep this energy going, man has consumed substances from Mother Nature that could enhance his sex life since the time immemorial. In other words the search for natural aphrodisiacs to enhance sexual desire, performance, and enjoyment is almost as old as the human race itself. While tracing back the history of sex, it was revealed that the human race throughout this universe had been using herbs and animal parts to enhance their sex life. Kings, princes, autocrats and the rich have been consuming philters comprised of exotic and costly ingredients such as pearl, gems, gold and silver to regain their lost sexual strength since antiquity. If it was within the control of mankind, he would have never given up his virility. Nature made him helpless and virility declines with age. For men, a rigid penis is not only for recreation but also for procreation that renders pleasure, reinforces pride, and allows one to nurture spiritual intimacy and ecstasy. It is not out of the way to mention that there is a saying *Man without erection is not a Man.*

Natural aphrodisiacs have been used for centuries on the basis of myths. Information available in the literature consists mainly of folklore, myths and anecdotal evidence as opposed to science. In fact, the United States Food and Drug Administration declared, *there is no scientific proof that any over-the-counter aphrodisiac works to treat sexual dysfunction*, so their consumption is on the basis of myths only.

193

Law of Similarity

According to one myth, the theories of "Therapeutic efficacy of signatures or the law of similarity," an object resembles genitalia may possess sexual powers. Food like bananas, asparagus, carrots, cucumber, and corn on the cob, eggplant, mushrooms, and horseradishes are sexually stimulating because of their phallic appearance. Similarly, mucilage plants such as jack fruit (*Artocarpus heterophyllus*); taro root (*Colocasia esculenta*); bombax *(Bombax malabaricum); okra (Hibiscus esculenta); mallows (Althea rosea); and orchis (Orchis maculata)* were used as aphrodisiacs on the myth that their mucilage would increase and strengthen semen *(Meyer 1993).*

Eating the genitals of the animals such as bulls, goat, and rams is considered to increase virility. Some of these are amusingly absurd. Similarly, Rhinoceros horn has been highly priced as an aphrodisiac for years because the horn has the same shape as an erect penis. It was considered to be the symbol of sexual power in many cultures. Likewise, oyster, clams, mussels, figs, peaches, and apricots have reputation to stimulate the sexual desire. It was believed that they are stimulating because they resemble female genitalia while peaches and apricots have a velvety erotic cleft. Some of the products can be rationalized on scientific concept for example oysters and clams are high in zinc, which is essential for the production of testosterone which could increase testosterone level and low libido. The remaining products are myth.

Myth or Reality

There is no scientific evidence that consumption of animal parts stimulates sex. This all comes from superstition, myth and culture. Survey of literature did not reveal any information that Casanova the lover man used such stimulants to retain his virility except more recently this myth came to some reality. Fisher at the department of chemistry, Barry University, Miami, Florida and Robert H. Shmerling, associate professor, Harvard Medical School, Boston, have discovered that mussels, clams and oysters contain compounds that have been shown to be effective in releasing sexual hormones such as testosterone and estrogen (Fisher and shmerling 2005). These compounds are D-aspartic acid and NMDA (N-methyl-D-aspartate). "We found there might be a scientific basis for the aphrodisiac properties of these mollusks," Fisher said. Not so fast, says food myth expert Dr. Robert H. Shmerling, an associate professor of medicine from Harvard Medical School. "The findings are certainly interesting, but we still have a ways to go before saying that there is scientific evidence that clams, oysters and scallops boost libido," he said. In addition, Shmerling wonders if animal studies linking D-aspartic acid and NMDA to the release of sex hormones is even relevant to humans. "This is a good example of the headlines getting well ahead of the science," Shmerling said. "It will take much more compelling evidence-with human subjects-to prove a link between seafood and libido."

Though some people do not believe in love potions, yet countless numbers of people have used them down the centuries. The majority of people with scientific background believe that sex boosters are mysterious and weird concoctions that contain bizarre ingredients. How come these formulations are still in use and are being sold over the counter as dietary supplements? There is a clear proof that people are assiduously buying them with hopes against hopes and spending lots of money to revive their lost libido.

Some of the myths can be explained on the basis of scientific evidences such as Ginseng. Ginseng root is shaped like a human with two legs and two arms; the myth is a spirit with human form lived in the root and eating ginseng to increase sexual power.

Panax ginseng is used in Traditional Chinese Medicine as an aphrodisiac. The extracts of the Chinese root *Panax ginseng* and the North American *Panax quinquefolius* contain a large group of ginsenosides, four-ring glycosides. Ginsenosides enhances nitric oxide synthesis in the endothelium of many organs, especially in the *corpora cavernosa* of the penis resulting erection according to Chen and Lee (1995) and Gillis (1997). Ginsenosides also enhance acetylcholine-induced and transmural nerve stimulation-activated relaxation. This increases cyclic guanosine monophosphate in the tissues and leads to stimulation of the low libido. It is more beneficial to women, since the biochemical process of erection is the same in the clitoris and morover, ginseng is also known to possess phytoestrogen activity. Since some of the studies are contradictory and methodological quality of the primary studies were too low to draw definitive conclusion.

Similarly, *Muira puama* have been clinically tested as an effective aphrodisiac on the basis of myth. *Muira puama* or *Liriosma ovata*, known as potency wood has been used in South American folk medicine for treating low libido and erectile dysfunction. French clinical study by Waynberg (1990) showed that *Muira puama* extract improved libido in 62% of patients with lack of sexual desire. Second study of Waynberg (1995) revealed very encouraging results. The herb enhanced libido in the test group, increased the frequency of intercourse and improved the ability to maintain erection. Studies conducted on *Muira puama* by Werbach *et al.* (1994) also demonstrated encouraging results. They claimed it to be the best herb for erectile dysfunction. However, additional controlled scientific data is required with double blind studies to prove the efficacy of the herb.

Bark of yohimbe is commonly used in aphrodisiac preparations such as Aphrodyne, Cobra, Erexol, Sexylady, Stayerect, Staminol, Passion Rx and Vipra. Yohimbe tree bark of the *Pausinystalia yohimbe* tree is grown in West Africa. Yohimbe bark contains indole alkaloids out of which yohimbine hydrochloride is the major ingredient used to treat ED. Yohimbine hydrochloride the active chemical in Yohimbe bark is only available legally in the US as a prescription drug. The indole alkaloids of thr bark affect the sympathetic system, dilating the blood vessel and lowering blood pressure. The flow of the blood in the pelvic area and the reflex excitability in the sacral area of the spinal cord are increased. On account of serious side effects of anxiety, panic attacks, hallucinations, fluctuation in blood pressure and heart rate, the FDA classifies yohimbine bark as an unsafe herb and due to the lack of effectiveness and the negative side effects of yohimbine; the FDA no longer recommends the prescription drug to treat ED (Murray 1995).

Damiana *(Turnera diffusa)* is another potential herb that contains essential oil, which stimulates the low libido of men and women. Native Mexicans have been using it to enhance their sex life. It contains several alkaloids that stimulate the sex organs, increase circulation, and relax tense muscles. It is mostly recommended in female sexual deficiency such as lack of sex interest in females. However, it has very limited clinical data. Only Arletti (1999) carried couple of clinical studies on the rats. The results were encouraging in support of the folk reputation. More studies are required to turn this myth into reality.

Saw palmetto or *Sabel serrulata* has gained increasing acceptance for the treatment of prostatic enlargement. It has estrogenic, anabolic constituents as well as other steroidal saponins. It has been used to enhance libido and stimulate breast enlargement. Supportive research, however, is lacking information as sex stimulant. In treating prostate enlargement *Saw palmetto* at least has potential as an alternative to Finasteride. Like Finasteride it doesn't tend to decrease libido and sexual performance (Carraro *et al.*1996). However, no scientific evidence is available for its use to enhance low libido, though it is being used as one of the ingredients in many aphrodisiac formulation.

Likewise, wild yam cream is very popular and sold over the Internet. Various species of the Genus Dioscorea are known as wild yam. Roots extracts of *Dioscorea deltoidea* or *Dioscorea villosa* are mostly used in creams. Many women are being encouraged to purchase and use this cream by luring advertisements. Diosgenin, present in the roots of *Dioscorea* species, is the precursor of hormones and can be converted into progesterone in the laboratory via several synthetic steps. The synthesis is not that simple. There is no evidence that diosgenin from the crude extract of wild yam can be converted into progesterone in the body. Though the human body is also a big laboratory, microflora and enzyme system can convert complicated compounds in couple of steps but there is no such evidence. Microbial degradation of diosgenin to progesterone is possible in the laboratory; however, it is not possible through the skin. Wild yam preparations are available in creams and oils containing the alcoholic extract of roots of wild yam. It is being advertised as "natural progesterone." It is a deceptive promotion. These are being promoted for premenopausal and postmenopausal symptoms along with breast enlargement. Lee reports (1996) that natural progesterone may improve sexual desire when used in physiologic doses. The cream contains wild yam root extract, aloe extract, and vitamin E in a creamy base along with some other preservatives. The preparation available in health food stores is quite variable for active ingredients. The yam extract contains many constituents along with diosgenin and it is not clear how much the yam extract called *natural progesterone* is actually being absorbed through the skin. Usually application on the soft part of the body is recommended such as breast, inner thigh, under the arms and pubic area. Vaginal creams are also available. That can be dangerous if the yam extract is of inferior quality or poorly standardized. It could be more harmful if there are traces of metals as contaminants in the herbal extracts. Traces of metals and pesticides are common contaminants in plant extracts.

Komesaroff *et al.* (2001) studied the wild yam extract for sexual activity. According to their findings, there was no evidence for an estrogenic effect in the yam cream formulation nor did it contain substances that were converted into progesterone in significant quantity. So claiming yam cream as elixir of life for women during low libido or menopause phase is a false claim by the manufacturers.

The use of chocolate as an aphrodisiac has been traced back to the seventeenth century by the Aztec and later on by the French. The myth is turning into reality. Recently, it has been justified that it contains phenylethylamine (PEA) the substance that triggers love. Further confirmations are required.

Soy products such as tofu and soy burger are also used as one of the ingredients in female aphrodisiac preparations. In female, vaginal dryness and atrophic vaginitis influence sexual activity during menopause (Albertazzi *et al.* 1998). Recent studies on soybeans indicated that dietary supplementation with phytoestrogens can be beneficial. Phytoestrogens mimic the effect of natural estrogens. Soy products containing isoflavones genistein and phytosterols are being recommended for relief of hot flashes, vaginal dryness and atrophic vaginitis (Baird *et al.* 1995). This could enhance sexual pleasure in the menopausal years. Moreover, women do not want to take hormone replacement therapy (HRT) because of its side effects which increase risk of breast cancer. Are soy products beneficial during menopause and do they provide the same benefits as HRT? There is no concrete scientific data to support this assumption. However, research is in progress. The best soy products are those that contain high contents of isoflavone such as whole soybean, soy flour and soy nuts. Since each woman reacts differently to soy products, women should try these themselves to see if soy in the diet reduces the symptoms. However, the FDA has not reviewed this product for safety and effectiveness.

Ginkgo biloba is another popular sex booster used in ED and has been recommended for women too. Some studies support the folkloric use as an aphrodisiac (Kang *et al.* 2002) where others are contradictory (Wheatly *et al.* 1999). More organized controlled studies are required to prove its efficacy as sex stimulant. It has been shown to have a Viagra like mechanism. It improves circulation via promoting the synthesis of nitric oxide and help in ED. It has a good future but needs extensive research.

Maca root (*Lepidium peruvianum*) is being used in many aphrodisiac preparations such as Intim X, MagnaRx, Male Boost and Enzyte. Meca root is the edible root of the Peruvian plant traditionally used to stimulate low libido. The myth is 2000 years old. Dr. Garry F. Gordon, President of International College of Advanced Longevity Medicine in Chicago, Illinois, calls it 'Natural Viagra.' Most of the studies have been carried out on rats and results are encouraging (Cicero *et al.* 2002). A couple of studies have been done on humans and the results are contradictory (Gonzalez *et al.* 2003). More recently, Bogani *et al.* (2006) found that Maca does not possess androgenic activities. Anyway, the plant has a bright future but needs more controlled clinical studies on humans to prove its folkloric use and to clear its contradictory studies.

Horny Goat Weed (*Epimedium sagittatum*) is one of the ingredients in many preparations of aphrodisiacs sold via the Internet such as Cobra, Chinese Virility pills, Stamanex, Avela, Enzyte and Vipra. Horny goat weed has been used for over 2000 years as an aphrodisiac. The herb received its more common name, horny goat weed, when goats grazing on the herb were observed to have significantly increased sexual desire. It is believed that horny goat weed has testosterone like effects that stimulate sexual activity in men and women (Chen 2006). Wang *et al.* (2004) found that the extract of the weed increased the production of androgens thus increasing testosterone level that increased in libido or sexual desire. Horny goat weed contains a prominent flavonoid called icariin. Icariin is a cGMP-specific PDE5 inhibitor like Viagra, Cialis and Levitra. However, there are no other scientific studies on humans to support the use of horny goat weed as an aphrodisiac. So it will remain as myth.

Tribulus terrestris is also used in many aphrodisiac preparations like Nirvana, Stamanex, Enzyte, Testrol, Vipra, Erect pills, Tribestan. The herb is available all over the world. It has been used as sex booster in Bulgaria (Zafar *et al.*, 1989). The Ayurvedi System of Medicine prescribed it for impotence and infertility in women. Traditional Chinese System of Medicine recommends it for premature ejaculation. Clinical studies by Gauthaman *et al.* (2002, 2003) showed encouraging results with mice. They concluded that tribulus extract appeared to possess aphrodisiac activity due to increase in androgen. Nechev and Mitev (2005) contradicted this study. Their findings suggest that steroid saponins of *Tribulus terrestris* do not possess direct or indirect androgen increasing activity. The plant has contradictory findings and more clinical studies on humans should be carried out to exploit its potential as sex stimulant.

Bois Bande *(Roupmala montana)* a native of the Caribbean is a hot aphrodisiac these days sold under the brand name Erectol. The bark of the tree is famous for its aphrodisiac property particularly erectile dysfunction. Seller claims that it is better than Viagra. It stimulates sexual appetite in both male and female by increasing blood flow to the sexual organs. Survey of literature reveals very limited information on the phytochemical investigations of the plant. No clinical data is available to support the myth. It means that the active substance remain elusive. The plant has a bright future for further investigations and findings could be very interesting.

Tongkat Ali (*Eurycoma longifolia*) is used as an aphrodisiac in Malaysia and Indonesia. Avela and Stamanex products contain Tongkat Ali as one of the constituents in their formulation. Clinical studies of the plant extract has been done on rats by Ang *et al.* (1997-2003) and showed increased sexual activity. All thesee studies are carried out in the same lab and require confirmation by other scientists. The plant contains interesting chemical constituents like indole alkaloids, quassioides and eurycomalactone and needs further research for their pharmacological action.

Dong Quai (*Alstonia scholaris*) found in 'Boom' preparation is considered to contain phytoestrogen activity. Few clinical studies have been done, and most of them have either been poorly designed or reported insignificant results. Most of the studies are done on animals and are conflicting. One human trial found no short-term estrogen like effects on the body (Hirata *et al.*1997). Thus the data is contradictory.

Zallou root (*Ferulis hermonis*) is another promising aphrodisiac found in the products Sexitivia A, Zallinex and Zamoreve. It is used in the Middle East to treat impotence in men and frigidity in women with amazing results. It is also used in healthy men and women to stimulate sexual pleasure for adding the extra zip to their life. In Beirut, the Lebanese Urological Society has sponsored clinical trials. To date, 7000 men have participated in this research and results have been very encouraging. Success has been achieved up to 80%. Zallouh root is rich in ferulic acid that increases blood flow to sexual organs. Nutranex is the first and only company to refine and standardize the Zallouh root extract under trademarked name Pur-Zall. The extract is combined with other ingredients and dispensed in capsule form; Zallinex is for men and Zamoreve for women. Hadidi *et al.* (2003) carried out a comparative study of ferula root extracts and sildenafil (Viagra) on copulatory behavior of rats. The results were not very encouraging. However, the systematic and controlled clinical humans' trial can make this herb the future Viagra from natural sources. It has a great potential.

Ashwagandha (*Withania somnifera*) is another herb that is being sold like hot cakes over the Internet. The herb has been used in the Ayurvedic System of Medicine in India for the last 4000 years. It is called Indian Ginseng. It stimulates the low libido in both men and women. Studies conducted by Iuvone and Esposito (2003) showed that ashwagandha can produce nitric oxide that is known to help erection in men. The roots of the plant contain steroidal alkaloids and steroidal lactones known as withanolides. Most of the pharmacological activity of ashwagandha has been attributed to two main withanolides, withferin A and withanolide D. Ashwagandha is considered to enhance sexual power, prevent impotence, infertility, low sperm count or seminal debility. In other studies by Ilayperuma *et al.* (2002) the root extract induced a marked impairment in libido, sexual performance, sexual vigor, and penile erectile dysfunction. These effects were partly reversible on cessation of treatment. Results are contradictory and need further investigation.

Safed Musli (*Chlorophytum borivilianum*) products are sold as Vita Ex and Musli Pak. It is extensively used in the Ayurvedic System of Medicine in India. Safed Musli is known as Divine herb in *Atharveda* (religious book of India). The tubers used for medicinal purposes contain about 27 steroidal alkaloids. It is extensively used to stimulate low libido. It is used in oligospermia and in increasing semen volume and total sperm count. However, no concrete scientific evidence is known to support this concept. Recently some encouraging results have been shown by Kenjale *et al.* (2008) on male Wistar albino rats. The herb has a great potential and requires extensive research to prove its potential. The herb is extensively consumed all over the world. The world annual demand is 50,000 tons, so there must be some reality in this myth.

Since antiquity, the saffron is used as an aphrodisiac. It has been cited in the *Kama Sutra* too for its aphrodisiac qualities. To the ancient Greeks and Persians, saffron was used to treat melancholia; a string of recent studies have shown that this herb is comparable to Prozac and impiramine for treating depression. No research supports the aphrodisiac properties of saffron, yet it has a great potential. The food with saffron can create a general state of wellness. Saffron is recommended to women, especially the ones with genital illnesses. It creates an atmosphere of serenity in intimate relationships. No other flower has a more ancient history than saffron crocus. However, there is no scientific data available to support its use as an aphrodisiac. Further investigations are required.

Licorice has been used as an aphrodisiac for vitality, and longevity, often known as an elixir of life. The rejuvenating and nutritive properties have made licorice one of the most versatile herbs widely used by herbalist of East and West. In India, licorice has an ancient reputation as an aphrodisiac in the *Kama Sutra* and *Ananga-Ranga* that comprise numerous recipes containing licorice for increasing sexual vigor Recent research by Armani *et al. (*2003) suggests that licorice may reduce testosterone levels in men After one week of treatment, testosterone levels decreased 26%. Interestingly, licorice is used in many aphrodisiacs preparations. The study is contradictory to its aphrodisiac use and needs further confirmation.

Vitamin E has always been promoted as an aphrodisiac but this is true only in rats. In the 1920s, scientists found that male rats suffering from vitamin E deficiency became sterile. Due to passage of time, this fact was twisted and vitamin E emerged as a sex vitamin for humans. Quacks have been raking in money on this inaccurate conception ever since much to the public ignorance. Vitamin E is generally non-toxic but there's a limit to this. Mega doses can cause stomach upset, diarrhea, dizziness, bleeding, and reduced sexual desire.

Likewise, Spanish fly constitutes small beetles found in France and Spain. Cantharidin is the active constituent isolated from beetles and has been promoted as an aphrodisiac for many years. There is hardly any truth in this. When applied to the penis, Spanish fly produces an itching, burning sensation that may lead to an erection. The same irritation and inflammation, however, can permanently damage the genitals. Death can occur with large doses, making Spanish fly a deadly aphrodisiac. So far Spanish fly will remain as a myth.

Conclusion

In view of the above discussion, the following questions arise:

Are aphrodisiacs myth or reality or both?
Do they have any future?
Can we find Viagra from natural products without side effects?

These questions need answers from a natural product chemist or herbalist and the answers do not come so easily. However, few of these aphrodisiac products work knowingly and have future where the majority of them are folkloric and superstitious. The nonscientific evidence and knowledge about aphrodisiacs has existed for thousands of years. There must be some truth in it. The mystery is still unsolved. Most likely, some of it is true and some of it is false. It is myth as well as reality. *Myth can be turn into reality with scientific evidences. Without myth there is no reality.* However, myth is a clue to reality. Myth gives us the way to explore and sometimes leads to reality. The herbs that were used as folkloric medicine in the past have revealed interesting results. Anticancer drugs such as vincalecoblastin from *vinca rosea*; taxol from yew trees *Taxus brevifolium;* antihypertensive reserpine from *Rauwolfia serpentina;* cardiotonic glycosides from *Digitalis purpurea;* precursor of hormones diosgenin and solasodine from *Dioscorea deltoidea* and *Solanum platanifolium (Puri 1974)* and so many others are the burning examples in support of this statement.

Likewise, if the systematic clinical research on the extracts of certain potential herbs as described in this chapter is carried out, it can definitely bring myth to reality and would give us the natural Viagra of tomorrow.

Research should be focussed on testing the efficacy of plant extracts for aphrodisiac activity. Isolation, purification and characterization of the active constituents from these extracts should be carried out by modern chromatographic and spectroscopic techniques. Finally the active constituents should be tested by clinical trials on animals and humans.

> *"What is a scientist after all? It is a curious man looking through a keyhole, the keyhole of nature, trying to know what's going on."*
> —*Jacques Yves Cousteau*

Chapter Twelve

Conclusion: What is a Good Aphrodisiac?

Sex is one of the most important aspects of life. Sound of three alphabets s e x imparts a tingling sensation down the spine and sends pleasant impulses to the mind. Sex is natural and an imperative biological phenomenon. It is an established fact in nature that when sperms mature, they go in search of eggs and vice versa. Thus, sex is an essential part of life. It is not only for procreation but also for recreation. However, when sex desire diminishes with age or any other physiological problems in the body, it can reflect on the daily life. On the other hand, unsatisfactory sex can cause mental frustration and ultimately leads to physical ailments. A majority of the people do not want to discuss their sexual problems even with their doctors. If the remedial action is not taken, problem can multiply. Thanks to Pfizer pharmaceuticals for the discovery of Viagra, at least people with some kind of sexual problems have started talking to their doctors.

Many people are embarrassed by talk of love and sexual desire. Some people believe that these sex topics are very personal and should not be discussed, while some advocate being free of social inhibitions that mask the true feelings and desires. *Communication, Sex and Intimacy Survey* done by SGI Research, a marketing and public opinion research company revealed, seven out of ten people said their partner has trouble discussing intimacy.

Sex is something that we should not talk about in open. That's what we learned from our parents in our childhood. Parents used to guide their offspring about most of the things of day to day life except sex. Sex was considered sinful act, however, the concept has changed with the passage of time. It is so sardonic that when you get into an all-male or all-female conversation, all topics percolate to sex-related discussions. No matter which way you look at it, sex gets involved in daily life. It has been said that sex is not everything but is a part of everything.

Sex lies at the root of life; we can never learn to reverence life until we know how to understand sex. In reality, many couples have an unsatisfactory sex life that becomes the root of problems. Not surprisingly, the majority of people put blame on their partners. No doubt, true source of pleasure lies within yourself, but you need a partner who believes in this ideology. What is the cause of unsatisfactory sex life? There could be many reasons. This could be due to physical problems or mental incompatibility of the partners. Discussion on domestic and workplace issues before, during or after the sexual act generally leads to arguments, fights and frustration. Most of the times, sex act is one sided. One of the partners is not turned on. Before one can take off, the other is already landed. Sometimes, female partners indulge in fake orgasm to perform their obligation to their spouses. In fact, their chemistry is not right. This one side act also results in frustration even after the act. Such type of sex is unsatisfactory and it creates a vacuum and hatred in both the partners. In many cases, it is a tragedy because they refuse to consult expert advice out of embarrassment and become victims of sex-boosters or aphrodisiac formulations easily available by mail orders. These aphrodisiac products do not mitigate their problems.

Why do people buy aphrodisiacs?

There are four categories of people who use aphrodisiacs.

1. People who have normal sexual desire but want to add some extra zip to their sexual life.

2. People who have low libido, premature ejaculation and want to stay longer in their action.

3. People who suffer from erectile dysfunction and need some help.

4. People who do not want to go to the physician to discuss their sexual problems. They do feel some kind of embarrassment in disclosing their sexual problems. It is easy to buy these products via the Internet. Tall claims by the sellers can tempt anybody who has low libido to buy these products.

Is all the "hype" about aphrodisiacs true?

Many enterprises that sell aphrodisiac formulations are marketing mostly over the Internet by mail orders that boast incredible claims of sexual arousal and enhancement such as "Enjoy *explosive sex tonight and every night . . . no matter how old you are"*. These unbelievable advertisements sell sex and people are buying them. Do they actually work? "The judicious answer is that we don't know," said Dr Barnaby Barratt, President of the American Association of Sexuality Educators, Counselors and Therapist, and Director of the Midwest Institute of Sexology, Detroit, Michigan.

Dr. Ira Sharlip, a Professor of Urology at the University of California-San Francisco and President of the Sexual Medicine Society of North America, says, "I don't know if they work-despite manufacturers claim that they do. I would like to see some evidence, but there is none."

Professor Stephen Myers, Director of the Australian Center for Complementary Medicine in Brisbane, said "the purported medicinal qualities of many herbs were based on tradition rather than research. This was particularly the case with herbal aphrodisiacs. I'm not aware of any rigorous scientific evidence for the therapeutic benefit of herbal aphrodisiacs," he said. "I don't think there is a culture that doesn't have a variety of things that are purported to be of benefit, but the area of aphrodisiacs is probably the area that's most exposed to myth."

A majority of the aphrodisiacs' preparations have nutrients and act as tonic. Some of them have a psychological or placebo effect for a short time. In fact, very little information is available regarding their benefit in erectile dysfunction or in any sexual dysfunction. According to FDA there is no scientific proof that these preparations treat sexual dysfunction. Unless it is scientifically proven, it is very difficult to believe that they really work and will remain as myth.

The human brain is more potent and powerful than any known aphrodisiac. The phrase *you have sex on the brain* is literally and scientifically correct. You may decide to use some type of aphrodisiac to rectify a sexual dysfunction, to enhance the sexual experience. However, no matter how powerful or potent the aphrodisiac is, it won't work without a meeting of the minds and hearts of the partners. Even Viagra would not produce the desired effect without including the input of the brain.

Naturopaths, family doctors, sex therapists, and psychologists agree: aphrodisiacs aren't going to help if an individual is not mentally healthy. Many experts think that aphrodisiacs only work by the power of mental suggestion anyway. "To be perfectly candid with you, aphrodisiacs don't work," says Sandra A. Davis, a psychotherapist and a board-certified sex therapist. "It's mostly placebo. They can shift your frame of mind."

The human body produces its own internal sexual stimulants such as neurotransmitters in the form of chemicals, electrical responses and glandular substances. Even external aphrodisiacs work by activating these internal stimulants, there is nothing more powerful than what already exist in our body. If aphrodisiacs truly work they are activating your own internal sexual mechanism

Are aphrodisiac preparations safe?

Exotically named herbal aphrodisiacs like Ashwagandha, Damiana, Dong Quai, Ginkgo, Ginseng, Horny Goat Weed, Kava-Kava, Potency wood, *Saw palmetto*, Tribulus and Yohimbine formulations are sold to stimulate low libido.

Aphrodisiac preparations are mostly from plants that can be toxic. There is a long list of poisonous and toxic plants that grow in tropical and subtropical countries. Aconitum, argemone, belladonna, bloodroot, buttercup, castor beans, croton, datura, digitalis, euphorbia, henbane, heliotropium, strophanthus, strychnos, poppy, yohimbine, and many more are very commonly found poisonous species. Socrates death penalty was carried out with a fatal dose of hemlock (*Conium maculatum*). These plants contain secondary metabolites such as alkaoids, glycosides, and related compounds that have definite pharmacological actions and can be toxic even in small doses. Some of these are used in small quantity as aphrodisiacs. Thre is a saying that *plants can kill as well as cure*. So, natural does not mean it is safe. There has been ample published data to support this statement (Marty 1999). Nevertheless, heavy metal contamination such as lead, mercury and arsenic have been reported during the production of herbal formulations and the amount above the detection limit of these contamiants can be fatal.

During February 1993-May 1995, the New York City Poison Control Center was informed about onset of illness in five previously healthy men after they ingested a substance marketed as a topical aphrodisiac; four of the men died. These cases were investigated by the New York City Department of Health. The decedents died from cardiac dysrhythmias, and all five patients had measurable levels of digoxin detected in their serum. Digoxin had not been prescribed for therapeutic purposes for any of these patients, and none had medical conditions associated with endogenous digoxin-like immunoreactive substances. The purported aphrodisiac contains bufadienolides, naturally occurring cardioactive steroids that have digoxin-like effects. Recently, New York City issued firm warning over fatal toad-venom/plant product aphrodisiac (Slatkin 2008). Another person died after ingestion of a product which is called love stone, Jamaican stone, black stone, and Chinese rock. It is available at certain stores and sex shops but is banned by the FDA. Health officials said that the harden resin made with toad venom contained chemical like bufadienolides that could disrupt heart rhythms.

Natural aphrodisiac preparations are comprised of several insects, animals, plant parts, and minerals that can have a deadly effect. Each plant or animal product in the preparation has many active constituents. They could be incompatible with each other or with the medicine an individual is taking. It could be harmful. We have discussed the interaction of herbal aphrodisiacs with modern medicine separately in chapter ten.

Side effects or toxicity studies of these aphrodisiac preparations are not reported or conducted. Moreover, standard guidelines for their Quality analysis and Quality control so far is not clearly spelled out. It is hard to accept them safe unless we have clinically proven data about their safety.

Do aphrodisiacs enlarge size of male organ?
Natural aphrodisiacs advertisements promote that the products increase size of the male organ. This information is invariably mentioned in most of the advertisement. It is a false perception and fictitious advertisements to lure people to buy their products. Contrary to the popular belief, erection size cannot be enhanced through exercises or supplementation of any sort. According to Mondaini and Gontero (2005) the penis is not a muscle; it is a complex organ consisting of erectile chambers. The size and hardness of erection is directly proportional to the volume of blood flows into the erect *corpus cavernosa*. If the volume of the blood is less or more then erection and thickness varies accordingly. Less volume of the blood in the chambers results in incomplete erection, small size, and thickness.

What is the best aphrodisiac?
The best aphrodisiac of all can't be bought online, in stores, off television, or ordered from a magazine. The best aphrodisiac is your partner who is gentle, caring, full of warmth and willing to explore new techniques. Everyone has his or her own desire to accomplish sexual pleasure. It is very pertinent for both the partners to care for each other needs that turn them on. Tender cares and communications if incorporated in lovemaking can stimulate sexual activity. Romantic encounter, right emotions, situation and environment stimulate low libido. Feelings and desires from the partners to make

love should be mutual. Both the partners should be willing for the sexual act. Both the partners must be in good moods. All other domestic or personal problems must be kept aside during the time of sexual act. Any communication except sexual fantasies during the sexual act can leads to unhealthy discussion, arguments, fight and frustration. If one of the partners is not feeling well, the act should be postponed, as one-sided sexual act will never give satisfaction and solace to ones mind.

Both the partners need to be mentally prepared. If any precaution is required, it should be taken before the act. The place must provide security, privacy, and serenity. Romantic music in the background will enhance the mood. The sexual act must be before a meal or at least 3-4 hours after a meal. This will enhance sexual performance. Pleasing flirtation, deep intimacy and nice compliments from both the partners will definitely act as a good aphrodisiac. Before the act, some kind of erotic fantasies will also work as an aphrodisiac. During the act, spontaneous verbal expression of ecstasy from both partners will tremendously elevate the pleasure.

> *"A woman is like a fruit which will only yield its fragrance when rubbed by the hands. Take for example, the basil; unless it is warmed by the fingers it emits no perfume. And do you not know that unless amber be warmed and manipulated it retains its aroma within? The same with woman, if you do not animate her with your frolics and kisses, with nibbling of her thighs and close embraces, you will not obtain what you desire."*
> - Cheik Nefzaoui (*The Perfume Garden*).

Gentle hugging, kissing and caressing the body are very essential in foreplay and during the sexual act and it should be very natural from both the partners. Faking and over acting can spoil the entire pleasure and throw the person out of the act. Most important thing is to care for each other after the sexual act. The majority of the couples neglect this very fundamental part of life. Some part of communication and body contact must be maintained after the act so that the partner does not feel abandoned after use. After the act, the tender loving care is very important part of lovemaking. If it is neglected, it leads to frustration and hatred. In other words, afterplay is rather more imperative than the foreplay.

Sexual fulfillment is an elixir for a harmonious married life. Thread of love with flowers of true emotions holds the garland of life forever. It can only happen if both the partners share their actual feelings. They have to open up to each other. In other words true to each other in every aspect of life. If the only contact is genital-penis, vagina and ejaculation, sex becomes mechanical and boring. One feels even unsatisfied and frustrated after this kind of act. Sex is more than physical sensation. It is beyond physical. It should be spiritual to get solace to mind. Physical and mental satisfaction of both the partners in a sexual relationship is very essential. When sex is performed according to the right way, it becomes an ecstasy, an inexhaustible source of energy like an ocean, rather than an exhausting ordeal. If used unwisely, it can destroy health, relationship, family values, future progeny and leads to insanity. With the right partner and environment, sex can prove wonderful. Sexuality does not exist in isolation; it is a spectrum of emotions, romantic responses full of sensuality and spirituality. Both men and women need the right environments to respond and often do not get them. Several studies have found that many men like it when their female partner initiates sex. A woman who initiates sex also often stimulates her partner's sex drive and his desire for her too and acts as an aphrodisiac.

Sex is a perplexing phenomenon and needs to be performed very carefully. Sometime there is a strong physical attraction or just lust or sex craving that leads to sex but the relationship doesn't last longer. This type of sex is devoid of love. It is just lust and has short performance that leads to frustration. The satisfaction is not there. The partners are not fully satisfied even after sex. This type of sex is considered a part-time activity. The relationship is short lived since it doesn't possess the right characteristics to grow into a healthy, happy and satisfying one. Something is lacking there. Well-known Arabian poet and writer Kahlil Gibran defines sex as "spiritual affinity." It is a spiritual chemistry that can only be felt in the heart and soul when two human beings meet very intimately on a deeper level. It is about trust, mutual respect, true love, sincerity, understanding, communication, and commitment. All these characteristics should be mutual from both partners, and then you can have *Great Sex* and do not need aphrodisiacs. Great Sex is more than just orgasms. Great Sex means that one has to establish a bridge between the body and the soul. Body to body, heart to heart and soul to soul communication between the partners is essential when bodies and souls resonate with the same frequency. There is no time and space. This type of sex leads to ecstasy. Great sex means combination of love, intimacy, bonding, communication, tenderness, and spirituality.

Some sexual problems have social and emotional roots and no medicine or aphrodisiac potion can fix it. One has to balance and discipline his or her life. Physical, mental and spiritual disciplines are the key to a happy and contented life. Regular exercise, good nutritional diet, basic supplements, and a disciplined life are required to revitalize our sex life.

The best aphrodisiac is a woman with a waist of fine shape, well endowed with beauty, charm, intellect, soft spoken, not caviling, and skillful in the art of making love and caring. In other words, a woman whose very presence delights her partner and arouses him sexually is the ultimate aphrodisiac. Likewise, the man can be a good aphrodisiac for woman who is keen to please her. He must possess good personality, handsome, healthy, intelligent, not autocratic, true, and sincere in his speech with women. He must keep his words. He must be generous, truthful, brave, and pleasant in conversation.

> "When you are together with the beloved, this union and partnership extend you into the depth world and helps bring the depth world into tune."
> — Jean Houston
> (The Search of the Beloved)

Chapter Thirteen

Future of Natural Aphrodisiacs

During recent years, the use of alternative medicines has become widespread. Among them natural products are in great demands. Out of approximately 250,000 to 350,000 higher plant species identified so far, about 35000 are used worldwide for medicinal purposes. Very limited investigations have been carried out on medicinal plants. The opportunity to explore medicines from natural sources is unlimited. During the mid-1990s, over 200 companies and research organizations worldwide are screening plant and animal compounds for medicinal properties. Taxol/ paclitaxel, vinblastine, vincristine, topotecan, irinotecan, etoposide, teniposide are the burning examples from plants (Kong *et al.* 2003). These products are being used to cure different types of cancer and fatal diseases even today.

According to WHO, 80% of the world population relies on traditional medicine mostly herbs for their primary health care. The international market of herbal products is estimated to be US $62 billion, which is poised to grow to $ 5 trillion by the year 2050.

The public's demand for herbal remedies has increased day by day, despite a lack of scientific evidence. Thus millions of desperate people suffering from horrendous diseases are turning to herbal remedies to prolong their lives. The Majority of them are being exploited; left uncured and governments in the U.S. and other countries are ignoring this fact. Aphrodisiacs fall in the same category.

Aphrodisiacs or sex boosters are sold like wildfire on the web. These products are sold on the basis of myth. It is comparatively easy to sell sex related products. Most of these preparations are consisted of 5-10 or more herbs and each of them is composed of complex constituents. A majority of them are not standardized and can be toxic even in small doses.

No doubt, the information on herbal aphrodisiacs in the literature is substantial but not very well respected in the scientific community due to several limitations as shown in Figure 18.

Limitations of Natural Aphrodisiacs Products
a. Shortage of Studies and Lack of Clinical Data
b. Studies Conducted on Animals
c. Misbranded and Spurious Herbal Aphrodisiacs
d. Interaction with Prescription Drugs
e. Inferior Products on the Market
f. Poor QA/QC & Standardization of Natural Aphrodisiacs
g. Picking the Right Product
h. Restriction on Licensing

a.   Shortage of Studies and Lack of Clinical Data
Aphrodisiacs are harried by several problems. The lack of reliable research data hurts the Aphrodisiac herbal industry. There is shortage of studies and always a lack of scientific data. General consensus feels that clinical research is not a preference. In their view, traditional knowledge and test of time are adequate proof (Wilks and Ferrette 1999, Dalen 1999). However, the *test of time* is not a proper guideline for implementing the safety of traditional therapy (Ernst *et al,* 1998). Aphrodisiacs apparently have been legging behind by keeping clinical trials of secondary importance. Unfortunately, few clinical trials have been carried out to investigate their efficacy. Reputed aphrodisiacs used on the basis of myth do not have positive scientific evidence to support their long traditional use.

"I haven't seen anything in the literature that says eating 10 oysters will get you hot and horny," says Linda Banner, a board certified sex therapist in San Jose, California, and a member of the American Association of Sex Educators, Counselors and Therapists. However, recent findings by Fischer and Shimerling (2005) are indicating some reality in eating oysters. They found that clams and oyster contain L-aspartic acid which can release sex hormones testosterone and estrogens in the body. Though the research is in a rudimentary stage, there is a bright hope to bring myth into reality.

It wasn't oysters that increased blood flow to the genitalia in research conducted by Alan Hirsch, MD, a neurologist and psychiatrist who is the neurological director at the Smell and Taste Treatment and Research Foundation in Chicago. Hirsch's findings suggest a way to a person's heart is not through the stomach, but through the nose-and the smell of oysters didn't put any of the study participants in the mood. Hirsch and his colleagues measured the effects certain aromas had on penile and vaginal blood flow and found them very encouraging (Hirsch 2004).

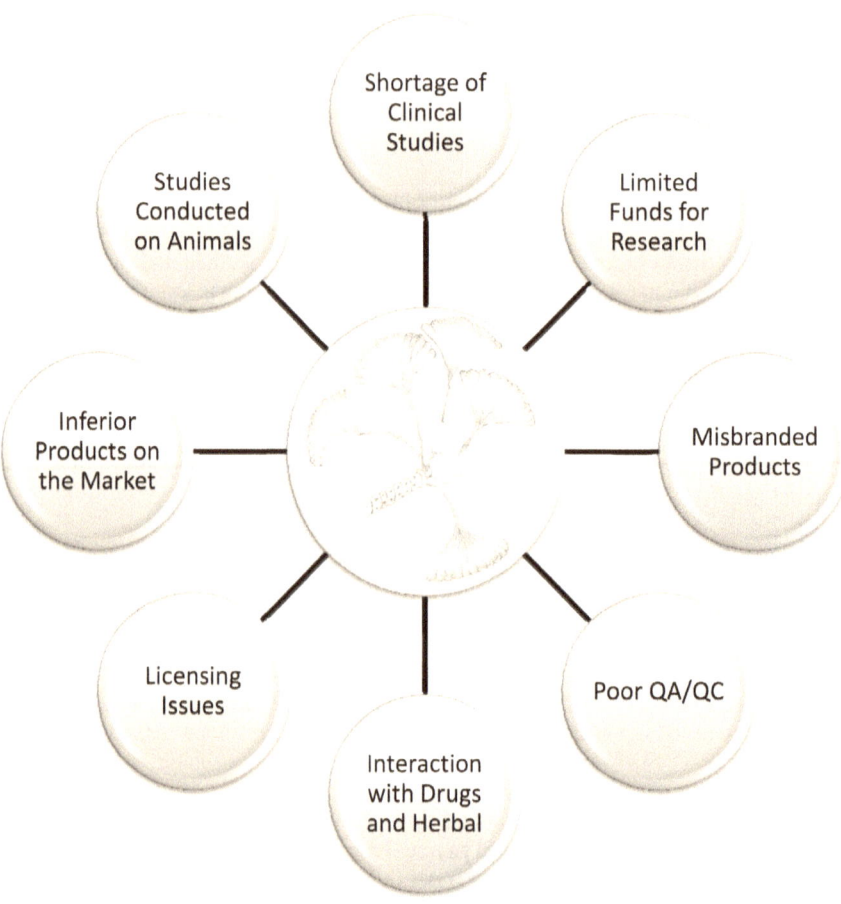

Figure 18: Limitations of Natural Aphrodisiacs

It is very difficult to accept the smell of food as aphrodisiacs since the clinical data vary from person to person. There aren't many scientific studies validating claims that certain foods or herbs actually increase sexual arousal, but oysters are one of those love potion foods with some scientific reasoning. It is believed that oysters contain large amount of zinc, a mineral that apparently helps the prostate gland, according to Chris Meletis, Chief Medical Officer at the National College of Naturopathic Medicine in Portland, Oregone. While the prostate is not directly linked to sexual performance, Meletis argues that good overall health is vital to sexual functioning.

Yohimbine appears to have the most scientific evidence directly linking it to an improvement in sexual function. The herb is a native to Africa and was used in African cultures to stimulate erections. Yohimbine does indeed help men who have problems getting and maintaining erections. The herb not only appears to increase blood flow, but also has side effects. Using more than 40 mg of yohimbine per day can cause dangerous side effects such as changes in heart rhythm, difficult breathing, potentially dangerous drop in blood pressure and temporary muscle paralysis in legs. It is one of monoamine oxidase inhibitors, a group of antidepressant drugs and can lead to many complications. A dose with 10 mg can induce mania in patients with bipolar depression. Till today, it is very difficult to accept this herb as an aphrodisiac on the basis of contradictory clinical data and side effects.

Some scientific evidence indicates ginseng, long used by many Asian societies as an aphrodisiac, improves blood circulation, and healthy blood flow is key to sexual arousal. Ginseng relaxes the artery walls, allowing blood to flow more easily. Viagra, the blockbuster anti-impotence drug, works in a similar way by increasing nitric oxide, which helps the arteries to relax. Ginseng has also been considered to increases sperm count, though there was nothing found published to substantiate this benefit. By far the most widely used herbal tonic and aphrodisiac is ginseng. Its overall export value worldwide exceeds $40 million annually and there are over six million users in the US alone. The roots of Asian ginseng (*Panax ginseng*) and American ginseng (*Panax quinquefolius*) have been used since ancient times. Chemical evidence supports the stimulatory effects of ginseng. Several steroids, triterpenoid D-glucosides (saponins) such as panaxsapogenin have been isolated from root extracts. Despite this, the US Food and Drug Administration found no evidence of enhanced sexual experience or potency resulting from its use.

Research has shown that ginkgo, like ginseng, can also increase blood flow to the pelvic. In a study conducted at the University of California in San Francisco, a group of men and women who suffered sexual problems as a side effect of taking antidepressants took 200 milligrams of ginkgo daily: 84% of them reported an improvement in sexual function. However, the study was flawed by the absence of a control group to account for possible placebo effects.

L-Arginine is an amino acid found naturally in many foods, including beef liver, pumpkin seeds, sesame seeds and raw almonds. L-Arginine triggers the production of nitric oxide, and increases the blood flow to the reproductive organ and may help treat erectile dysfunction but in a clinical trial in humans, no difference was established between L-arginine 500 mg 3 times daily and placebo (Klotz *et al.* 1999). However, in combination with yohimbine hydrochloride and pycnogenol (Pine pinaster) it showed improvement in men suffering with ED (Lebret *et al.* 2002; Stanislavov (2003). The effects of L-Arginine alone are unknown.and more research is required with L-arginine alone.

Spanish fly is another example derived from beetles of the *Lytta vesicatoria* species. It is the most notorious aphrodisiac of all. Its active ingredient is cantharidin that irritates the bladder and urethra, causing increased blood flow to the genitals and stimulating libido. It can permanently scar urethral tissue and infect the genitourinary tract. It may leads to an abnormally prolonged or constant erection (priapism) or an engorged vulva and vagina, both of which are often painful. It can be poisonous or even fatal. This Product needs to be rgulated.

Damiana, *Muira puama*, *Saw palmetto*, Maca root, Horny Goat Weed, Tribulus, Tongkat Ali, Dong Quai, Catuaba, Zallouh roots, and Ashwagandha, Licorice and Safed Musli are extensively used in many aphrodisiac products under their proprietary name as shown in Appendix A & B. Some of them are advertised to have great potential; however, these herbs need more scientific evidence to prove their effectiveness.

According to the U.S. Food and Drug Administration (FDA), no aphrodisiac has been scientifically proven to be effective at meeting its claims. It's also important to know that aphrodisiacs are similar to other herbal supplements and are not regulated by the FDA.

b. Studies Conducted on Animals
Most of the data of aphrodisiacs is obtained from animals testing. It is not appropriate to use that data to establish the safety of these herbs on human beings. A study conducted was on the Damiana species *Turnera diffuse* and *Pfaffia paniculata* extracts on the sexual behavior of male rats in Pharmacology at the Department of Biomedical Sciences, Univ. of Modena, Italy. It was reported that sexually potent and sexually sluggish/impotent male rats were treated orally with different amounts of *Turnera diffusa and Pfaffia paniculata* fluid extracts. Extracts had no effect on the copulatory behavior of sexually potent rats. However, both plant extracts-singly or in combination-improved the copulatory performance of sexually sluggish/impotent rats.

Similarly, Zheng *et al.* (2000) reported Maca's effect on sexual function on mice and rats. The results were encouraging. Further studies were also done with rats by Cicero *et al.* (2002) and indicated that sexual activity was tripled in mice. Gonzalez *et al.* (2003) carried out some human trials with maca and reported that it increased sexual desire in men within 8 weeks of treatment. However, the findings were contradicted by Bogani *et al.* (2006) who found that Maca did not exert androgenic activities.

There have been quite a few studies of Tongkat Ali for aphrodisiac activities on rats by Ang *et al.* (1997-2003). The effects of Tongkat Ali were studied on sexually experienced rats, castrated rats, sexually inexperienced rats and middle-aged rats. All these studies indicated an increase in the rats' sexual activities. No human clinical trials were done.

Many examples of herbs as aphrodisiacs can be cited that showed encouraging results with the preliminary randomized clinical trials but further controlled studies were not done. Thus, the clinicians are caught between encouraging results of randomized clinical trials and the relative lack of controlled long-term data. The latter information is required to make right therapeutic decisions.

c. Misbranded and Spurious Herbal Aphrodisiacs
Studies reveal that herbal products may contain ingredients, sometimes toxic, not listed on the label. These unidentified ingredients may be unintentionally included in the product eg, misidentification of a toxic plant as a desired nontoxic plant or adulterated for increased effect such as addition of a pharmaceutical agent to an herbal preparation.

On July 12, 2006, the Federal Drug Administration (FDA) warned consumers not to purchase or consume Zimaxx, Libidus, Neophase, Nasutra, Vigor-25, Actra-Rx, or 4EVERON. These products are promoted and sold on web sites as "dietary supplements" for treating erectile dysfunction (ED) and enhancing sexual performance, but they are in fact illegal drugs that contain potentially harmful undeclared ingredients. FDA has not approved these products and there is no guarantee of their safety and effectiveness or of the purity of their ingredients (Date Posted: 7/13/2006).

FDA advises consumers who have used any of these products to discontinue use and to consult their health care provider. FDA encourages anyone experiencing ED to seek guidance from a health care provider before purchasing a product to treat this medical condition.

"These products threaten the public health because they contain undeclared chemicals that are similar or identical to the active ingredients used in several FDA-approved prescription drug products. This risk is even more serious because consumers may not know that these ingredients can interact with medications and dangerously lower their blood pressure," said Dr. Steven Galson, Director of FDA's Center for Drug Evaluation and Research. Chemical analysis by FDA revealed that Zimaxx contains sildenafil, which is the active pharmaceutical ingredient in Viagra, a prescription drug approved in the United States to treat ED. The other products contain chemical ingredients that are analogues of either sildenafil or a pharmaceutical ingredient called vardenafil. Vardenafil is the active ingredient in Levitra, a prescription drug that, like Viagra, is approved in the United States to treat ED. There is no mention of any of these ingredients in any of the illegal products' labeling.

This deception poses a threat to consumers because the undeclared ingredients may interact with nitrates found in some prescription drugs (such as nitroglycerin) and lower blood pressure to dangerous levels. Consumers with diabetes, high blood pressure, high cholesterol, or heart disease often take nitrates. ED is a common problem in men with these conditions, and they may seek products like the ones noted above because these products claim that they are *all natural* or that they do not contain the active ingredients used in FDA-approved ED drugs. Nevertheless, the manufacturing source of the active ingredients in these dietary supplements is unknown; there is no assurance that the ingredients are safe, effective, or pure."

FDA warning letters to the firms marketing these products state that the products are illegal drugs based on claims made for the products or their ingredients. The letters also state that the products' labeling is false and misleading because it fails to disclose the presence of the chemical ingredients or the potential side-effects associated with the products' consumption. FDA instructed agency staff to stop the importation of Libidus, and the agency recently stopped a shipment of 4 EVERON from entering the United States.

In another FDA survey, the agency analyzed 17 dietary supplements marketed on the internet to treat ED and to enhance sexual performance in men. "Our survey found that many of the so-called *dietary supplements* marketed as treatments for erectile dysfunction actually contain non-dietary chemicals, including chemicals used as active ingredients in FDA-approved drugs. The claims made for these products were in fact claims made for the undeclared non-dietary chemicals they contain, which rendered them illegal drugs. FDA is committed to protecting the public health by removing such illegal and dangerous products from the market," said Margaret O'K. Glavin, FDA's Associate Commissioner for Regulatory Affairs.

During the past years, the FDA has found numbers of sex enhancers' herbal pills that contained the active ingredients found in Viagra, Cialis, or Levitra. One of the recalls formulations was called Liviro3 that contained tadalafil the main constituent of Cialis.

Spiked prepararations of natural aphrodisiacs have been found in Thailand, Taiwan, Canada, Australia, New Zealand, Hong Komg, Malaysia, UK, and US. Sildenafil analogues such as N-ethylpiperazine moiety instead of an N-methylpiperazine, and an acetyl group instead of the sulfonyl group, named acetildenafil, an N-ethylpiperazine moiety instead of an N-methylpiperazine (homosildenafil), and an N-hydroxylethylpiperazine moiety instead of an N-methylpiperazine, named hydroxyhomosildenafil were found by Blok-Tip *et al.* (2004) in herbal aphrodisiac products. Several undeclared drugs were identified in herbal remedies, such as sildenafil, tadalafil, testosterone, or glibenclamide by Bogusz *et al.* (2006). Very recently, detection of sildenafil analogues such as homosildenafil, hydroxyhomosildenafil, and acetildenafil in herbal products for erectile dysfunction has been demonstrated by Oh *et al.* (2006).

Quantities of ingredients listed on the label of aphrodisiac preparations also vary greatly causing toxic effects and unsafe products for consumption. Sometimes the concentration of the active ingredients varies substantially or these are negligible in some formulations. In a recent study a test of 25 ginseng products found that some of them did not even contain ginseng (Harkey *et al.* 2001). To determine the variability in a range of ginseng herbal products available in the U. S., they identified and measured the concentration of marker compounds by using HPLC and liquid chromatography-tandem mass spectrometry. Twenty-five commercial ginseng preparations from the genera *Panax or Eleutherococcus* were obtained from a local health food store and analyzed for 7 ginsenosides (marker compounds for Panax species, which include Asian and American ginseng) and 2 eleutherosides (marker compounds for *Eleutherococcus senticosus,* also known as Siberian ginseng). All plant products were correctly identified by botanical plant species (ie, Panax species or *E. senticosus*); however, concentrations of marker compounds differed significantly from labeled amounts. There was also significant product-to-product variability: concentrations of ginsenosides varied by 15-and 36-fold in capsules and liquids, respectively, and concentrations of eleutherosides varied by 43-and 200-fold in capsules and liquids, respectively. The data suggest that US ginseng products are correctly labeled as to plant genus; however, variability in concentrations of marker compounds suggests that standardization may be necessary for quality assurance and that characterization of herbal products should be considered in the design and evaluation of studies on herbal products.

Labeling recommendations exist for products containing *Larrea tridentata* (chaparral) comfrey (Symphytum spp.); *Piper methysticum*; *Serenoa repens (Saw Palmetto)*; and *Hypericum perforatum (*St. John's Wort) and so on. In addition, AHPA (American Herbal Product Association) published an entire volume of information related to establish herb safety concerns, entitled *Botanical Safety Handbook* (Mcguffin *et al.*1997). This reference classifies over 500 herbs with safety categories that can assist both manufacturers in their labeling and consumers in making informed choices in their use of herbs. A general rule for assuring responsible use of an herbal product is to follow all of the labeled directions. If the product bears a caution that suggests that the product is inappropriate for your use, you should take that message seriously.

Garlic supplements are one of the most popular herbal products used as aphrodisiac formulations. Garlic may help lower cholesterol and high blood pressure, slow the development of atherosclerosis, and aid other conditions such as low libido. But testing by ConsumerLab.com found one product without any of a key garlic compound and another had less than 1% of the expected amount of this compound. Other products claiming high potency were actually low potency. Two products were contaminated with lead. In all, eight products failed testing.

Ginseng is the common ingredient of many sexual enhancement products. But ConsumerLab.com has found problems in many ginseng supplements over the years. In this newest review, six products failed to pass testing due to lead contamination, lack of ingredient, or inadequate labeling. One product had less than 10% of its claimed amount of ginsenosides despite its "EXTRA STRENGTH" label. A major store brand product was contaminated with lead. Quality and safety assessment of ginseng was carried out by Durgant *et al.* (2005). 47 samples from 20 suppliers were collected by them comprising American and Asian ginseng. The results showed that 24 samples contain organochlorine pesticides above the detection limit. Toxic metals such as arsenic, mercury, lead, and cadmium were also present in most of the samples above the detection limit.

ConsumerLab.com tested supplements used to enhance male and female sexual performance made with L-arginine, yohimbe, and epimedium (horny goat weed). But only six of eleven different supplements were found to contain key listed ingredients and meet other quality criteria. Lead contamination and/or low levels of expected plant compounds caused five products to fail testing. The lab tested another batch of some sexual enhancement products. This report comprises 35 different ingredients commonly used in sexual enhancement supplements. Evidence of potential benefit was found to exist for only 14. The report covers products such as *ArginMax for Women, Excite Male Performance, Libido-Max,* and *Source Naturals Male Response.* Significant amounts of lead in supplements made with a form of ginkgo, *saw palmetto* was also found. Several ginkgo supplements were also low in key compounds.

Adulterants were also found in Asian Patent Medicine (TCM) sold in the USA. These formulations comprise multiple products including herbs, animal parts, and minerls. 260 Asian herbal products collected from California retail herbal stores by California Health Department and were analyzed by GC-MS and atomic absorption methods (KO 1998). Most of the products were contaminated with lead, arsenic, and mercury. 83 contained undeclared pharmaceuticals or heavy metals. 23 have more than one adulterant. The most common undeclared were ephedrine, chlorpheniramine, methyltestosterone, and phenacetine. The remaining products cannot be assumed to be safe of toxic ingredients in view of their batch to batch inconsistency. The studies were under taken by the health department to educate the public, the herbal industry, and the medical community about the potential danger of Asian patent medicine (Traditional Chinese Medicines). Likewise, contamination of mercury above the detection limit was found in extensively used aphrodisiac herbal Tongkat Ali in Malaysia (Ang and Lee 2006).

d. Interactions with Prescription drugs

One issue that needs to be addressed is the interaction between aphrodisiac formulations and prescription drugs (Ernst 1998, 2000a-c). Several complications have been found in patients taking these aphrodisiacs comprising herbs with their prescription drugs. Instead of stimulating their sexual activity, they could be ended in the hospital. Some information is available as reported by Cupp (2000); however, inadequate data in this field is available. More research is required to investigate the interactions between herbal medicines constituting aphrodisiac preparations and prescription drugs. Severe and systematic evaluation of all aphrodisiac herbal products is required to face the challenge. Drug-herbal aphrodisiacs interaction has been summarized in Appendix K. The subject has been discussed in details in chapter ten.

e. Inferior Products on the Market

Many products on the market do not use standardized extract, because these are costly. However, these products do not contain enough active ingredients to offer its benefit. It is very pertinent that the product is manufactured at a facility that follows strict GMP compliance. These are the strictest manufacturing standards, the same standards that pharmaceutical drugs follow to make the highest quality products. It's important to note that under the Dietary Supplement Health and Education Act of 1994, herbal aphrodisiac products are considered dietary supplements, so manufacturers do not have to guarantee the content or effectiveness of their products. As a result many of these products contain very little or none of the ingredients they claim to or too much ingredient (dangerous!). Still others contain harmful contaminants that could cause serious damage to the body. In a study by Ziglar (1979), a test of 54 ginseng products found that 25% of them did not even contain ginseng. Likewise, an analysis of commercial yohimbe products sold in the United States revealed that few products were found with any appreciable levels of yohimbine, raising concerns about the quality control of some of these products (Betz *et al.* 1995).

During recent years there have been more similar reports as discussed in the previous section. It is amazing what consumers will do when they don't have the ingredients in their products! That's why it is very pertinent to find a high quality supplement, one with standardized herbal extract, manufactured under GMP compliance by a well highly-trained, highly-credential product formulator.

f. Poor QA/QC and Standardization of Natural Aphrodisiacs

The present lack of quality control and standardization of herbal medicinal products is of great concern. Whenever such products are independently analyzed, the results show that pesticides, herbicides, or heavy metals contaminate most of them (Consumer lab 2001). This is very dangerous. It can be fatal and ultimately hurts the reputation of the herbal industry. The herbal industry should have strict quality control and quality standards to save this disaster. Meanwhile, consumers and health care professionals are in a fix while trying to decide which brands of herbal medicines to buy or recommend. It is always advisable to buy standard extracts with analytical values or to buy brands and extracts that have been tested for their active constituents and these constituents meet the cocentration standards. The FDA does not regulate QA/QC and standardization of aphrodisiac herbal products since these are sold over the counter as food supplements. These products are not regulated as drugs, so no legal standards exist for their processing, harvesting or packaging. Thus, manufacturers are not required to prove the efficacy, toxicity, side effects and interaction of these products. Consumers are depending upon the integrity of the manufacturers. In spite of the good intentions of majority of the manufactures, it is very difficult to control the quality of the aphrodisiac herbal products. Quality control depends upon many factors; some of them are as follow:

f.1. Identification of the Raw Material

Proper identification of the plant at the time of collecting the plant parts from wild source is very important. Improper identification of the herb is detrimental to the quality of the products. The manufacturer, most of the time, is not even aware of this problem. The herb suppliers, who supply the raw material, are not devoid of the identification's problems especially in the case of expensive herbs, less expensive substitute is used. Sometimes substituted plant parts contain toxic substances and can be injurious to health. Proper authentication of the raw material is very pertinent. Those companies who are cultivating their own species are better off. They can control these factors. Most of these factors can be controlled by implementing standard operating proceedures leading to Good Agricultural Practices (GAPs).

f.2. Ecological Conditions

Variation of the active constituents due to ecological conditions is also a matter of great concern. Soil, temperature, and climate also effect the concentration of the active constituents present in the plants. Ginseng growing at two different places will show variations in ginsenosides contents. To avoid the variation in chemical constituents, it is pertinent to collect the material from the same place or cultivate them under similar ecological conditions.

f.3. Drying and storage

Proper drying and storage of the collected plant parts follow it. Improper drying and storage will definitely impact the active principles. Drying at low temperature will save the thermolabile constituents. It will also protect the enzymes to activate. During growth and storage, pesticides residues, microorganism, aflatoxins, radioactive substances, and heavy metals can contaminate crude plant materials.

f.4. Extraction

Extraction of the material with pure solvents and concentrating them under proper conditions of temperature affect the quality of the active principles. Organic solvents from plastic containers used for the extraction of herbs often contaminate the final extracts with organic phthalates, which can be toxic. Solvents for extraction should be distilled and stored in clean glass containers.

f.5. Analysis of the Active Constituents

Standardization is another important and logical issue but not an easy one. Owing to lack of proper standard parameters for the standardization of an aphrodisiac preparation, herbal industry is in question. There have been several instances where substandard herbs are reported. Many products containing feverfew had no detectable levels of the active constituent parthenolide-responsible for antimigraine activity. Instead of curing migraine, it became a headache for the consumer (Awang 1993). A *Consumer Reports study* found that level of ginsenosides the active component in ginseng varied significantly in ten different products brands (Consumer Report 1995). Ziglar (1979) reported that 60% of the 54 ginseng tested contained less amount of active constituent and 20 % did not show any trace of the active constituent. A Good Housekeeping Institute analysis (1998) of six widely used St John's Wort supplements capsule and four liquid extracts revealed variation in concentration of the active constituent hypericin and pseudohypericin. The study showed a 17 fold difference between the capsules containing the smallest amount of hypericin and those containing the largest amount, based on manufacture's maximum recommended dosage. A 7 to 8 fold difference from the highest to lowest levels was found in liquid extracts (Consumer safety symposium 1998). Likewise, twenty products containing *Ephedra* were tested at the University of Arkansas. Researchers found many differences in alkaloid content from product to product and between two lots of the same product. One product was devoid of alkaloidal contents (Gurley 2000). Another recent report indicated that a leading laboratory had tested five brands of ginger, ginkgo, ginseng, melatonin, *Saw palmetto,* and St. John's Wort and milk thistle purchased at five stores in the Dallas area. Majority of the products did not reveal the stated concentration of the active constituents (Roffman 2000). There are many examples of herbal medicines that have been adulterated with other more toxic herbs; potent drugs like phenylbutazone, synthetic corticosteroids and other prescription drugs or heavy metals (Segasothy 1991, Chan 1993). Aphrodisiac herbal products consisted of synthetic compound like sildenafil to obtain the desired effect. James Neal-Kababick, director of Oregon-based Flora Research Laboratories, said about 90 percent of the hundreds of samples he has analyzed contained forms of patented pharmaceuticals. Some of the samples have doses more than twice that of prescription erectile dysfunction medicine.

Thus standardization of product is a difficult process but very essential. Aphrodisiac preparation is a combination of several herbs and each herb comprises of several chemical constituents. In many cases, the overall pharmacological effects are not due to single compound but several compounds causing synergistic effects. This indicates that standardization should calibrate more than one component. It is difficult to quantitate all these constituents in the formulation. Hence a fingerprint chromatogram of the extracts or preparation obtained by HPLC, gas chromatography (GC) or thin layer chromatography (TLC) is very essential. This fingerprint should represent identity, purity, and therapeutic efficacy of extracts from the same herb. There is no problem to quantify the single principal active constituent. If a preparation contains more than four herb extracts, each extract may be characterized before mixing and after mixing to establish the identity of the product. Ginseng has been used as a powder or a crude extract of the plant roots. The quality control of commercial ginseng preparations is difficult due to the diverse compounds present. Most previous quality control methods using TLC or HPLC-UV (or-MS) cannot be expected to cover a wide range of compounds in the commercial ginseng preparations. The metabolic fingerprinting of ginseng preparations was performed by (1) H-NMR spectroscopy. Although (1)H-NMR spectroscopy could provide information about the total profile of the compounds present, low resolution and overlapping signals make it difficult for further identification of each compound. For overcoming the problem two-dimensional J-resolved NMR spectra and multivariate data analysis techniques were applied for the analysis. Principal component analysis (PCA) of projected J-resolved NMR spectra shows a clear discrimination among those samples by principal component 1 and principal component 3. The loading plot of PC values obtained from all NMR signals indicates that alanine, arginine, choline, fumaric acid, inositol, sucrose, as well as ginsenosides are important metabolites to differentiate the preparations from each other. This method allows an efficient discrimination of a ginseng preparation in less than 15 minutes without any pre-purification steps (Yang *et al.,* 2006). Catuaba is another example where it is difficult to classify the commercial samples. Catuaba commercial samples can also be identified by this technique.

Secondly, toxicity studies are very pertinent to indicate the safety of the product. Simple bioassay for biological standardization of herbal drugs should be incorporated to develop animal models for toxicity and safety evaluation. Finally, clinical trials should also be performed to establish the therapeutic efficacy of the final product. NCCAM (National Center for Complementary and Alternative Medicine) recently announced its intention to fund the development of standardized preparation for use in clinical trials of herbs. These include feverfew, valerian root, echinacea and many more.

New standardization technology is available. Whitehall-Robins, a division of American Home Products Corporation is launching a new herbal supplement line of six products under the Centrum umbrella. They are utilizing a new standardization technology known as **PharmPrint**™ that identifies the key constituents in the herb. This unique technology is necessary to (a) Identify, quantify, and control the active components of herb compound. (b) Determine the activity of each component in a specific bioassay and (c) Ensure the components are present in predetermined quantities for a given manufactured batch. The **PharmPrint**™ process was initially developed at the University of Southern California in the 1970s. After years of testing this process has received its first patent in 1996 (*http://nccam.nih.gov/).*

g. Picking the Right Products

Picking the most suitable product is another hard task. The right herbal remedy is there, with the right dose and under the proper guidance of the herbalist or physician. Now it is very pertinent to find a manufacturer who creates a high-quality product that would be therapeutically effective. A high quality product with good laboratory practice, quality control, quality analysis, and standardizations should be picked up from a reputable manufacturer; however, still it is a big challenge. Anyway, the following suggestions are given to help the consumers. Check the background profile of the manufacturers, their reputation in the market, quality, and reliability of the product etc. Examine the manufacture and expiry date along with the batch number of the product. Verify the active ingredients and their quantities mentioned on the label. Check the name and address of the manufacturer and distributor, if any. Are they implementing Standard Operating Proceedures (SOPs), Good Agricultural Practices (GPAs), Good Laboratory Practices (GLPs), Good Supply Practices (GSPs), and Good Manufacturing Practice (GMPs)?

h. Restriction on Licensing

There is no restriction on the licensing of aphrodisiac products. These are being sold on the basis of dietary supplements. There is a great exploitation. Adults ingest these products for self-treatment of low libido and health maintenance considering them safe. The active constituents of an aphrodisiac product are not properly defined. Some contain toxic ingredients that may not be identified on the label. These unidentified ingredients may be intentionally or unintentionally included in the product e.g., misidentification of a toxic plant as a desired nontoxic plant or adulterated for increased effect e.g., addition of a pharmaceutical agent to an herbal preparation.

In our opinion, aphrodisiac formulations, or any herbal remedies should be controlled under some standards. Caution is urged with the trend towards licensing of all herbal remedies not only herbal aphrodisiacs. Only those remedies should be licensed that have clinically proved data, less side effects, cost effective and long history of safe use. The FDA also may prohibit sale of an herbal product proven to have serious or unreasonable risk under conditions of use on the label or as commonly consumed. Prohibition of an herbal product generally occurs after marketing and extensive distribution to the public.

We must fight against the unproven and potentially harmful products sold freely as potent aphrodisiacs. We must demand proper diagnosis and conventional proven effective treatment for the various sexual problems. We must not allow these companies to take advantage of an individual's despair or frustration for benefit of profit.

Limitations of Research in Natural Aphrodisiacs

1. Limited Funds for Sexual Dysfunction Research

Most of the funds in the research and development industry are allocated to life saving drugs for the cure of various cancers, heart diseases, hypertension, rheumatic arthritis, and gastrointestinal disorders. Sexual dysfunction is the last one to be considered. It is a general consensus that sex diminishes with age so sexual dysfunction subject is not taken seriously. It is considered as natural decline of sex. The herbal industry cannot support the high costs of long-term studies because of the limitations that herbal medicinal products can hardly be patented. The encouragement for research investments is much lower as compared to the pharmaceutical sector. The figures for retail sales of herbs show that there should be plenty of funds. However, the challenge is to allocate some money to research and development section, particularly for herbs that show promising primary results. Limited funds available to the industry are not enough to carry out research in herbal aphrodisiacs to enable to separate reality from the myth.

2. Lack of Scientific Interest

Due to the sensitive nature of the subject there is lack of scientific interest. Very few people would like to discuss sexual dysfunction. If some scientist shows interest, he is not taken seriously by his colleague or superior, rather they make fun of him. However, lack of funds and encouragement could not attract scientist to carry out research in the field of atural aphrodisiacs. Similarly, there are so many scientists who want to chase myth to discover reality but are discouraged. Ample patience and interest is required in research particularly in natural product industry. If one screens 10,000 plants, there may be few that will show confirmed biological activity and to isolate the active constituent is another *Herculean task*.

221

3. Very Few Human Volunteers

Due to the sensitivity of the subject; very few volunteers are available. People feel some kind of embarrassment to discuss about their sex problems particularly men. Thanks to Pfizer for Viagra, at least people started to discuss with their doctors. Very few people take prescription from their doctors. Most of the people suffering from any kind of sexual dysfunction do self-treatment to avoid embarrassment. That is the answer that herbal sexual enhancers are still in vogue. To offer as a volunteer for aphrodisiac testing would not be easy for them. Likewise, women do not feel comfortable going through the trials, measuring physiological responses using a variety of special instruments: an ultrasound device that measures blood flow, a pH probe that measures alkalinity, another probe to measure the lengthening and widening of uterous muscle tissues, and a biothesiometer to determine sensitivity to applied pressure etc. Time is not long enough to come when there won't be any hesitation by the people to volunteer for aphrodisiac research too.

4. Limited Information on the Female Sexual Dysfunction

In the past years, nobody cared for the female sexual problems. The issue remained dormant until recently Viagra revolution ignited fire in this subject. Generally, women keep silent about their problems unless these come to the surface. Now they have started discussing their sexual dysfunction with their physicians. Herbal industry has also come in action and numerous herbal preparations are being sold for female sexual dysfunction on the basis of myths. Some of the myths are under investigations where the other still need to be investigated.

5. Complicated Research on Female Sexual Dysfunction

Research in the area of FSD is very complicated and depends upon various aspects such as emotional, physical, and socioeconomic. Many pharmaceuticals companies such as Proctor Gamble, Pfizer, Nastech, and Vivus are involved in the research of female sexual dysfunction; however, progress is very slow.

Methods for Evaluating Aphrodisiac Activity

Research on aphrodisiac activity from plant should follow a comprehensive and systematic methodology. The plant extract of various concentrations is first tested in healthy and active animal models and later on in humans. A brief methodology is summarized for determining the aphrodisiac activity from plants, using male/female albino rats as model. The systematic approach is illustrated in Figure 19. This methodology is a good start for screening the plants extracts for aphrodisiac activity and has been followed by many phytochemists (Yakubu *et al*. 2007).

Conclusion

The future of aphrodisiac herbal products is very bright but depends upon many factors. Some of the problems have been described and discussed above. We have to overcome these limitations that are hindering in the progress of research in the herbal industry. History reveals that life saving drugs have been obtained from the nature in the past

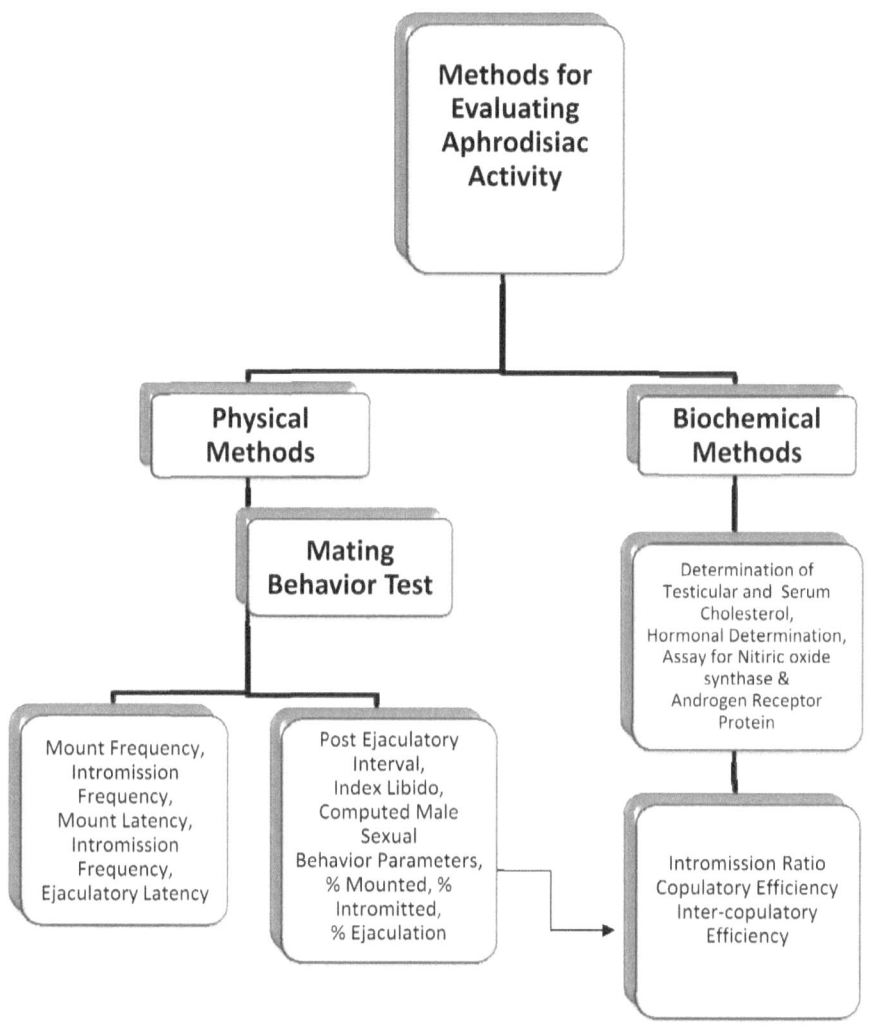

Figure 19: Methods for Evaluating Aphrodisiac Activity

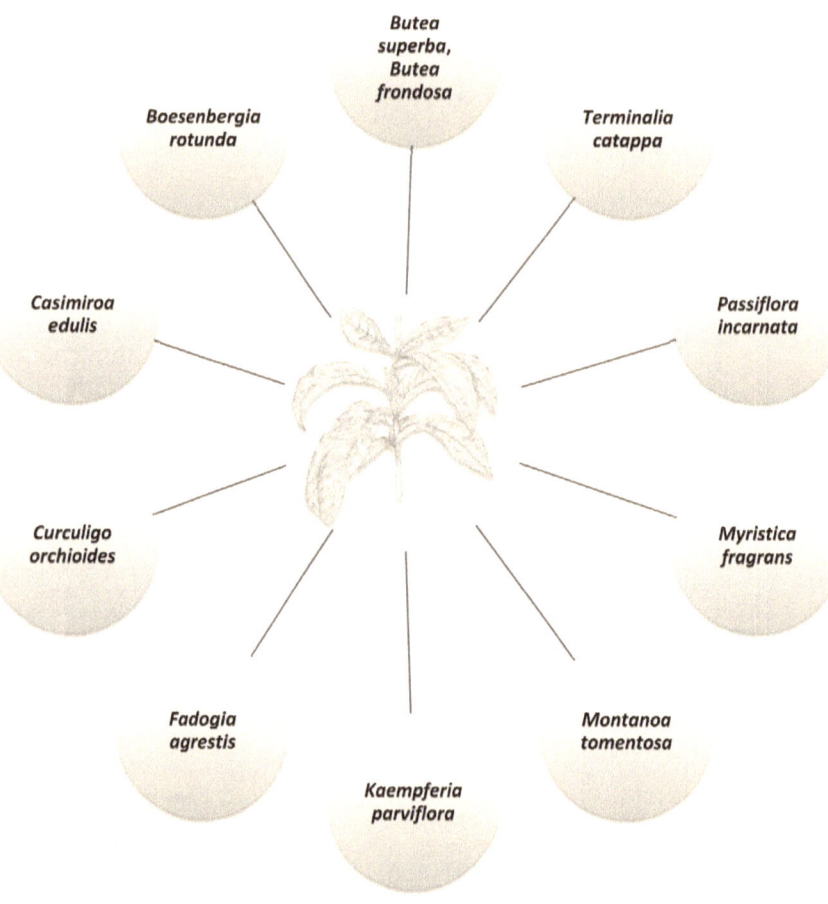

Figure 20: Natural Aphrodisiacs of the Future

starting from cardiotonics, antihypertensives, analgesics, antibiotics to anticancer drugs; and who knows that the Viagra of future for male and female may be from natural sources? On account of the serious side effects of Viagra, natural product chemists are seriously involved in the screening of various potential plant species with strong folkloric background for their aphrodisiac activity in different parts of the world. Appendix Z and Figure 20, show the various plants being assessed for aphrodisiac activity throughout the world. The Natural products industry in different parts of the world should get serious and allocate significant funds towards research and development to investigate these herbal aphrodisiacs that reveal strong history of myths and show promising results during their preliminary clinical studies. An organized and systematic research is required with a special emphasis on reliable clinical trials. There is a possibility to bring myth to reality by isolating the future Viagra from the natural sources.

> *"The average orgasm is only ten seconds long. The average frequency of intercourse is one or twice a week. That's twenty seconds a week, about one and a half minute a month, about eighteen minutes a year. In fifty years, that's about fifteen hours. For fifteen hours of ectasy we devote how many thousands and thousands of hours to thinking about sex, worrying about sex; wishing for sex and planning for sex?"*
> — *Alan P. Brauer and Donna J. Brauer ESO*

References

The list of the citations compiled here is by no means exhaustive. The authors have tried to keep the text references to a minimum. Some of the significant references have been given for further reading or information. The list below comprises references from each chapter.

Chapter One
Introduction

Eisenberg DM, Davis RB, Ettner SL, Appel S, Wilkey S, Van RM (1998) Trends in alternative medicine use in the United States, 1990-1997: results of a follow-up national survey. *JAMA* 280: 1569-75.

Ernst E, White A (2000) The BBC survey of complementary medicine use in the UK. *Complement Ther Med* 8: 32-6.

Heller A (1996) Herbs & Pharmacy: Natural Alliance. *Pharm Times* 62: 75-76.

Pritchard J (2007) Herbal Sex Pills Pose Dangers. *Associated Press,* Posted Nov., 13.

Puri RK, Puri R (2001) Alternative Medicine-A Review. *Indian J. Nat. Prouct* 17: 3-10.

*Texts available on the subject:

Bonnard M (1999) The Viagra Alternative: The Complete Guide to Overcoming Erectile Dysfunction Naturally, Healing Art Press, Rochester, Vermont.

Camphausen RC (1999) The Encyclopedia of Sacred Sexuality from Aphrodisiac and Ecstasy to Yoni Worship Zap Lam Yoga, InnerTraditions, Rochester, Vermont.

Choueke E (1998) Aphrodisiacs: A guide to What Really Works, Carol Publishing Group, New Jersey.

Dash B, Ramaswamy S (2001) Indian Aphrodisiacs, Roli Books Pvt. Ltd., New Delhi, India.

Kilham C (2004) Hot Plants: Nature's Proven Sex Boosters for Men and Women, St Martin Griffin Press, New York.

Lee WH, Lee L (1994) The Encyclopedia of Concentrated Aphrodisiacs, Instant Improvement, Inc, New York.

Levene P (1985) Aphrodisiacs: Fact and Fiction, Javelin Books, Poole, Dorset, UK.

Luca D (1998) Botanica Erotica: Arousing body, Mind and Spirit, Healing Art Press, Rochester, Vermont.

Mcgilvary M, Hodgson L (1988) Aphrodisiacs, Kyle Cathie Limited, London, U.K.

Meyer C (1986) Herbal Aphrodisiacs from World Sources, Meyer books Publisher, Glenwood, Illinois.

Miller RA (1985) The Magical and Ritual Use of Aphrodisiacs, Destiny Books, Rochester, Vermont.

Mitchell D (1999) Nature's Aphrodisiacs, Dell Publishing, a division of Random House Inc, New York.

Mitchell D (2005) Sexual Vigor for life-The Natural Way, ibooks Inc, New York.

Morgenthaler J, Joy D (1994) Better Sex through Chemistry, Smart Publications, Petaluma, CA.

Nickell N (1999) Nature's Aphrodisiacs, The Crossing Press, Freedom, California.

Ratsch C (1997) Plants of Love: The History of Aphrodisiacs and A Guide to Their Identification and Use, Ten Speed Press, Berkeley, California.

Roybal BA, Skowronki G (2002) Sex Herbs: Nature's Sexual Enhancers for both Men and Women, Gramercy Books, 2nd ed., NY.

Sahelian R (2003) Natural Sex Boosters, www.RaySahelian.com., USA.

Stark R (1981) The Book of Aphrodisiacs, A Starborough Book, Stein and Day Publishers, NY.

Taberner PV (1985) Aphrodisiacs: The Science and the Myth, Croom Helm Ltd., UK.

Tisserand M (1993) Essence of Love, Fragrance, Aphrodisiaca and Aromatherapy for Lovers, HarperCollins Publishers, NY.

Walton AH (1958) Aphrodisiacs: from Legend to Prescription, Associated Booksellers, Westport, Connecticut, USA.

Watson CM, and Hynes A (1993) Love Potions, A Guide to Aphrodisiacs and Sexual Pleasures, Jeremy P. Tarcher, Inc, CA, USA.

Chapter Two
History of Aphrodisiacs

Andrew L, Nicola G (2000) Homeopathy, the Principles and Practice of Treatment, D. K. Publishing, NY.

Charles Fowkes (1990) Ed., "The Illustrated Kama Sutra, Ananga-Ranga-Perfumed, Garden" trld, by Sir Richard Burton and F.F. Arbuthnot, The Hamlyn Publishing Group, London, UK.

Chattoe J (2003) CNN's, Friday, May 30, 2003, posted: 8:52 AM.

Lockie A (2000) Encyclopedia of Homeopathy, D.K. Publishing, NY.

Lovell MS (1999) *Introduced,* "The Perfume Garden of the Cheikh Nefzoui-A manual of Arabian Erotology', Translated by Sir Richard Burton, A Signet Classic, Penguin group, NY.

Moore BP (1965) translated, Ovid (Naso Ovidius Publius)*, Ars Amatoria*, Folio Society, London, UK.

Shang A, Huwiler-Muntener K, Nartey L, Juni P, Dorig S, Stern J, Pewsner D, Egger M (2005) Are the clinical effects of homoeopathic placebo effects? Comparative study of placebo-controlled trials of homoeopathy and allopathy. *The Lancet* 366: 726-732.

Tiquia R (1999) The University of Melbourne Proceedings of the Second Australian Symposium on Traditional Medicine and Wildlife, Melbourne, Australia, March.

Walton AH (1958) Aphrodisiacs: from Legend to Prescription, Associated Booksellers, Westport, Connecticut, p. 101.

William JR (1931) Treatment of Sexual Impotence in Men and Women, Eugenics Publishing Company, New York.

Wyke VB, Wink M (2005) Medicinal Plants of the World, Timber Press, Inc, Portland, Oregon, USA, p. 10.

Chapter Three
Male Sexual Dysfunction

Ayta IA, Mckinlay JB, Krane RJ (1999) The likely worldwide increase in erectile dysfunction between 1995 and 2025 and some possible policy consequences. *BJU Int* 84: 50-56.

Caretta N, Ferlin A, Palego PF, Foresta C (2005) Erectile dysfunction in aging men: testosterone role in therapeutic protocols. *J. Endocrinolo Invest* 28(11): 108-11.

Feldman HA, Goldstein I, Hatzichtistou DG, Krane RJ, Mckinlay JB (1994) Impotence and its medical and pschychological correlates results of Massachusetts male aging study. *J. Urol* 151: 54-61.

Goldstein I (2000) Male Sexual Circuitry Working Group for the Study of Central Mechanisms in Erectile Dysfunction. *Sci Am* 283: 70-75.

Guay AT, Spark RF, Bansal S, Cunningham GR, Goodman NF, Nankin SM, Perez JB (2003) American Association of Clinical Endocrinologist Medical Guidelines for Clinical Practice for the evaluation and treatment of male sexual dysfunction. A couple's Problem. *Endocrinology Pract* 9(1): 78-95.

Ji-Kan R, Sun US, *et al.* (2004) Plasma transforming growth factor-β levels in patients with erectile dysfunction. *Asian J Androl* 6: 349-353.

Laumann EO, Paik A, Rosen RC (1999) Sexual Dysfunction in the United States: Prevalence Predictors. *Jour Amer Med Assoc* 281: 537.

Rogers K (1996) Sexual dysfunction can be harmful effect of many drugs. *Drug Topics* June 10:71.

Selvin E, Burnett AL, Platz EA (2007) Prevalence of and Risk factors for Erectile Dysfunction in US Men. *Am. J. med* 120: 151-157.

World Health Organization, *Education and treatment in Human Sexuality: The Training of Health Professional.* Geneva, Switzerland: World Health Organization 1975, Technical Report Series, and No 572.

Chapter Four
Female Sexual Dysfunction

Besson R, McInnes R, Smith MD, Hodgson G, Koppiker N (2002) Efficacy and safety of sildenafil citrate in women with sexual dysfunction associated with female sexual arousal disorder. *J. Women Health Grad Based Med* 11: 367-77.

Cohen AJ (1998) Long term safety and efficacy of ginkgo biloba for antidepressant induced sexual dysfunction. Psychiatry On-Line; *www.priory.com/pharmol/ginkgo. htm*

Cohen AJ (1996) Treatment of Antidepressant-induced Sexual Dysfunction with Ginkgo biloba Extract. *Health Watch* 6: 1.

Cohen AJ, Bartlik B (1998) Ginkgo biloba for antidepressant-induced sexual dysfunction. *J Sex Marital Therapy* 24(2): 139-143.

Couper-Smartt JD, Rodham R (1973) A technique for surviving side-effects of tricyclic drugs with reference to reported side effects. *J Int Med Res* 1: 473-6.

da Silva AL (2002) Anxiogenic properties of *Ptychopetalum olacoides* Benth. (Marapuama) *Phytother Res* 16(3): 223-6.

Dash B and Ramaswamy S (2001) Indian Aphrodisiacs, Roli Books Pvt. Ltd., New Delhi, India.

Ellison JM, Deluca P (1998) Fluoxetine-induced genital anesthesia relieved by Ginkgo biloba extract (letter). *J Clin Psychiatr* 59(4): 199-200.

Fugh-Berman A, Kronenberg F (2001) Red clover (*Trifolium pratense*) for menopausal women: current state of knowledge. *Menopause* 8(5): 333-337.

Gillis CN (1997) Panax Ginseng Pharmacology: a NO link? *Biochem Pharmacol* 54(1): 1-8.

Gitlin M (1997) Sexual Side Effects of Psychotropic Medications. *Psychiatric Clinics of North America Annual of Drug Therapy* 4: 61-89.

Grimes JB, Labbate LA, Hines AH (1996) Sexual dysfunction induced by SSRIs. Presented at the American Psychiatric Association 149th Annual Meeting, New York, May 4-9 (Abstract 95).

Kessler RC, McGonagle KA, Zhao S, Nelson CB, Hughes M, Eshleman S, Wittchen HU, Kendler, KS (1994) Lifetime and 12-month prevalence of DSM-III-R psychiatric disorders in the United States: Results from the National Comorbidity Survey. *Arch Gen Psychiatry* 51(1): 8-19.

Kleijen J, Knipschild P (1992) Ginkgo biloba. *The Lancet* 340: 1136-1139.

Jacobsen FM (1992) Fluoxetine-induced sexual dysfunction and an open trial of Yohimbine. *J Clin Psychiatr* 53(4):119-22.

Laucella L (1997) Hormone Replacement Therapy, Conventional Medicines and Natural Alternatives, Your Guide to Menopausal Health Care Choices. Lowell House, Los Angeles.

Laumann EO, Paik A, Rosen RC (1999) Sexual Dysfunction in the United States: Prevalence and Predictors. *Jour Amer Med Assoc* 281: 537.

Meston CM, Bradford A (2007) Sexual Dysfunction in Women. *Ann. Rev. Clin Psychol* 3: 233-256.

Montejo AL, Llorce G, Izquierdo JA (1996) Sexual Dysfunction with SSRIs: a comparative analysis. Presented at the American Psychiatric Association 149th Annual Meeting, New York, May 4-9 (Abstract 717).

Newton KM *et al*. (2006) Treatment of symptoms of menopause. *Annals of Internal Medicine* 145: 869-879.

Price J, Grunhaus LJ (1990) Treatment of Clomipramine-induced anorgasmia with Yohimbine: a case report. *J Clin Psychiatr* 51(1): 32-3.

Raffa R (1998) Screen of receptor and uptake-site activity of hypericin component of St. John's wort reveals alpha receptor binding. *Life Sciences* 62: 265-270.

Siegel RK (1979) Ginseng abuse syndrome. *JAMA* 241: 1614-1615.

Stolze H (1982) An Alternative to Treat Menopausal Complaints. *Gyne* 3: 14-16.

Suzuki O *et al.* (1984) Inhibition of Monoamine Oxidase by Hypericin. *Planta Medica* 50: 272-74.

Tice JA, Ettinger B, Ensrud K, *et al.* (2003) Phytoestrogen supplements for the treatment of hot flashes: the Isoflavone Clover Extract (ICE) study. *Journal of the American Medical Association* 290(2): 207-214.

Vermeulen A (1976) The Hormonal Activity of the Postmenopausal Ovary. *J Clin Endocrinol Metab* 42: 246-253.

Chapter Five
Neurotransmitters as Natural Aphrodisiacs

Koshland DE (1992) The Molecule of the Year. *Science* 18: 1861-1865.

Meston CM, Frohlich PF (2000) The neurobiology of sexual function. *Arch Gen Psychiatry* 57: 1012-1030.

Stahl SM (1997a) Estrogen makes the brain a sex organ [Brainstorms]. *J Clin Psychiatr* 58: 421-422.

Stahl SM (1997b) Sex therapy in psychiatric treatment has a new partner: reproductive hormones [Brainstorms]. *J Clin Psychiatr* 58: 468-469.

Stahl SM (1998a) Brain fumes: yes, we have NO brain gas [Brainstorms]. *J Clin Psychiat* 59: 6-7.

Stahl SM (1998b) How psychiatrists can build new therapies for impotence. [Brainstorms]. *J Clin Psychiatr* 59: 47-48.

Stahl SM (1998c) Nitric oxide physiology and pharmacology [Brainstorms]. *J Clin Psychiat* 59: 101-102.

Stahl SM (2000) Essential Psychopharmacology, 2nd ed., New York, Cambridge University Press, NY.

Chapter Six
Alstonia scholaris (Saptparn)

Arulmozhi S, Mazumdar PM, Ashok PN, Sathiya Narayanan, L (2007) Pharmacological Activities of Alstonia scholaris Lin (Apocynaceae). *Pharmacognosy Review* 1: 163-170.

Baliga MS, Jagetia GC, Ulloor JN, Baliga MP *et al.* (2004) The Evaluation of the acute toxicity and long term safety of hydroalcoholic extract of Sapthatparna (Alstonia scholaris) in mice and rats. *Toxicology Letters* 151: 317-36.

Cai XH, Du ZZ, Luo XD (2007) Unique monoterpenoids indole alkaloids from *Alstonia scholaris*. *Org Lett* 9: 1817-20.

Cai XH, Tan QG, Liu YP, Feng T, Du ZZ, Li WQ, Luo XD (2008) A cage -monoterpene indole alkaloids from Alstonia scholaris. *Org Lett* 10(4): 577-80.

Channa S, Dar A, Ahmed S, Atta-ur-Rahman (2005) Evaluation of Alstonia scholaris leaves for broncho-vasodilatory activity. *J. Ethnopharmacol* 97(3): 469-76.

Chopra RN (1992) Indigenous Drugs of India, Academic Publishers, Calcutta, India.

Elisabetsky E, Costa-Campos L (2006) The alkaloid Alstonine: A Review of its Pharmacological Properties. *Evi. Based Comp Alt Med 3: 39-48.*

Feng T, Cai XH, Zharo PJ, Du ZZ, Li WQ, Lua XD (2009) Monoterpenoid indole Alkaloids from the Bark of Alstonia scholaris. *Planta Medica July 16, Aheaad of print.*

Gosh R, Roychowdhry DC, Litaka Y (1988) The structure of an alkaloid picrinine from Alstonia scholaris. *Acta Crysta* 44: 2151-2154.

Jagitia GC, Baliga MS (2003) Induction of developmental toxicity in mice treated with Alstonia scholaris in utero. *Repro Toxicol* 68(6): 472-78.

Kam TS, Nyeoh KT, Sim KM, Yoganathan K (1997) Alkaloids from Alstonia scholaris. *Phytochemistry* 45(6): 1303-05.

Macabeo AP, Krohn K, Gehle D, Read RW, Brophy JJ, Cordell GA, Franzblau SG, Aguinaldo AM (2005) Indole alkaloids from the leaves of Philippine Alstonia scholaris. *Phytochemistry* 66: 1158-62.

Miller RA (1993) The Magical and Ritual Use of Aphrodisiacs, Destiny Books, Rochester, Vermont.

Nadkarni K (1982) The Indian Materia Medica, Popular Prakashan, Vol 1, p. 80.

Rahman AU, Alvi KA, Abbas SA, Voelter W (1987) Isolation of 19, 20-Z -Vallesamine and 19, 20-E-Vallesamine from Alstonia scholaris. *Heterocycles* 26(2): 413-419.

Rahman AU, Ghazala MA, Fatima J, Alvi KA (1985) Scholaricine, an alkaloid from *Alstonia scholaris*. *Phytochemistry* 24(11): 2771-73.

Salim AA, Garson MJ, Craik DJ (2004) New indole alkaloids from the bark of Alstonia scholaris. *J. Nat Product* 67: 1591-4.

Williamson EM (1998) Potter's Herbal Cyclopedia, The C.W. Daniel Company limited, Saffron Walden, Essex, UK, p. 18.

Yamauchi T, Abe F, Chen RF, Nonaka GI, Santisuk T, and Padolina WG (1990b) Alkaloids from leaves of *Alstonia scholaris* in Taiwan, Thailand, Indonesia and the Philippines. *Phytochemistry* 29(11): 3547-52.

Yamauchi T, Abe F, Padolina WG, Dayrit FM (1990a) Alkaloids from leaves and bark of Alstonia scholaris in the Philippines. *Phytochemistry* 29: 3321-25.

Wang F, Ren FC, Liu JK (2009) Alstonic acids A and B, unusual 2,3-secofernane from Alstonia scholaris. *Phytochemistry* 70(5): 650-4.

Angelica Sinensis (Dong Quai)

Amato P, Christophe S, Mellon PL (2002) Estrogenic activity of herbs commonly used as remedies for menopausal symptoms. *Menopause* 9(2): 145-150.

Chen QC, Lee J, Jin W, Youn U, Kim H, Lee IS, Zhang X, Song K, Seong Y, Bae K (2007) Cytotoxic constituents from Angelica sinensis radix. *Arch Pharm Res* 30(5): 565-9.

Circosta C, Pasquale RD, Palumbo DR, Samperi S, Occhiuto F (2006) Estrogenic activity of standardized extract of Angelica sinensis. *Phtother Res* 20(8): 665-9.

Deng S, Chen S, Yao P, Nikolic D, Breemen RB, Bolton JL, Fong HS, Farnsworth NR, Pauli GF (2006) Serotonergic Activity-Guided Phytochemical Investigation of the Roots of Angelica sinensis. *J. Nat. Prod* 69 (4): 536-541.

Fu YF (1988) Treatment of 34 cases of infertility due to tubal occlusion with compound Dang guai injection by irrigation. *Jiangsu J. Trad Med* 9: 15-16.

Hann SK, Park YK, Im S, *et al.* (1991) Angelica-induced phytophotodermatitis. *Photodermatol Photoimmunol Photomed* 8(2): 84-85.

Hirata JD, Swiersz LM, Zell B, *et al.* (1997) Does Dong quai have estrogenic effects in postmenopausal women? A double-blind, placebo-controlled trial. *Fertil Steril* 68(6): 981-986.

Hon P, Lee C, Choang TF, Chui K, Wong HNC, (1990) A ligustilide dimer from Angelica sinensis. *Phytochemistry* 29: 1189.

Huntley AL, Ernst E (2003) A systematic review of herbal medicinal products for the treatment of menopausal symptoms. *Menopause* 10(5): 465-476.

Kim AR, EL-Aty AM, Choi JH, Lee KB, Shim JH (2006) Identification of volatile components in Angelica species using supercritical-CO (2) fluid extraction and solid phase microextraction coupled to gas chromatography-mass spectrometry. *Biomed Chromatogr 20*: 1267-73.

Kotani N, Oyama T, Sakai I, *et al.* (1997) Analgesic effect of an herbal medicine for treatment of primary dysmenorrhea-a double-blind study. *Am J Chin Med* 25(2): 205-212.

Lao SC, Li SP, Kelvin KWK, P Li, Wan JB, Wang YT, Dong , TT and Tsim WK (2004) Identification and quantification of 13 components in Angelica sinensis (Dang gui) by gas chromatography-mass spectrometry coupled with pressurized liquid extraction. *Analyitica Chimica Acta* 526: 131-137.

Lau CB, Ho TC, Chan TW, Kim SC (2005) Use of dong quai (Angelica sinesis) to treatperi or post menopausal symptoms in women with breast cancer: Is it appropraite? *Menopaus* 12(6): 734-740.

Lu XH, Liang H, Zhao YY (2003) Isolation and identification of the ligustilide compounds from the root of Angelica sinensis. *Zhongguo Zhong Yao Za Zhi* 28(5): 423-5.

Murray MT (1996) The Healing Power of Herbs, Prima Publishing, Rocklin, CA, pp. 43-49.

Newall CA, Anderson LA, Phillipson JD (1996) Herbal Medicines: A Guide for Health-Care Professionals, Pharmaceutical Press, London, pp. 28-9.

Page RL, Lawrence JD (1999) Potentiation of warfarin by Dong quai. *Pharmacotherapy* 19(7): 870-876.

Pizzorno JE, MT Murray (1996) "Angelica species," A Textbook of Natural Medicine, Bastyr University Publications, Bothell, WA, Vol. 1.

Powell CB, Dibble SL, *et al*. (2002) Use of Herbs in Women Diagnosed Ovarian cancer. *International J. Gynecologic Cancer* 12, 214-217.

Scott, GN, Elmer, GW (2002) Update of Natural Product-Drug Interaction. *Amer J. Health System Pharmacist* 59: 339-347.

Su DM, Ho TC, Chan TW, Kim SC (2005) New dimeric phthalide derivative from Angelica sinensis. *Yao Xue Xue Bao* 40(2): 141-4 (Article in Chinese).

Sun Y, Li S, Song H, Tian S (2006) Extraction of ferulic acid from Angelica sinensis with supercritical CO_2. *Nat Prod Res* 20(9): 835-41.

Wu M, Sun X, Dai Y, Guo F, Huang L, Liang Y (2005) Determination of constituents of essential oil from Angelica sinensis by gas chromatography-mass spectrometry. *J. Central South Univ Technology* 12: 430-436.

Zhiping H, Dazeng W, Lingyi S, *et al*. (2002) Treating amenorrhea in vital energy-deficient patients with Angelica sinensis-astragalus membranaceus menstruation-regulating decoction. *J Trad Chin Med* 6(3): 187-190.

Zschocke S, Liu J, Stuppner H, *et al*. (1998) Comparative study of roots of Angelica sinensis and related Umbelliferous drugs by thin layer chromatography, high-performance liquid chromatography, and liquid chromatography-mass spectrometry. *Phytochemical Analysis* 9(6): 283-290.

Avena Sativa (Oat)

Butt SM, Tahir-Nadeem M, Khan MK, Shabir R, Butt MS (2008) Oat: unique among the cereals. *Eur J Nutr* 47(2): 68-79.

Bye C, Fowle ASE, Letley E, Wilkenson S (1974) Lack of effect of *Avena sativa* on cigarette smoking. *Nature* 252: 580-581.

Dimberg LH, Gissen C, Nisson J (2005) Phenolic Compounds in oat grains (Avena sativa) grown in conventional and organic system. *Ambio* 34: 331-7.

Fukushima M, Watanabe S, Kushima K (1976) Extraction and purification of a substance with luteinizing hormone releasing activity from the leaves of Avena sativa. *Tohoku J Exp Med* 119(2): 115-22.

Gibson L, Garren B (2002) Origin, History, and Uses of Oat (*Avena sativa*) and Wheat (*Triticum aestivum*). Iowa State University, Department of Agronomy. Available from http://www.agron.iastate.edu/courses/agron212/Readings/Oat_wheat_history.htm.

Karmally W, Montez MG, Palmas W (2005) Cholesterol bearing benefits of oat-containing cereal in Hispanic American. *J.Am Diet Assoc* 105: 967-970.

Kurtz ES, Wallo W (2007) Colloidal oatmeal: history, chemistry and clinical properties. *J.Drug dermatol* 6: 167-70.

Okazaki Y, Ishizuka A, Ishihara A, Nishioka T, Iwamura H (2007) New Dimeric compounds of avenanthramide phytoalexin in oats. *J.org Chem* 72: 3830-9.

Tapola N, Karvonen H, Niskanen L, Mikola M, Sarkkinen E (2005) Glycemic responses of oat bran products in type 2 diabetic patients. *Nutr Metab Cardiovasc Dis* 15(4): 255-61.

Wenzig E, Kunert O, Ferreira D, Schmid M, Schuhly W, Bauer R, Hiermann A (2005) Flavonolignans from Avena sativa. *J Nat Product* 68: 289-92.

Capsicum annuum (Cayenne)

Abdel S, Moszik O, Szolcsanyi J (1995) Studies on the effect of intragastric capsaicin on gastric ulcer and on the prostacyclin-induced cytoprotection in rats. *Pharmacology Research* 32: 209-215.

BBC News (2006) March 15, 06:00 GMT

Buiatti E, Palli D, Decaril A, *et al.* (1989) A case-control study of gastric cancer and diet in Italy. *Int. J. Cancer* 44: 611-6.

Barnes J, Anderson LA, Phillipson JD (2002) Herbal Medicines, London, Pharmaceutical Press, UK.

Chu YF, Sun J, Wu X, Liu RH (2002) Antioxidant and Antiproliferative Activities of Common Vegetables. *J. Agric Food Chem* 50: 6910-6.

Fugh-Berman A (2000) Herb-drug interactions. *Lancet* 355: 134-38.

Kawaguchi Y, Ochi T, Takaishi Y, Kawazoe K, Lee KH (2004) New sesquiterpenes from Capsicum annuum. *J. Nat Product* 67: 1893-6.

Lee IO, Lee KH, Pyo JH, Kim JH, Choi YJ, Lee YC (2007) Anti-inflammatory effect of capsaicin in Helicobacter pylori-infected gastric epithelial cells. *Helicobacter* 12 (5): 510-7.

Lopez-Carrillo L, Avila M, Dubrow R (1994) Chili pepper consumption and gastric cancer in Mexico: A case study. *Amer J Epiderm* 139: 263-71.

Yoshioka M, St-Pierre S, Drapeau V, *et al* (1999) Effects of red pepper on appetite and energy intake. *Br J Nutr* 82: 115-23.

Yoshioka M, St-Pierre S, Suzuki M, Tremblay A (1998) Effects of red pepper added to high-fat and high-carbohydrate meals on energy metabolism and substrate utilization in Japanese women. *Br J Nutr* 80: 503-10.

Chlorophytum borivilianum (Safed Musli)

Acharya D, Mitaine-Offer AC, Kaushik N, Miyamoto T, Mirjolet JF, Duchamp O, Lacaille-Dubois MA (2009) Cytotoxic spirostane saponins from the roots of Chorophytum borivilianum. *J.Nat Product* 72: 177-81.

Handa SS (1996) Rasayana Drugs. In: Handa SS, Kaul MK (Eds). Supplement to Cultivation and Utilization of Medicinal Plants, Vol. 1. Jammu Tawi: Regional Research Laboratory, India, pp. 509-10.

Kenjale R, Shah R, Sathaye S (2008) Effects of Chlorophytum borivilianum on sexual behaviour and sperm count in male rats. *Phytother Res* 22(6): 796-801.

Kothari SK (2004) Safed Musli (Chlorophytum borivilianum) revisited. *J Med Arom Plant Sci* 26: 60--3.

Puri HS (2003) Rasayana-Ayurvedic Herbs for Longevity and Rejuvenation. Taylor and Francis, London, pp. 212-24.

Thakur M, Bhargava S, Dixit VK (2006) Immunomodulatory Activity of *Chlorophytum borivilianum* Sant. F. *eCAM*, December 13. P.1-5

Thakur M, Chauhan NS, Bhargava S, Dixit VK (2009) A Comparative Study on Aphrodisiac activity of Some Ayurvedic herbs in Male Albino Rats. *Arch Sex Behav Jan 13(ahead of print)*.

Crocus sativus (Saffron)

Abdullaev-Jafarova F, Espinosa-Aguirre JJ (2004) Biomedical properties of saffron and its potential use in cancer therapy and chemoprevention trials. *Cancer Detection and Prevention* 28(6): 430-436.

Abe K, Saito H (2000) Effects of saffron extract and its constituent crocin on learning behavior and long-term Potentiation. *Phytother Research* 14: 149-52.

Akhondzadeh S, *et al.* (2004) Comparison of Crocus sativus and impiramine in the treatment of mild to moderate depression: a pilot double-blind randomized trial. *Biomed Central Complementary and Alternative Medicine* 4(1): 12.

Akhondzadeh S, *et al.* (2005) Crocus sativus in the treatment of mild to moderate depression: a double-blind, randomized and placebo-controlled trial. *Phytotherapy Research* 19(2): 148-151.

Chevalier A (2000) Encyclopedia of Herbal Medicine, Dorling Kindersley, NY.

Escribano J, *et al.* (1996) Crocin, safranal and picrocrocin from saffron (Crocus sativus) inhibit the growth of human cancer cells in vitro. *Cancer Letters* 100 (1-2): 23-30.

Garcia-Olmo DC (1999) Effects of long-term treatment of colon adenocarcinoma with crocin, a carotenoid from saffron (Crocus sativus): an experimental study in the rat. *Nutrition and Cancer* 35(2): 120-126.

Heidary M, Reza NJ, Delfan B, Birjandi M, Kaviani H, Givrad S (2008) Effect of saffron on semen parameters of infertile men. *Urol J* 5(4): 255-9.

Hosseinzadeh H, Khosravan V (2002) Anticonvulsant effects of aqueous and ethanolic extracts of Crocus sativus stigmas in mice. *Archives of Iranian Medicine* 5: 44-47.

Hosseinzadeh H, Ziaee T, Sadeghi A (2008) The effect of saffron, Crocus sativus, extract and its constituents, safranal and crocin on sexual behaviors in normal male rats. *Phytomedicine* 15(6-7): 491-5.

Noorbala AA, Akhondzadeh S, Tamacebi-Pour N, Jamashidi A-H (2005) Hydro-alcoholic extract of Crocus sativus L. verus fluoxetine in the treatment of mild to moderate depression: a double blind randomized pilot trial. *J. Ethanopharmacol* 97: 281-284.

Straubinger M, Jezussek M, Waibel R, Winterhalter P (1997) Novel glycosidic constituents of saffron. *J. Agric. Food Chem* 45: 1678-1681.

Sujata V, Ravishankar GA, Venkataraman, LV (1992) Methods for the analysis of the saffron metabolites crocin, crocetins, picrocrocin and safranal for the determination of the quality of the spice using thin-layer chromatography, high-performance liquid chromatography and gas chromatography. *J. Chromatogr* 624: 497-502.

Tarantilis PA, Polissiou M, Manfait M (1994a) Separation of picrocrocin, cis-trans crocins and safranal of saffron using high-performance liquid chromatography with photodiode-array detection. *J. Chromatogr* 664: 55-61.

Tarantilis PA, *et al*. (1994b) Inhibition of growth and induction of differentiation of promyelocytic leukemia (HL-60) by carotenoids from Crocus sativus. *Anticancer Research* 14(5A): 1913-1918.

Willan A (2003) The Washington Post, March 12.

Willard P (2001) Secrets of Saffron, Beacom Press, Boston, Massachusetts.

Zarghami, NS, Heinz, DE (1971a) Monoterpene aldehydes and isophorone-related compounds of saffron. *Phytochemisty* 10: 2755-2761.

Zarghami NS, Heinz DE (1971b) The volatile constituents of saffron. *Lebensm. Wiss. Technol* 4 (2): 43-45.

Zhang Y, Shoyama Y, Sugiura M, Saito H (1994) Effect of Crocus sativus on the ethanol-induced impairment of passive avoidance performances in mice. *Biological and Pharmaceutical Bulletin* 17: 217-221.

Epimedium sagittatum (Horny Goat Weed)

Bensky D, Gamble A, Kaptchuk T (1992) Chinese Herbal Medicine Materia Medica, Revised Edition, Eastland Press, Seattle.

Bown D (1995) Encyclopedia of Herbs and their Uses, Dorling Kindersley, London.

Chen CC, Huang YL, Sun CM, Shen CC, Tang CM (1996) New Prenylflavones from the leaves of Epimedium sagittatum. *J Nat Prod* 59(4): 412-4.

Chen JK, Chen TT (2003) Chinese Medical Herbology and Pharmacology, City of Industry, Art of Medicine Press, Inc, CA.

Chen KK, Chiu JH (2006) Effect of Epimedium brevicornum Maxim extract on elicitation of penile erection in the rat. *Urology* 67(3): 631-5.

Chiu JH, Chen KK, Chein TM, Chio WF, Chen CC, Wang JY, Lui WY, Wu, CW (2006) Epimedium brevicornum Maxim, extract relaxes rabbit EC50 corpus cavernosum through multitargets on nitric oxide /cyclic guanosinmonoposphate signaling pathway. *Int J. Impot Res* 18: 335-42.

De Naeyer A, Pocock V, Milligan S, De Keukeleire D (2005) Estrogenic activity of a polyphenolic extract of the leaves of Epimedium brevicornum. *Fitoterapia* 76: 35-40.

Dell AM, Galli GV, Dal CE, Belluti F, Matera R, Zironi E, Pagliuca G, Bossio E (2008) Potent Inhibition of phosphodiestrase-5 by Icarin derivatives. *J. Natural Prod* 71: 1513-1517.

Jin XL (1981) Flavonoids from Epimedium sagittatum. *Zhong Yao Tong Bao* 6: 20-22.

Kuroda M, Mimaki Y, Sashida Y, Umegaki E, Yamazaki M, Chiba K, Mohri T, Kitahara M, Yasuda A, Naoi N, Xu ZW, Li MR (2000) Flavonol glycosides from Epimedium sagittatum and their neurite outgrowth activity on PC12h cells. *Planta Med* 66(6): 575-7.

Liang HR, Vuorela P. Vuorela H. Hiltunen R (1997) Isolation and immunomodulatory effect of flavonol glycosides from Epimedium hunanense. *Planta Medica* 63: 316-319.

Lin CC, Ng LT, Hsu FF, *et al.* (2004). Cytotoxic effects of Coptis chinensis and Epimedium sagittatum extracts and their major constituents (berberine, coptisine and icariin) on hepatoma and leukaemia cell growth. *Clin Exp Pharmacol Physiol* 31: 65-9.

Makarova MN, Pozharitskaya ON, Shikov AN, Tesakova SV, Makarov VG, Tikhonov VP (2007) Effect of lipid-based suspension of Epimedium koreanum Nakai extract on sexual behavior in rats. *J Ethnopharmacol* 114(3): 412-6.

Ning H, Xin Z, Lin G, Banie L, Lhe TF, Lin C (2006) Effects of Icarin on Phosphodiesterase-5 activity in vitro and cyclic guanosine monophosphates level in cavernous smooth muscle cells. *Urology* 68: 1350-4.

Oshima Y, Okamoto M, Hikino H (1989) Sagittatins A and B Flavonoid Glycosides of Epimedium sagittatum Herbs. *Planta Med* 55: 309-11.

Partin JF, Pushkin YR (2004) Tachyarrhythmia and hypomania with horny goat weed. *Psychosomatics* 45(6): 536-7.

Shen P, Guo BL, Gong Y, Hong DY, Hong Y, Yong EL (2007) Taxonomic, genetic, chemical and estrogenic characteristics of Epimedium species. *Phytochemistry* 68: 1448-58.

Wang GJ, Teal TH, Lin LC (2007) Prenylfavonol acylated flavonol glycosides and related compounds from Epimedium sagittatum. *Phytochemistry* 68: 2455-64.

Wang S, Zheng Z, Weng Y, *et al.* (2004) Angiogenesis and anti-angiogenesis activity of Chinese medicinal herbal extracts. *Life Sci* 74: 2467-78.

Wu C, Sheng Y, Zhang Y, Zhang J, Guo B (2008) Identification and characterization of active compounds and their metabolites by high performance liquid chromatography/Fourier transform ion cyclotron resonance mass spectrometry after oral administration of a herbal extract of Epimedium koreanum nakai to rats. *Phytotherapy Res 22(18): 2813-2824.*

Wu H, Lien EJ, Lien LL, *et al.* (2003) Chemical and pharmacological investigations of *Epimedium* species: a survey. *Prog Drug Res* 60: 1-57.

Yap SP, Shen P, Butler MS, *et al.* (2005) New estrogenic prenylflavone from *Epimedium brevicornum* inhibits the growth of breast cancer cells. *Planta Med* 71: 114-9.

Yap SP, Shen P, Li J, Lee LS, Yong FL (2007) Molecular and pharmacodynamic properties of estrogenic extracts from traditional Chinese medicinal herb Epimedium. *J.Ethnopharmacol* 113: 218-24.

Yong EL, Wong SP, Shen P, Gong YH, Li J, Hong Y (2007) Standardization and evaluation of botanical mixtures; lessons from a traditional Chinese herb, Epimedium, with oestrogenic properties. *Novatis Found Symp* 282: 173-88. Discussions 188-91: 212-8

Zhang HF, Yang YS, Li ZZ, Wang Y (2008) Simultaneous extraction of epimedine A,B, C and Icarin from Herba Epimedii by ultra sonic technique. *Ultrason Sonochem* 5: 376-85.

Erythroxylum catuaba (Catuaba)

Antunes E, Gordo WM, deOliveira JF, Teixeira CE, Hyslop S, De Nucci G (2001) The relaxation of isolated rabbit corpus by the herbal medicine Catuama and its constituents. *Phytother Res* 15: 416-21.

Bartram T (1995) Encyclopedia of Herbal Medicine, Ed Grace Publishers. UK.

Bernardes A (1984) A Pocket book of Brazilian Herbs, Editora e arta Ltd Brazil.

Campos MM, Fernandes ES, Ferreira J, Santos AR, Calixto JB (2005) Antidepressants-like effects of Trichilia catigua (Catuaba) extract: evidence for dopaminergic-mediated mechanism. *Psychopharmacology (Berl)* 182 (1): 45-53.

Daolio C, Beltrame FL, Ferreira AG, Cass QB, Cortez DA, Ferreira MM (2008) Classification of commercial Catuaba samples by NMR, HPLC and chemometrics. *Phytochem Anal* 19: 218-28.

Grafe E, Lude W (1978) Alkaoids from Erythroxylum vacciniifolium .The structure of Catuabine A, B and C. *Arch Pharm* (weinheim) 311(2): 139-52.

Kilham C (2006) Herb Sex Boosters, Discovery Health. (http:health.discovery.com/centers/sex/libido/amazon.html)

Kletter C, Glasi S, Presser A, Werner I, Reznicek G, Narantuya S, Cellek S, Haslinger E, Jurentsch J (2004) Morphological, chemical and functional analysis of catuaba preparations. *Planta Med* 70: 993-1000.

Manabe H, Sakagami H, Kusano H, Fujimaki, M, Wada C, Komatsu N (1992) Effects of Catuaba Extracts on Microbial and Hiv infection, Horiuchi Itaro & Co Tokyo, Japan, *In Vivo* 6(2): 161-165.

Van-Straten M (1994) Guarana: The Energy Seeds and Herbs of the Amazon Rainforest, C.W. Daniel Company, Ltd. Essex, England.

Zanolari B *et al.* (2003) Tropane alkaloids from the bark of Erythroxylum vacciniifolium. *J. Nat Prod* 66: 497-502.

Zanolari B, Guilet D, Marston A, Queroz EF, Paulo Mde Q, Hostettmann K (2005) Methylpyrrole tropane alkaloids from the bark of Erythroxylum vacciniifolium. *J. Nat Prod* 68: 1153-8.

Eurycoma Longifolia (Tongkat Ali)

Ang HH, Hitotsuyanagi Y, Fukava H, Takeya K (2002) Quassinoids from Eurycoma longifolia. *Phytochemistry* 59(8): 833-7.

Ang HH, Cheang HS (1999) Studies on the anxiolytic activity of Eurycoma longifolia Jack roots in mice. *Jpn J Pharmacol* 79(4): 497-500.

Ang HH, Cheang HS (2001) Effects of Eurycoma longifolia jack on laevator ani muscle in both uncastrated and testosterone-stimulated castrated intact male rats. *Arch Pharm Res* 24(5): 437-40.

Ang HH, Cheang HS, Yusof AP (2000) Effects of Eurycoma longifolia Jack (Tongkat Ali) on the initiation of sexual performance of inexperienced castrated male rats. *Exp Anim* 49(1): 35-8.

Ang HH, Ikeda S, Gan, EK (2001) Evaluation of the potency activity of aphrodisiac in Eurycoma logifolia Jack. *Phytother Res* 15: 435-6.

Ang HH, Lee KL (2002) Effect of Eurycoma longifolia Jack on libido in middle-aged male rats. *J Basic Clin Physiol Pharmacol* 13(3): 249-54.

Ang HH, Lee KL (2002) Effect of Eurycoma longifolia Jack on orientation activities in middle-aged male rats. *Fundam Clin Pharmacol* 16(6): 479-83.

Ang HH, Lee KL, Kiyoshi M (2004) Sexual arousal in sexually sluggish old male rats after oral administration of Eurycoma longifolia Jack. *J Basic Clin Physiol Pharmacol* 15 (3-4): 303-9.

Ang HH, Ngai TH (2001) Aphrodisiac evaluation in non-copulator male rats after chronic administration of Eurycoma longifolia Jack. *Fundam Clin Pharmacol* 15(4): 265-8.

Ang HH, Ngai TH, Tan TH (2003) Effects of Eurycoma longifolia Jack on sexual qualities in middle aged male rats. *Phytomedicine* 10 (6-7): 590-3.

Ang HH, Sim MK (1997) Eurycoma longifolia Jack enhances libido in sexually experienced male rats. *Exp Anim* 46(4): 287-90.

Ang HH, Sim MK (1998) Eurycoma longifolia increases sexual motivation in sexually naive male rats. *Arch Pharm Res* 21(6): 779-81.

Ang HH, Sim MK (1998) Eurycoma longifolia JACK and orientation activities in sexually experienced male rats. *Biol Pharm Bull* 21(2): 153-5.

Bedir E, Abou-Gazar H, Ngwendson JN, Khan IA (2003) Eurycomaoside: a new quassinoid-type glycoside from the roots of Eurycoma longifolia. *Chem Pharm Bull* (Tokyo) 1(11): 1301-3.

Cyranoski D (2005) Malaysia researchers bet big on home grown Viagra. *Nature* 11: 912.

Farouk AE, Benafri A (2007) Antibacterial activity of Eurycoma longifolia Jack, A Malayasian medicinal plants. *Saudi Med J* 28 (9): 1422-4.

Guo Z, *et al.* (2005) Biologically Active Quassinoids and their Chemistry. *Curr Med Chem* 12: 173-190.

Jiwajinda S, Santisopasri V, Murakami A, Sugiyama H, Gasquet M, Riad E, Balansard G, Ohigashi H (2002) In vitro anti-tumor promoting and anti-parasitic activities of the quassinoids from Eurycoma longifolia, a medicinal plant in Southeast. Asia. *J. Ethnopharmacol* 82(1): 55-8.

Kuo PC, Damu AG, Lee KH, Wu TS (2004) Cytotoxic and antimalarial constituents from the roots of Eurycoma longifolia. *Bioorg Med Chem* 12: 57-44.

Kuo PC, Shi LS, Damu AG, Su CR, Huang CH, Ke CH, Wu JB, Lin AJ, Bastow KF, Lee KH, Wu TS (2003) Cytotoxic and antimalarial beta-carboline alkaloids from the roots of Eurycoma longifolia. *J Nat Prod* 66 (10): 1324-7.

Le-Van-Thoi, Nguyen-Ngot-Suong (1970) Constituents of Eurycoma longifolia, Jack. *J Org Chem* 35(4): 1104-9.

Tjahjadi RV (2003) "Big Interest in a Small Tree, Bioresearch of *Eurycoma longifolia*, Pasak Bumi (Eurycoma longifolia) in South East Asia." *BioTani Indonesia Foundation/ PAN Indonesia*, Issue 6-20, May 18.

Ferula hermonis (Zallouh Root)

Ahmedhodjaieva A, *et al*. (1999) Influence of Ferutinine on the secretion of Leutnizing Hormone. *Pharmacol Toxicol* (Moscow) 4: 37-38.

El-Thaher TS, Matalka KZ, Taha HA, Badwan AA (2001) Ferula hermonis 'Zallouh" and enhacing erectile function in rats: efficacy and toxicity study. *Int J. Impot Res* 13: 247-51.

Galal A (2000) Sesquiterpenes from Ferula hermonis Boiss, *Pharmazie* 55: 961-2.

Hadidi KA, Aburjai T, Battah AK (2003) A comparative study of Ferula root extracts and sildenafil on copulatory behaviour of male rats. *Fitoterapia* 74: 242-246.

Khleifat K, Homady MH, Tarawneh KA, Shakhanbeh J (2001) Effect of Ferula hormonis extract on social aggression, fertility and some physiological parameters in prepubertal male mice. *Endocr J* 48(4): 473-82.

Kilham C (2004) Hot Plants, Natures Proven Sex Boosters for Men and Women, St Martin S. Griffin, NY.

Lhuillier A, Fabre N, Cheble E, Oueida F, Maurel S, Valentine A, Fouraste I, Moulis C (2005) Daucane sesquiterpenes from Ferula hermonis. *J. Nat Prod* 68: 468-71.

Zanoli P, Rivasi M, Zavatti M, Brusiani F, Vezzalini F, Baraldi M (2005b) Activity of single components of ferulis hermonis on male rat sexual behavior. *Int J. Res 17: 513-18.*

Zanoli P, Zavatti M, Rivasi M, Baraldi M (2005a) Ferula hermonis impairs sexual behavior in hormone-primed female rats. *Physiol Behav* 86: 69-74.

Zavatti M, Montanari C, Zanoli P (2006) Role of ferutinin in the impairment of female sexual function induced by Ferula hermonis. *Physiol Behav* 89: 656-661.

Ginkgo biloba (Maidenhair Tree)

Ashton AK, Ahrens K, Gupta S, Masand PS (2000) Antidepressant-induced sexual dysfunction and Ginkgo biloba. *Am J Psychiatry* 157: 836-8375.

Auguet M, Clostre F (1983) Effects of an extract of Ginkgo biloba and diverse substances on the phasic and tonic components of the contraction of an isolated rabbit aorta. *Gen Pharmacol* 14: 277-280.

Auguet M, DeFeudis FV, Clostre F, Deghengh R (1982) Effects of an extract of Ginkgo biloba on rabbit isolated aorta. *Gen Pharmacol* 13: 225-230.

Bedir E, Tatli II, Khan RA, Zhao J, Takamatsu S, Walker LA, Goldman P, Khan IA (2002) Biologically active secondary metabolites from *Ginkgo biloba. J Agric Food Chem* 50(11): 3150-5.

Braquet P, Ed (1988) Ginkgolides-Chemistry, Biology, Pharmacology and Clinical Perspectives, Vol., I, J. R. Prous, Barcelona, Spain. 1988.

Brown D (2001) The herbal: essential guide to herbs for living, Pavilion, London.

Cohen AJ, Bartlik B (1998) Ginkgo biloba for antidepressant-induced sexual dysfunction. *J Sex Marital Ther* 24: 139-143.

Evans JR (2000) Ginkgo biloba for age related macular degeneration. *Cohran Databse Syst Rev* 2: CD001775.

Huh H, Staba EJ (1992) The Botany and Chemistry of *Ginkgo biloba* L. *Journal of Herbs, Spices & Medicinal Plants* 1(1/2): 91-124.

Hyun SK, Kang SS, Son KH, Chung HY, Choi JS (2005) Biflavone glucosides from Ginkgo biloba yellow leaves. *Chem Pharm Bull (Tokyo)* 53(9): 1200-1.

Kang BJ, Lee SJ, Kim MD, Cho MJ (2002) A placebo-controlled double-blind trial of Ginkgo biloba for antidepressant-induced sexual dysfunction. *Hum Psychopharmacol* 17: 279-284.

Lai SM, Chen IW, Tsai MJ (2005) Preparative isolation of terpene trilactones from Ginkgo biloba leaves. *J Chromatogr A* 1092(1): 125-34.

Lebuisson DA, Leroy L, Rigat G (1986) Treatment of senile macular degeneration with Ginkgo biloba extract. A preliminary double blind drug vs placebo study. *Presse Med* 15(31) 1556-8.

Mahadevan S, Park Y (2008) Multifaceted therapeutic benefits of Ginkgo biloba L.: chemistry, efficacy, safety and uses. *J. Food Sci* 73(1): R14-9.

Marcocci L, Maguire JJ, Droy-Lefaix MT, Packer L (1994) The nitric oxide-scavenging properties of Ginkgo biloba extract EGb 761. *Biochem Biophys Res Commun* 201: 748-755.

Paick JS, Lee JH (1996) An experimental study of the effect of Ginkgo biloba extract on the human and rabbit corpus cavernosum tissue. *J Urol* 156: 1876-1880.

Perry EK, Pickering AT, Wang WW, Houghton PJ, Perry NS (1999) Medicinal plants and Alzheimer's disease: from ethnobotany to phytotherapy. *Pharm. Pharmacol* 51: 27-54.

Singh B, Kaur P, Gopichand, Singh RD, Ahuja PS (2008) Biology and chemistry of Ginkgo biloba. *Fitoterapia* 79: 401-18.

Sohn M, Sikora R (1991a) *Ginkgo biloba* extract in the Huh, H. and E. J. Staba (1992). The Botany and Chemistry of *Ginkgo biloba* L. *Journal of Herbs, Spices & Medicinal Plants* 1(1/2): 91-124.

Sohn M, Sikora R (1991b) *Ginkgo biloba* extract in the therapy of erectile dysfunction. *J. Sc Educ. Ther* 17: 53-61.

Stromgaard K, Nakanishi K (2004) Chemistry and biology of terpens trilactones from Ginkgo biloba. *Angew Chem Int Ed Engl* 43: 1640-58.

Tang Y, Lou P, Wang J, Li Y, Zhuang S. (2001) Coumaroyl flavonol glycosides from the leaves of Ginkgo biloba. *Phytochemistry* 58: 1251-6.

Waynber J, Brewer S. (2000) Effects of Herbal vX on libido and sexual activity in premenopausal and postmenopausal women. *Adv Ther* 17: 255-62.

Wheatley D (1999) Ginkgo biloba in the treatment of sexual function due to antidepressant drugs. *Human Psychopharmacol* 14: 512-513.

Yang JF, Zhou DY, Liang ZY (2009) A new polysaccharide from leaf of Ginkgo biloba. *Fitoterapia* 80 (1): 43-47.

Glycyrrhiza glabra (Licorice)

Anon (2005) Glycyrrhiza glabra. Monograph. *Alternative Medicine Review* 10(3): 230-237.

Armanini D, Bonanni G, Mattarello MJ, Fiore C, Sartorato P, Palermo M (2003) Licorice consumption and serum testosterone in healthy man. *Experimental and Clinical Endocrinology and Diabetes* 111(6): 341-343.

Armanini D, Bonanni G, Palermo M, *et al.* (1999) Reduction of serum testosterone in men by licorice. *New England Journal of Medicine* 341: 1158.

Davis EA, Morris DJ (1991) Medicinal uses of licorice through the millennia: the good and plenty of it. *Mol Cell Endocrinol* 78(1-2): 1-6.

Fiore C, Eisenhut M, Ragazzi E, Zanchin G, Armanini D (2005) A history of the therapeutic use of liquorice in Europe. *J Ethnopharmacol* 99(3): 317-24.

Kitagawa I, Chen WZ, Hori K, *et al.* (1994) Chemical studies of Chinese licorice-roots.I. Elucidation of five new flavonoid constituents from the roots of Glycyrrhiza glabra L. collected in Xinjiang. *Chem Pharm Bull (Tokyo)* 42(5): 1056-62.

Li W, Asada Y, Yoshikawa T (2000) Flavonoid constituents from Glycyrrhiza glabra hairy root cultures. *Phytochemistry* 55(5): 447-56.

Li ZR, Wong YQ, Deng ZZ (2005) Two new compounds from *Glycerrihiza glabera,* Asian. *Nat Prod Res* 7 (4): 677-680.

Nassiri M, Hosseinzadeh H (2008) Review of Pharmacological Effects of Glycyrrhiza sp and its Bioactive Compounds. *Phytotherapy Res 22 (18): 709-724.*

Hypericum perforatum (St. John's Wort)

Cannon-Smith TW, Kaufman JH (2007) Improved Ejaculatory Control and Sexual Satisfaction in Pilot Study of Men Taking Hypericum perforatum Extract. *The internet Journal of Nutrition and Wellness* 3: 2.

Chatterjee SS, Bhattacharya SK, Wonnemann M (1998) Hyperforin as a possible antidepressant component of hypericum extracts. *Life Sci* 63(6): 499-510.

Ernst E, Rand JI, Barnes J (1998) Adverse effects profile of the herbal antidepressant St. John's Wort (Hypericum perforatum). *Eur J Clin Pharmacol* 54(8): 589-94.

Gaster B, Holroyd J (2000) St John's wort for depression: a systematic review. *Arch Intern Med* 160: 152-6.

Lantz MS *et al.* (1999) St. John's Wort and antidepressant drug interactions in the elderly. *J Geriatr Psychiatry Neurol* 12(1): 7-10.

Miller LG (1998) Herbal Medicinals: selected clinical considerations focusing on known or potential drug-herb interactions. *Arch Intern Med* 158(20): 2200-11.

Parker V, Wong AH, Boon HS, Seeman MV (2001) Adverse reactions to St. John's Wort. *Can J Psychiatry* 46(1): 77-9.

Sleno L, Daneshfar R, Echert GF, Muller WF, Volmer DA (2006) Mass spectral characterization of phloroglucinol derivatives hyperforin and adhyperforin. *Rapid Commun Mass Spectrom* 20(18): 2641-8.

Smelcerovic A, Spiteller M (2006) Phytochemical analysis of nine *Hypericum* L. species from Serbia and the F.Y.R. Macedonia. *Pharmazie* 61(3): 251-2.

Stevinson C, Ernst E (1999) Hypericum for depression. An update of the clinical evidence. *Eur Neuropsychopharmacol* 9: 501-5.

Tanaka N, Takaishi Y (2006) Xanthones from Hypericum chinense. *Phytochemistry* 67(23): 2568-2572.

Lepidium meyenii (Maca)

Bogani P, Simonini F, Iriti M, Rossoni, M, Faora F, Poletti A, *Visioli F* (2006) Lepidium meyenii (Maca) does not exert direct androgenic activities. *J Ethnopharmacol* 104 (3): 415-7.

Brooks NA, Wilcox G, Walker KZ, Ashton JF, Cox MB, Stojanovska L (2008) Beneficial effects of Lepidium meyenii (Maca) on psychological symptoms and measures of sexual dysfunction in postmenopausal women are not related to estrogen or androgen content. *Menopause* 15(6): 1157-62.

Cicero AF, Bandieri E, Arletti R (2001) Lepidium meyenii Walp improves sexual behaviour in male rats independently from its action on spontaneous locomotor activity. *J Ethnopharmacol* 75: 225-229.

Cicero AF, Piacente S, Plaza A, *et al.* (2002) Hexanic maca extract improves rat sexual performance more effectively than methanolic and chloroformic maca extracts. *Andrologia* 34: 177-179.

Comhaire FH, Mahmoud A (2003) The role of food supplements in the treatment of the Infertile man. *Reprod Biomed Online* 7(4): 385-391.

Cui B, Zheng BL, He K, Zheng QY (2003) Imidazole alkaloids from Lepidium meyenii. *J Nat Prod* 66(8): 1101-3.

Dording CM, Fisher L, Papakostas G, Farabaugh A, Sonawalla S, Fava M, Mischoulon D (2008) A double-blind, randomized, pilot dose-finding study of maca root (L. meyenii) for the management of SSRI-induced sexual dysfunction. *CNS Neurosci Ther 14*(3): 182-91.

Gonzales GF, Cordova A, Vega K, Chung A, Villena A, Gonez CJ (2003) Effect of Lepidium meyenii (Maca), a root with aphrodisiac and fertility-enhancing properties, on serum reproductive hormone levels in adult healthy men J. *Endocrinol* 176(1): 163-8.

Gonzales GF, Cordova A, Gonzales C (2001) *Lepidium meyenii* (Maca) improved semen parameters in adult men. *Asian J. Androl* 3(4): 301-3.

Gonzales GF, Cordova A, Vega K (2002) "Effect of *Lepidium meyenii* (MACA) on sexual desire and its absent relationship with serum testosterone levels in adult healthy men." *Andrologia* 34(6): 367-72.

Gonzales GF, Gasco M, Cordova A, Chung A, Rubio J, Villegas L (2004) Effect of Lepidium meyenii (Maca) on spermatogenesis in male rats acutely exposed to high altitude (4340 m). *J Endocrinol* 180(1): 87-95.

Gonzales GF, Nieto J, Rubio J, Gasco M (2006) Effect of Black maca (Lepidium meyenii) on one spermatogenic cycle in rats. *Andrologia* 38(5): 166-72.

Gonzales GF, Ruiz A, Gonzales C (2001) "Effect of *Lepidium meyenii* (maca) roots on spermatogenesis of male rats." *Asian J. Androl* 3(3): 231-3.

Gonzalez GF, Gonzalez C (2008) The Methyltetrahydro-{beta}-Carbolines in Maca (Lepidium meyenii). *Evid Based Complement Alternat Med*, June 19.

Muhamad I, Zhao J, Dunbar DC, Mustafa, *Khan IA* (2005) New alkamides from maca (Lepidium meyenii). *J. Agric Food Chem* 53: 690-3.

Muhammad I, Zhao J, Dunbar DC, Khan IA (2002) Constituents of Lepidium meyenii 'Macca. *Phytochemistry* 59(1): 105-10.

Piacente S, Carbone V, Plaza A, Zampelli A, Pizza C (2005) Investigation of the tuber constituents of maca (Lepidium meyenii Walp). *J Chem Agric Food* 53(3): 690-693.

Popenoe H, *et al.* (1990) *Lost Crops of the Incas,* National Academy Press, Washington, DC.

Rowland DL, Tai W (2003) A review of plant-derived and herbal approaches to the treatment of sexual dysfunctions. *J Sex Marital Ther* 29: 185-205.

Tellez, MR, Khan IA, Kobaisy M, Schrader KK, Dayan FE, Osbrink W (2002) Composition of the essential oil of Lepidium (walp). *Phytochemistry* 61(2): 149-55.

Zhao J, Muhammad I, Dunbar DC, Mustafa J, Khan I (2005) A New alkamides from Maca (Lepidium meyenii). *J Agric Food Chem* 53(3): 690-3.

Zheng, BL *et al.* (2000) Effect of a lipidic extract from lepidium meyenii on sexual behavior in mice and rats. *Urology* 55(4): 598-602.

Zenico T, Cicero AF, Valmorri L, Mercuriali M, Bercovich E (2009) Subjective effects of Lepidium meyenii (Maca) extract on well-being and sexual performance in patients with mild erectile dysfunction: a randomised, double blind clinical trial. *Andrologia* 41(2): 95-9.

Mandragora officinarum (Mandrake Root)

Bare D (2004) The Mysterious Mandrake. *Winston -Salem Journal* October, 30, Salem, New York.

Carter AJ (1996) Narcosis and nightshade. *British Medical Journal* 313: 1630-1632.

Fleisher A, Fleisher Z (1994) The Fragrance of Biblical Mandrake. *Economic Botany* 48: 243-51.

Grieve M (2003) Mandrake In: A Modern Herbal, Dover Publishers, 1971, New York, Available at: http://www.botanical.com/botanical/mgmh/mgmh.html Posted 1995, Accessed, June 19.

Mendelson J, Mello N (1986) The Encyclopedia of psychoactive Drugs: Flowering Plants, Magic in Bloom, Chelsea House Publishers, New York.

Thompson CJS (1968) The Mystic Mandrake, M.B.E University Books, New Hyde Park, NY.

Vandaveer C (2003) Why have mandrake on the Mantle? Killerplants.com assessed, Dec 15, 2003.

Mucuna pruriens (Velvet-Beans)

Amin KMY, Khan MN, Zillur-Rehman S, *et al.* (1996) *"Sexual function improving effect of Mucuna pruriens in sexually normal male rats"*. *Fitoterapia* 67 (1): 53-58.

Damodaran M, Ramaswamy R (1937) Isolation of 3, 4 dihydroxyphenylalanine from the seeds of *M. pruriens. Biochem J* 31: 2149-52.

Donati D, Lampariello LR, Pagani R, Guerranti R, Cinci G, Marinello E (2005) Antidiabetic oligocyclitols in seed of Mucuna pruriens. *Phytother Res* 19: 1057-60.

Ghosal S, Singh S, Bhattacharya SK (1971) Chemistry and Pharmacology. *Planta Medica* 19: 280-84.

Giuliano F, Allard J (2001a) Dopamine and male sexual function. *Eur Urol* 40:601-608.

Giuliano F, Allard J (2001b) Dopamine and sexual function. *Int J Impot Res* 13 (3): S18-S28.

Manyam BV, Dhanasekaran M, Hare TA (2004) Neuroprotective effects of the antiparkinson drug Mucuna pruriens. *Phytother Res* 18(9): 706-12.

Misra L, Wagner H (2004) Alkaloidal constituents of Mucuna pruriens seeds. *Phytochemistry* 65(18):2565-7.

Misra L, Wagner H (2007) Extraction of bioactive princilples from mucuna pruriens seeds. *Indian J. Biophysics* 44: 56-60.

Modi KP, Patel NM, Goyal RK (2008) Estimation of L-dopa from Mucuna pruriens LINN and formulations containing M. pruriens by HPTLC method. *Chem Pharm Bull* (Tokyo) 56(3): 357-9.

Shukla KK, Mahdi AA, Ahmad MK, Shankhwar SN, Rajender S, Jaiswar SP (2008) Mucuna pruriens improve male fertility by its action on the hypothalamus-pituitary-gonadal axix. *Fertil Ster Oct 28 (accepted for publication).*

Tan NH, Fung SY, Sim SM, Marinello E, Guerranti R, Aguiyi JC (2009) The protective effect of Mucuna pruriens seeds against snake venom poisoning. *J Ethnopharmacol* 123 (2): 356-8.

Muira Puama (Potency Wood)

Bartram T (1995) Encyclopedia of Herbal Medicine, Grace Publishers Dorset, England. p. 299.

Bernardes A (1984) *A Pocketbook of Brazilian Herbs,* Editora e, Arta Ltda, Brazil.

Brazilian Pharmacopeia (1956) Muira puama. Ptychopetalum olacoides, Rio de Janeiro, Brazil.

British Herbal Pharmacopoeia (1983) British Herbal Medicine Association, West York, England, pp. 132-133.

Bucek E, Fournier G, Dadoun H (1987) Volatile constituents of Ptychopetalum olacoides root oil. *Planta Med* 53(2): 231.

Cherksey BD (1996) Method of preparing Muira puama extract and its use for decreasing body fat percentage and increasing lean muscle mass. United States Patent No. 5516516.

Duke JA (1985) *CRC Handbook of Medicinal Herbs*, CRC Press, Inc, Florida.

Jayasuriya H, Herath KB, Ondeyka JG, Guan Z, Borris RP, Tiwari S, de Jong W, Chavez F, Moss J, Stevenson DW, Beck HT, Slattery M, Zamora N, Schulman M, Ali A, Sharma N, MacNaul K, Hayes N, Menke JG, Singh SB (2005) Diterpenoid, steroid, and triterpenoid agonists of liver X receptors from diversified terrestrial plants and marine sources. *J. Nat. Prod* 68(8): 1247-52.

Mowrey DB (1993) *Herbal Tonic Therapies*, Keats Publishing Inc, Connecticut.

Piato Al, Detanico BC, Jesus JF, Lhullier FL, Nunes DS, Elisabetsky E (2008). Effects of marapuama in the chronic mild stress model: Further investigations of antidepressant properties. *J. Ethnopharmacol* 118: 300-304.

Rowland D L, Tai W (2003) A review of plant-derived and herbal approaches to the treatment of sexual dysfunctions. *J. Sex. Marital Ther* 29(3): 185-205.

Schultes RE, Raffauf RF (1990) *The Healing Fores-Medicinal and Toxic Plants of the Northwest Amazonia*, R.F. Dioscorides Press, Portland OR.

Schwontkowski D (1993) Herbs of the Amazon, Traditional and Common Uses, Science Student BrainTrust Publishing, Utah.

Siqueira IR, Fochesatto C, Torres IL, da Silva Al, Nunes DS, Elisabetsky E, Netto, CA (2007) Antioxidant activities of Ptychopetalum olcaoides (muira puama) in mice brain. *Phytomedicine* 14, 763-9.

Steinmetz E (1971) Muira puama. *Quart. J. Crude Drug Res* 11(3): 1787-9.

Tang W, Hioki H, Harada K, Kubo M, Fukuyama Y (2008) Clerodane diterpenoids with NGF-potentiating activity from Ptychopetalum olacoides. *J.Nat Prod* 7: 1760-3.

Waynberg J (1995) Male Sexual Asthenia-Interest in a Traditional Plant-Derived Medication, *Ethnopharmacology Conference, Strasbourg, France*, March.

Waynberg J, Brewer S (2000) Effects of Herbal vX on libido and sexual activity in premenopausal and postmenopausal women. *Adv. Ther* 17: 255-62.

Waynberg, J (1990) Contributions to the Clinical Validation of the Traditional Use of Ptychopetalum guyanna, Presented at the First International Congress on Ethnopharmacology, Strasbourg, France, June 5-9.

Werbach MR, Murray MD, Michael ND (1994) *Botanical Influences on Illness, A Sourcebook of clinical research*, Third Line Press, Tarzana, California p. 200.

Panax quinquefolius (American Ginseng)

Bensky D, Gamble A, Kaptchuk T (1993) *Chinese Herbal Medicine: Materia Medica*, Eastland Press, Seattle, p 358-9.

Beveridge TH, Li TS, Drover JC (2002) Phytosterol content in American ginseng seed oil. *J Agric Food Chem* 50(4): 744-50.

Drewes SE, George J, Kahn F (2003) Recent findings on natural products with erectile-dysfunction activity. *Phytochemistry* 62: 1019-1025.

Duke J (1989) *Ginseng: A Concise Handbook*, Reference Publications, Algonac, MI, p 36.

Foster S (1996) *Herbs for Health*, Interweave Press, Loveland, CO. p 48-9.

Hong B, Ji YH, Hong JH, *et al.* (2002) A double-blind crossover study evaluating the efficacy of Korean red ginseng in patients with erectile dysfunction: a preliminary report. *J Urology* 168: 2070-2073.

Ligor T, Ludwiczuk A, Wolski T, Buszewski B (2005) Isolation and determination of ginsenosides in American ginseng leaves and root extracts by LC-MS. *Anal Bioanal Chem* 383: 1098-105.

Morris AC, Jacobs I, McLellan TM, *et al.* (1996) No estrogogenic effect on ginseng ingestion. *Int J Sport Nutr* 6: 263-71.

Nakamura S, Yoshikawa M, *et al.* (2007) Medicinal Flowers. XVII. New dammarane type triterpene glycosides from flower buds of American ginseng, Panax quinquefolium, L. *Chem Pharm Bull* (Tokyo) 55: 1342-8.

Shibata S, Tanaka O, Shoji J, Saito H. (1985) Chemistry and pharmacology of *Panax. Econ Med Plant Res* 1: 218-84. Vol 1, Wagner H, Farnsworth NR (Eds). Academic Press, London.

Siegel RK (1979) Ginseng abuse syndrome. *JAMA* 241: 1614-1615.

Su J, Li HZ and Yang CR (2003) Studies on saponin constituents in roots of Panax quinquefolium, *Zhongguo Zhong Yao Za Zhi* 28(9): 830-3.

Wang CZ, Zhang B, Song WX, Wang A, Ni M, Luo X, Aung HH, Xie JT, Tong R, He TC Yuan, CS (2006) Steamed American ginseng berry: ginsenosides analyses and anticancer activities. *J.Agric Food Chem* 54: 9936-42.

Zhou Y, Song F, Lin S (1998) Water-soluble ginsenosides in American ginseng, *LiX. Zhongguo Zhong Yao Za Zhi* 23(9): 551-2.

Panax ginseng (Koren Ginseng)

Brown DJ (1996) *Herbal Prescriptions for Better Health,* Prima Publishing, Rocklin CA, p. 129-38.

Choi HK, Seong DH, Rha KH (1995) Clinical efficacy of Korean red ginseng for erectile dysfunction. *Int J Impotence Res* 7: 181-6.

Choi T (2008) Botanical characteristics, pharmacological effects and medicinal components of Korean Panax ginseng CA Meyer. *Acta Pharmacol Sin* 29(9): 1109-18.

Choi YD, Rha KH, Choi HK (1999) In vitro and in vivo experimental effect of Korean red ginseng on erection. *J Urol* 162: 1508-1511.

Choi YD, Xin ZC, Choi HK (1998) Effect of Korean red ginseng on the rabbit corpus cavernosal smooth muscle. *Int J Impot Res* 10: 37.

de Andrade E, de Mesquita AA, de Almeida Claro J, de Andrade PM, Ortiz V, Paranhos M, Srougi M, Kiter G (2007) Study of the efficacy of Korean Red Ginseng in the treatment of erectile dysfunction. *Asian J Androl* 9 (2); 241-244.

Dou DQ, Chen YJ, Liang LH, Pang FG, *Shimizu N* (2001) Six new dammarane-type triterpene saponins from the leaves of Panax ginseng. *Chem Pharm Bull (Tokyo)* 49(4): 442-6.

Gillis CN (1997) Panax ginseng pharmacology: a nitric oxide link*? Biochem Pharmacol* 54: 1-8.

Hong B, Ji YH, Hong JH, *et al.* (2002) A double-blind crossover study evaluating the efficacy of Korean red ginseng in patients with erectile dysfunction: a preliminary report. *J Urol* 168: 2070-2073.

Hou JP (1977) The chemical constituents of ginseng plants. *Comp Med East West* 5(2): 123-45.

Jang DJ, Lee MS, Shin BC, Lee YC, Ernst E (2008) Red ginseng for treating erectile dysfunction: a systematic review. *Br J Clin Pharmacol* 66(4): 444-50.

Newall CA, Anderson LA, Phillipson JD (1996) *Herbal Medicines: A Guide for Healthcare Professionals,* Pharmaceutical Press, London, p.1145-50.

Price A, Gazewood J (2003) Koren red ginseng for treatment of erectile dysfunction. *J Farm Pract* 52: 20-1.

Qiu Y, Lu X, Pang T, Ma C, Li X, Xu G (2008) Determination of radix ginseng volatile oils at different ages by comprehensive two-dimensional gas chromatography/ time-of-flight mass spectrometry. *J Sep Sci* 31(19): 3451-7.

Rowland DL, Tai W (2003) A review of plant-derived and herbal approaches to the treatment of sexual dysfunctions. *J Sex Marital Ther* 29: 185-205.

Salvati G, Genovesi G, Marcellini L, *et al.* (1996) Effects of *Panax ginseng* CA Meyer saponins on male fertility. *Panmineva Med* 38: 249-54.

Shibata S, Tanaka O, Shoji J, Saito H (1985) Chemistry and pharmacology of Panax. In *Economic and Medicinal Plant Research,* vol 1, Wagner H, Hikino H, Farnsworth NR (Eds), Academic Press, London, 217-84.

Tamaoki J, Nakata J, Kawatani K, *et al.* (2000) Ginsenoside-induced relaxation of human bronchial smooth muscle via release of nitric oxide. *Br J Pharmacol* 130: 1859-1864.

Tomoda M, Hirabayashi K, Shimizu N, *et al.* (1993) Characterization of two novel polysaccharides having immunological activities from the root of *Panax ginseng. Biol Pharm Bull* 16: 1087-90.

Vazquez I, Aguera-Ortiz LF (2002) Herbal products and serious side effects: a case of ginseng-induced manic episode. *Acta Psychiatr Scand* 105: 76-77.

Eleutherococcus senticosus (Siberian Ginseng)

Asano K, Takahashi T, Miyashita M, *et al*. (1986) Effect of *Eleutherococcus senticosus* extract on human physical working capacity. *Planta Med* 48: 175-7.

Ben-Hur E, Fulder S (1981) Effect of *P. ginseng saponins* and *Eleutherococcus senticosus* on survival of cultured mammalian cells after ionizing radiation. *Am J Chin Med* 9: 48-56.

Bohn B, Nebe CT, Birr C (1987) Flow cytometric studies with *Eleutherococcus senticosus* extract as an immunomodulating agent. *Arzneim-Forsch Drug Res* 37: 1193-6.

Brown DJ (1996) *Herbal Prescriptions for Better Health,* Prima Publishing, Rocklin CA, 69-77.

Collisson RJ (1991) Siberian ginseng *(Eleutherococcus senticosus)*. *Brit J Phytother* 2: 61-71 [review].

Dowling EA, Redondo DR, Branch JD, Jones S, McNabb G, Williams MH (1996) Effect of *Eleutherococcus senticosus* on submaximal and maximal exercise performance. *Med Sci Sports Exerc* 28: 482-9.

Farnsworth NR, Kinghorn AD, Soejarto DD, Waller DP (1985) Siberian ginseng *(Eleutherococcus senticosus)*: Current status as an adaptogen. In *Economic and Medicinal Plant Research,* vol 1, Wagner H, Hikino HZ, Farnsworth NR (Eds), Academic Press, London, pp.155-215 [review].

Hikino H, Takahashi M, Otake K, Konno C (1986) Isolation and hypoglycemic activity of eleutherans A, B, C, D, E, F and G: glycans of *Eleutherococcus senticosus* roots. *J Natural Prod* 49: 293-7.

Kelly GS (1997) Sports nutrition: A review of selected nutritional supplements for endurance athletes. *Alt Med Rev* 2: 282-95.

Kupin VI, Polevaia EB (1986) Stimulation of the immunological reactivity of cancer patients by eleutherococcus extract. *Vopr Onkol* 32: 21-6 [in Russian].

McGuffin M, Hobbs C, Upton R, Goldberg A (1997) *American Herbal Products Association's Botanical Safety Handbook,* CRC Press, Boca Raton, FL, p. 45.

McNaughton L (1989) A comparison of Chinese and Russian ginseng as ergogenic aids to improve various facets of physical fitness. *Int Clin Nutr Rev* 9: 32-5.

McRae S (1996) Elevated serum digoxin levels in a patient taking digoxin and Siberian ginseng. *Can Med Assoc J* 155: 293-5.

Wagner H, Nörr H, Winterhoff H (1994) Plant adaptogens. *Phytomed* 1: 63-76 (review).

Pausinystalia yohimbe (Yohimbe)

Betz JM, White, KD, Der Marderosian A (1995) Gas chromatographic determination of yohimbine in commercial yohimbe products. *J AOAC Int.* 78(5): 1189-1194.

Bown D (1995) *Encyclopedia of Herbs and Their Uses*, DK Publishing, Inc, New York, 322.

Bruneton J (1995) *Pharmacognosy, Phytochemistry, Medicinal Plants,* Lavoisier Publishing, Paris.

Budavari S, Ed. (1996) *The Merck Index: An Encyclopedia of Chemicals, Drugs, and Biologicals,* 12th ed. Merck & Co, Inc. Whitehouse Station, NJ.

Chan Q, Li P, Zhang Z, Li K, Liu J, Li Q (2008) Analysis of yohimbine alkaloids from Pausinystalia yohimbe by non-aqueous capillary electrophoresis and gas chromatography-mass spectrometry. *J. Sep Sci* 31: 2211-8.

Duke JA (1997) *The Green Pharmacy,* Rodale Press, Emmaus, PA, pp. 176, 188, 191-192, 288.

Ernst E and Pittler MH (1998) Yohimbine for erectile dysfunction: a systematic review and meta-analysis of randomized clinical trials. *J Urol* 159(2): 433-436.

Galitzky J, Taouis M, Berlan M, *et al.* (1988) Alpha 2-antagonist compounds and lipid mobilization: evidence for a lipid mobilizing effect of oral yohimbine in healthy male volunteers. *Eur J Clin Invest* 18: 587-94.

Goldberg MR, Robertson D (1983) Yohimbine: a pharmacological probe for the study of the alpha 2-adrenoceptor. *Pharmacol Rev* 35: 143-180.

Grasing K, *et al.* (1996) Effects of yohimbine on autonomic measures are determined by individual values for area under the concentration-time curve. *J Clin Pharmacol* 36(9): 814-822.

Guirguis WR (1998) Oral treatment of erectile dysfunction: from herbal remedies to designer drugs. *J Sex Marital Ther* 24: 69-73.

Mackey D (2004) Nutrients and Botanicals for Erectile Dysfunction, *Alternative Medicine Review* 9(1): 4.

McGuffin M, Hobbs C, Upton R, Goldberg A (1997) American Herbal Product Association's *Botanical Safety Handbook,* CRC Press, Boca Raton, Florida.

Morales A, *et al.* (1987) Is yohimbine effective in the treatment of organic impotence? Results of a controlled trial. *J Urol* 137(6): 1168-1172.

Owen JA, et al. (1987) The pharmacokinetics of yohimbine in man. *Eur J Clin Pharmacol* 32(6): 577-582.

Reichert R (1997) Yohimbine Pharmacokinetics. *Quarterly Rev Nat Med* Spring 17-18.

Reid K, *et al.* (1987) Double-blind trial of yohimbine in treatment of psychogenic impotence. *Lancet* 2: 421-423.

Riley AJ (1994) Yohimbine in the treatment of erectile disorder. *Br J Clin Pract* 48(3): 133-136.

Siddiqui MA, More-O'Ferrall D, Hammod RS, Baime RV, Staddon AP (1996) Agranulocytosis associated with yohimbine use. *Arch Intern Med* 156 (11): 1235-6, 1238.

Susset JG, Tessier CD, Wincze J, *et al.* (1989) Effect of yohimbine hydrochloride on erectile impotence: a double-blind study. *J Urol* 141: 1360-1363.

Vogt HJ, Brandl P, Kockott G, *et al.* (1997) Double blind, placebo controlled safety and efficacy trial with yohimbine hydrochloride in the treatment of nonorganic erectile dysfunction. *Int J Impot Res* 9: 155-161.

Piper methysticum (Kava-Kava)

Almeida JC, Grimsley EW (1996) Coma from the health food store: interaction between kava and alprazolam. *Ann Intern Med.* 125(11): 940-941 [letter].

Bloomfield HH (1998) Healing Anxiety with Herbs, HarperCollins Publishers, New York, pp 97-98.

Bone K (1993/1994) Kava: a safe herbal treatment for anxiety. *British Journal of Phytotherapy* 3(4): 147-153.

Chevallier A (1996) The Encyclopedia of Medicinal Plants, Ist American Edition, Boston, DK Publication, Dist. Houghton Mifflin.

Chienthavorn O, Smith RM, Wilson ID, Wright B (2005) Superheated water chromatography-nuclear magnetic resonance spectroscopy of kava lactones. *Phytochem Anal* 16(3): 217-21.

Dharmaratne HR, Nanayakkara NP, Khan IA (2002) Kavalactones from *Piper methysticum,* and their 13C NMR spectroscopic analyses. *Phytochemistry* 59(4): 429 33.

Dragull K, Yoshida WY, Tang S (2003) Piperidine alkaloids from Piper methysticum. *Phytochemistry* 63(2): 193-8

Gessner B, Cnota P, Steinbach TS (1994) Extract of kava-kava rhizome in comparison with diazepam and placebo. *Zeitschrift fýr Phytotherapie* 15: 30-37.

Head KA, Miller AL, Eds. (1998) *Piper methysticum* (kava-kava). *Altern Med Rev* 3(6): 458-460.

Humberston CL, Akhtar J, Krenzelok EP (2003) Acute hepatitis induced by Kava-Kava. *J. Toxicolo Clinical Toxicol* 41: 109-113.

Keville K, Korn P (1996) Herbs for Health and Healing, Rodale Press Inc., Emmaus, Pa, p33.

Kumar V (2006) Potential Medicinal Plants for CNS disorders: an overview. *Phytother Res* 20 (12): 1023-1035.

Lehmann E, Kinzler E, Friedermann J (1996) Efficacy of a special kava extract (*Piper methysticum*) in patients with states of anxiety, tension and excitedness of non-mental origin-a double-blind placebo-controlled study of four weeks treatment. *Phytomedicine* 3(2): 113-119.

Linderberg D, Pitule-Schodel H (1990) D, L-kavain in comparison with oxazepam in anxiety states. *Forthsch Med* 108(2): 49-54.

MacGregor FB, Abernethy VE, Dahabra S, *et al.* (1989) Hepatotoxicity of herbal remedies. *Br Med* 299: 1156-1157.

Mayell, M (1998) Natural energy: a consumer's guide to legal, mind-altering and mood-brightening herbs and supplements. New York, Three Rivers Press, p 198.

Newall C, Anderson LA, Phillipson JD (1996) Herbal Medicines: A Guide for Health Care Professionals, Pharmaceutical Press, London, p 296.

Norton SA (1998) Herbal medicines in Hawaii from traditional to convention. *Hawaii Med* 57(1): 382-386.

Reader's Digest (1999) The Healing Power of Vitamins, Minerals and Herbs, Reader's Digest Association Inc, Pleasantville, NY, pp 320-321.

Smith MW (1998) Herbal medicine and psychiatry: potential for toxicity, Presented at the 151st annual meeting of the American Psychiatric Association, June 3, Toronto.

Sticker F, Baumuller HM, Seitz K, Vas Lakis D, Seitz G, Seitz HK, Schuppan D (2003) Hepatitis Induced by Kava/Piper methysticum Rhizomes. *J. Hepatol* 39: 62-7.

Whiton PA, Lau A, Saisbury A, Whitehouse J, Evans CE (2003) Kava Lactones and the kava-kava controversy. *Phytochemistry* 64: 673-679.

Wu D, Nair MG, Dewitt DL (2002) Novel compounds from Piper methysticum Forst (Kava Kava) roots and their effect on cyclooxygenase enzyme. *J Agric Food Chem* 50(4): 701-5.

Xuan TD, Fukuta M, Wei AC, Elzaawely AA, Khanh TD, Tawata S (2008) Efficacy of extracting solvents to chemical components of kava (Piper methysticum). *Nat Med (Tokyo)* 62(2): 188-94.

Polygonum multiflorum (Fo-Ti)

Bone K (1996) *Clinical Applications of Ayurvedic and Chinese Herbs,* Phytotherapy Press, Warwick, Australia, 49-51.

Cárdenas A, Restrepo JC, Sierra F, Correa G (2006) Acute hepatitis due to shen-min: a herbal product derived from Ploygonum multiflorum. *J Clin Gastroenterol* 40(7):629-32.

Foster S (1996) *Herbs for Your Health,* Interweave Press, Loveland, CO, pp. 40-1.
Foster S, Yue CX (1992) *Herbal Emissaries: Bringing Chinese Herbs to the West,* Healing Arts Press, Rochester, VT, pp.79-85.

Kim HK, Choi YH, Choi JS, Choi SU, Kim YS, Lee KR, Kim YK, Ryu SY (2008) A new stilbene glucoside gallate from the roots of Polygonum multiflorum. *Arch Pharm Res* 31: 1225-9.

Oerter KK, Janfaza M, Wong JA, Chang RJ (2003) Estrogen bioactivity in Fo-Ti and othre herbs used for their estrogen-like effects as determined by a recombinant cell bioassay. *J. Clin Endocrinol Metab* 88(9): 4075-6.

Yao S, Li Y, Kong (2006) Preparative isolation and purification of chemical constituents form the root of Polygonum multiflorum by high speed counter current chromatography. *J.Chromatgr A* 19: 1115(1-2): 64-71.

Yim TK, Wu WK, Mak DH, Ko KM (1998) Myocardial protective effect of an antraquinone containing extract of *Polygonum multiflorum* ex vivo. *Planta Medica* 64(7): 607-11.

Yong C, Mingfu W, Robert TR, Ching-Tang Ho (1999) 2,2-Diphenyl-1-picrylhydrazyl radical scavenging active components from *Polygonum multiflorum* Thunb. *J. Agric Food Chem* 47: (6) 2226-2228.

Yoshizaki M, Fujino H, Arise A, Ohmura K, Arisawa M, Morita N (1987) Polygoacetophenoside, A new acetophenone glucoside from Polygonum multiflorum *Planta Med* 53(3): 273-5.

Roupala montana (Bois Bande)

Cutler HG (1998) Natural Products and their potential in agriculture: A personal overview. In: Biologically Active Natural Products: Potential Use in Agriculture. ACS Symposium Series 380, American Chemical Society, Washington DC pp 1-22.

Lans C (2007) Ethnomedicines used in Trinidad and Tobago for reproductive problems. *Journal of Ethnobiology and Ethnomedicine* 3: 13.

Omega Alpha Pharmaceuticals place, Toranto, Ontario, Canada. (Omega sells *Erectol* comprising this herb).

Serenoa repens (Saw palmatto)

Bennett BC, Hicklin JR (1998) Uses of saw palmetto (Serenoa repens, Arecaceae) in Florida. *Economic Botany* 52(4): 381-393.

Bruneton J (1995) Pharmacognosy, Phytochemistry, Medicinal Plants, Lavoisier Publishing, Paris.

Duke JA (1985) Handbook *of Medicinal Herbs,* CRC Press, Boca Raton, Florida.

Honsel R, Keller K, Rimpler H, Schneider G, Eds (1994) Hagers Handbuch der Pharmazeutischen Praxis, 5th ed. Vol. 6, Springer Verlag, Berlin-Heidelberg, pp.680-687.

Marks LS, Tyler VE (1999) *Saw palmetto* extract: newest (and oldest) treatment alternative for men with symptomatic benign prostatic hyperplasia. *Urology* 53(3): 457-461.

Newall CA, Anderson LA and Phillipson JD (1996) *Herbal Medicines: A Guide for Health-Care Professionals,* The Pharmaceutical Press, London.

Pytel YA, Vinaroy A, Lopatkin N, Sivkov A, Gorilovsky L, Raynaud JP (2002) Long Term clinical and biological effects of the liidosterol extract of Serenoa repens in patients with symptomatic benign prostatic hyperplasia. *Adv Ther* 19: 297-306.

Sorenson WR, Sullivan D (2006) Determination of Compesterol, Stigmasterol in *Saw palmato*. *JAOAC* 89: 22-34.

Tacklind J, MacDonald R, Rutks I, Wilt TJ (2009) Serenoa repens for benign prostatic hyperplasia. *Cochrane Database Syst Rev* April 15(2): CD001423.

Vogel VJ (1970) *American Indian Medicine*, University of Oklahoma Press, Norman, OK, pp.365-366.

Wilt T, Ishani A, Stark G, MacDonald R, Mulrow C, Lau J (2000) Serenoa repens for benign prostatic hyperplasia (Cochrane Review), In: The Cochrane Library, Issue 1, Oxford, Update Software.

Wilt TJ, *et al.* (1998) Saw palmetto extracts for treatment of benign prostatic hyperplasia, A systematic review. *JAMA* 280: 1604-9.

Tribulus terrestris (Gokshura)

Adimoelja A (2000) Phytochemicals and the break-through of traditional herbs in the management of sexual dysfunctions. *Int J Androl* 23: 82-84.

Adimoelja A, Adaikan PG (1997) Protodioscin from herbal plant Tribulus terrestris L. improves male sexual functions possibly via DHEA. *Int J Impot Res* 9: S64.

Antonio J, Uelmen J, Rodriguez R, Earnest C (2000) The effects of Tribulus terrestris on body composition and exercise performance in resistance-trained males. *Int J Sport Nutr Exerc Metab* 10(2): 208-15.

Arcasoy HB, Erenmemisoglu A, Tekol Y, Kurucu S, Kartal M (1998) Effect of Tribulus terrestris L. saponin mixture on some smooth muscle preparations: a preliminary study. *Bull Chim Farm* 137(11): 473-5.

Bardin CW, Swerdloff RS, Santen RJ (1991) Androgens: risks and benefits. *J Clin Endocrinol Metab 73: 4-7.*

Bedir E, Khan IA (2000) New steroidal glycosides from the fruits of Tribulus terrestris *J Nat Prod* 63(12): 1699.

Bourke CA (1983) Hepatopathy in sheep associated with Tribulus terrestris. *Aust Vet J* 60(6): 189.

Bourke CA (1984) Staggers in sheep associated with the ingestion of Tribulus terrestris. *Aust Vet J* 61(11): 360-3.

Brown GA, *et al.* (2000) "Effects of anabolic precursors on serum testosterone concentrations and adaptations to resistance training in young men." *International Journal of Sport Nutrition and Exercise Metabolism* 10 (3): 340-359.

Conrad J, Dinchex D, Klaiber I, Milka S, Kostova I, Kraus W (2004) A novel furostanol saponin from Tribulus terrestris of Bulgarian origin. *Fitoterapi.* 75(2): 117-22.

De Combarieu F, Fuzzati N, Lovati N, Mercalli F (2003) Furostanol saponins from Tribulus terrestris. *Fitoterapi* 74(6): 583-91.

Gauthaman K, Adaikan PG, Prasad RN (2002) Aphrodisiac properties of Tribulus Terrestris extract (Protodioscin) in normal and castrated rats. *Life Sci* 71: 1385.

Gauthaman K, Ganesan AP, Prasad RN (2003) "Sexual effects of puncturevine (*Tribulus terrestris*) extract (protodioscin): an evaluation using a rat model." *Journal of Alternative and Complementary Medicine* 9 (2): 257-265.

Gauthaman K, Ganesan AP (2008) The hormonal effects of Tribulus terrestris and its role in the management of male erectile dysfunction-an evaluation using primates, rabbit and rats. *Phytomedicine* 15: 44-54.

Huang JW, Tan CH, Jiang SH, Zhu DY (2003) Terrestrinins A and B, two new steroid saponins from Tribulus terrestris. *J Asian Nat Prod Res* 5(4): 285-90.

Kostova I, Dinchex D, Rentsch GH, Dimitrox V (2002) Two new sulfated furostanol saponins from Tribulus terrestris. *Z Naturforsch* 57(1-2): 33-8.

Lv AL, Zhang N, Sun MG, Huang YF, Sun Y, Ma HY, Hua HM, Pei YH (2008) One new cinnamic imide dervative from the fruits of Tribulus terrestris. *Nat Prod Res* 22(11): 1013-6.

Neychev VK, Mitev VI (2005) The aphrodisiac herb *Tribulus terrestris* does not influence the androgen production in young men. *Journal of Ethnopharmacology* 101 (1-3): 319-323.

Su L, Chen G, Feng SG, Wang W, Li ZF, Chen H, Liu YX, Pei YH (2009) Steroidal saponins from Tribulus terrestris. *Steroids* 74(4-5): 399-403.

Sun W, Gan J, Tu G, Guo Z, Zhang Y (2002) A new steroidal saponin from Tribulus terrestris Linn. *Nat Prod Lett* 16(4): 243-7.

Temraz A, Gindi OD, Kadry HA, De Temmasi N, Braca A (2006) Steroidal saponins from the aerial parts of Tribulus alatus Del. *Phytochemistry* 67(10): 1011-8.

Wu G, Jiang S, Jiang F, Zhu D, Wu H, Jiang S. (1996) Steroidal glycosides from Tribulus terrestris. *Phytochemistry* 42(6): 1677-81.

Xu T, Xu Y, Liu Y, Xie S, Si Y, Xu D (2009) Two new furostanol saponins from Tribulus terrestris L. *Fitoterapia.* May 11(Accepted).

Xu Th, Xu YJ, Xie SX, Zhao HF, Han D, Li Y, Niu JZ, Xu DM (2008) Two new furostanol saponins from Tribulus terrestris L. *J Asian Nat Prod Res* 10(5-6): 419-23.

Xu Y, Xie SX, Zhao HF, Xu TH, Xu, DM (2001) Studies on the chemical constituents from Tribulus terrestris, *Yao Xue Xue Bao* 36(10): 750-3 (In Chinese language).

Xu YX, Chen HS, Liang HQ, Gu ZB, Liu WY, Leung WN, Li TJ (2000) Three new saponins from Tribulus terrestris. *Planta Med* 66(6): 545-50.

Yan W, Ohtani K, Kasai R, Yamasaki K (1996) Steroidal saponins from fruits of Tribulus terrestris. *Phytochemistry* 42(5): 1417-22.

Zafar R, Lalwani M (1989) Tribulus terrestris Linn-a review of the current knowledge. *Indian Drugs* 27(3):148-153.

Turnera diffusa (Damiana)

Alarcon-Aguilara FJ, *et al*. (2002) Investigation on the hypoglycemic effects of extracts of four Mexican medicinal plants in normal and alloxan-diabetic mice. *Phythother Res* 16(4): 383-86.

Alcaraz-Melendez L, Delgado-Rodriguez J, Real-Cosio S (2004) Analysis of essential oils from wild and micropropagated plants of damiana (Turnera diffusa). *Fitoterapia* 75: 696-701.

Arletti R, *et al*. (1999) Stimulating property of *Turnera diffusa* and *Pfaffia paniculata* extracts on the sexual-behavior of male rats. *Psychopharmacology* 143(1): 15-19.

Blumenthal M, Busse WR, Goldberg A, *et al*. (Eds) 1998 *The Complete Commission E Monographs: Therapeutic Guide to Herbal Medicines,* Integrative Medicine Communications, Boston, MA, 325-6.

Bradley PR, Ed (1992) *British Herbal Compendium,* Vol 1, British Herbal Medicine Association, Bournemouth, Dorset, UK, 71-2.

Dominguez XA, Hinojasa M (1976) Mexican medicinal plants xxviii.Isolation of 5-hydroxy-7,'3,-4' trimethoxy-flavone from Tuenera diffusa.

Duke JA (1985) *CRC Handbook of Medicinal Herb,* CRC Press, Boca Raton, FL, p.492.

Jiu J (1966) A survey of some medicinal plants of Mexico for selected biological activity. *Lloydia* 29: 250-59.

Kumar S, Taneja R, Sharma A (2006) Pharmacognostic standardization of *Turnera aphrodisiaca* Ward. *J Med Food* 9(2): 254-60.

Newell C, Anderson LA, Phillipson JD (1996) Herbal Medicines: A Guide for Health-Care Professional. England, Pharmaceutical Press, London.

Piacente S, Camargo EE, Zampelli A, Garcioso JS, Souza Brito AR, Pizza C, Vilegas W (2002) Flavonoides and arbutin from Turnera diffusa. *Z. Naturforsch* (c) 57(11-12): 983-5.

Rowland DL, Tai W (2003) A review of plant-derived and herbal approaches to the treatment of sexual dysfunctions. *J. Sex Marital Ther* 29: 185-205.

Spencer KC, Seigler DS (1981) Teteraphylline B from Turnera diffusa. *Planta Medica* 43: 175-8.

Willard T (1991) The Wild Rose Scientific Herbal, Canada, Wild Rose College of Natural Healing Ltd, 104-105, Jun; 29(3): 185-205.

Zava DT, Dollbaum CM, Blen M (1998) Estrogen and progestin bioactivity of foods, herbs and spices. *Proc Soc Exp Biol Med* 217: 369-78.

Zhao J, Dasmahapatra AK, Khan SI, Khan IA (2008) Anti-aromatase activity of the constituents from damiana (Turnera diffusa). *J. Ethnopharmacol* 120(3): 387-393.

Zhao J, Pawar RS, Ali Z, Khan IA (2007) Phytochemical Investigations of Turnera diffusa. *J.Nat Product* 70: 289-92.

Withania somnifera (Ashwaghandha)

Abou-Douh AM (2002) New withanolides and other constituents from the fruit of *Withania somnifera. Arch Pharm (Weinheim)* 335 (6): 267-76.

Anon (2004) *Withania somnifera* (ashwagandha)-monograph. *Altern Med Rev* 9(2): 211-214.

Bone K (1996) Clinical Applications of Ayurvedic and Chinese Herbs, Monographs for the Western Herbal Practitioner, Phytotherapy Press, Australia, pp.137-141.

Brekhman, II (1980) Man and Biologically Active Substances, The Effect of Drugs, Diet and Pollution on Health. Pergamon Press, UK, p.1-89

Gupta GL, Rana AC (2007) *Withania somnifera* (Ashwagandha): *Pharmacognosy Review 1: 129.*

Ilayperuma I, Ratnasooriya WD, Weerasooriya TR (2002) Effect of Withania somnifera root extract on the sexual behavior of male rats. *Asian J Androl* 4(4): 295-8.

Iuvone T, Esposito G (2003) Induction of nitric oxide synthase expression by *Withania somnifera* in macrophages. *Life Sci* 72(14): 1617-25.

Khajuria RK, Suri KA, Gupta RK, Satti NK, Amina M, Suri OP, Qazi GN (2004) Separation, identification, and quantification of selected withanolides in plant extracts of Withania somnifera by HPLC-UV(DAD)-positive ion electrospray ionisation-mass spectrometry. *J Sep Sci* 27(7-8): 541-6.

Kulkarni SK, Dhir A, (2008) Withania somnifera: an Indian ginseng. *Prog Neuropsychopharmacol Biol Psychiatry* 1: 32(5):1093-105.

Matsuda H, Murakami T, Kishi A, Yoshikawa M (2001) Structures of withanosides I, II, III, IV, V, VI, and VII, new withanolide glycosides, from the roots of Indian Withania somnifera DUNAL and inhibitory activity for tachyphylaxis to clonidine in isolated guinea-pig ileum. *Bioorg Med Chem* 9(6): 1499-507.

Mishra LC, Singh BB, Dagenais S (2000) scientific basis for the therapeutic use of Withania somnifera (ashwagandha): a review. *Altern Med Rev* 5(4): 334-46.

Misra L, Mishra P, Pandey A, Sangwan RS, Tuli R (2008) Withanolides from Withania somnifera roots. *Phytochemistry* 69(4): 1000-4.

Nadkarani AK (1976) Indian Materia Medica, Bombay Popular Prakashan, Vol 1, pp. 1292-94

Paramanick S, Roy A, Gosh S, Majumdar HK, Mukopadhayay (2008) Withanolide Z, a new chlorinated withanolide from Withania somnifera. *Planta Med* 74(14):1745-8.

Rastogi RP, Mehrotra, BN (1998) Compendium of Indian Medicinal Plants. Vol. 6 Central Drug Research Institute, New Delhi.India.

Chapter Six B
Truffles as Aphrodisiacs

Buzzini P, Gasparetti C, Turchetti B, Cramarossa MR, Vaughan-Martini A, Martini A, Pagnoni UM, Forti L (2005) Production of volatile organic compounds (VOCs) by yeasts isolated from the ascocarps of black (*Tuber melanosporum*Vitt.) and white (*Tuber magnatum* Pico) truffles. *Arch Microbiol* 184(3): 187-93.

Gioacchini AM, Menotta M, Bertini L, Rossi I, Zeppa S, Zambonelli A, Piccoli G, Stocchi V (2005) Solid-phase microextraction as chromatography/mass spectrometery identification of truffles. *Rappid Commun Mass Spectro* 19: 2365-70.

Harki E, Klaeba A, Talou T, Dargent R (1996) Identification and quantification of Tuber melanosporum Vitt.Sterols. *Steroids* 61: 609-12.

Jinming G, Lin H, Jikai L (2001) A novel sterol from Chinese truffles *Tuber indicum. Steroids* 66: 771-5.

Mauriello G, Marino R, D'Auria M, Cerone G, Rana GL (2004) Determination of the Volatile organic compounds from truffles via SPME-GC-MS. *J Chromatogr Sci* 42(6): 299-305.

McPartland JM, Vilgalys RJ, Cubeta MA (1997) Mushroom poisoning. *Am Fam Physician* 55(5): 1797-800; 1805-9; 1811-2.

Murat C, Martin F (2008) Sex and truffles; first evidence of Perigord black truffle outcrosses. *New Phytol* 180(2): 260-3.

Pinillos MA, Gomez J, Elizald J, Duefias A (2003) Poisoning by foodstuffs, plants and mushrooms. *An sist Sanit Navar* 6: 243-63.

Chapter Six C
Aphrodisiacs from Animal Kingdom
Deer Antler

Bensky D, Gamble A, Kaptchuk T (1993) Eds, Chinese Herbal Medicine: Materia Medica, Revised edition, Eastland Press Inc. Seattle, WA, p. 336-338.

Dalefield RR, Oehme FW (1999) Controversies in toxicology; Deer velvet antler: some unanswered questions on toxicology. *Vet Human Toxicol* 41(1): 39-41.

Falloon J, Wellington P (1992) *The Deer Farmer*, Trevor Walton, Editor, New Zealand, p 2.

Ewashkiw C, Marion A (2001) *Velvet Antler-a Gift from Nature,* Antler Farm, Alberta, Canada.

Zheng Z (1997) Analysis on the therapeutic effect of combined acupuncture and medication in 297 cases of male sterility. *J. Traditional Chinese Medicine.* 17: 190-93.

Spanisf Fly

Karras DJ, Farrell SE, Harrigan RA, Henretig FM, Gealt L (1996) Poisoning from "Spanish fly" (cantharidin). *Am J Emerg Med* 14(5):478-83.

Marcovigi P, Leoni S, Calbi G, Valtancoli E, Ravaglia G (1995) Acute poisoning caused by cantharidin ingestion for aphrodisiac purposes: A clinical case. *Minerva Anestesiol* 61(3): 105-7

Rowland B (2005) Gale Encyclopedia of Alternative Medicine, The Gale Group Inc. Farmington Hills, Michigan.

Tagwireyi D, Ball DE, Loga PJ, Moyo S (2000) Cantharidin poisoning due to "Blister beetle" ingestion. *Toxicon* 38(12): 1865-9

Miscellaneous Animal Products Sold as Aphrodisiacs

Gay J (2002) Anchorage Daily News (Viagra affect reduced animal sales). December, 27, 2002

Kweli T (2007) Aphrodisiacs and Potency Restorers Drugs-X news. Blog Pharmacy, Sun, 11/04/2007.

McCarthy S (1999) Can Viagra save the tigers? *Health & Body*, Sept 22:

Ratsch C (1997) Plants of Love; The History of Aphrodisiacs and A Guide to Their Identification and Use, Ten Speed Press, Berkeley, California

Chapter Six D
Minerals as Natural Aphrodisiacs

Prasad AS (1993) Biochemistry of Zinc, Plenum Press, New York.

Scott R, *et al.* (1998) The effect of oral selenium supplementation on human sperm motility. *Br J Urol* 82: 76-80.

Nielsen FH, Hunt CD, Mullen LM, Hunt JR (1987) Effect of dietary boron on mineral, estrogen, and testosterone metabolism in postmenopausal women. *FASEB J* 1(5): 394-7.

Winston D, Maimes S (2007) "Adaptogens: Herbs for Strength, Stamina, and Stress Relief," Healing Arts Press, Rochester, Vermont, USA

Chapter Seven
Hormones as Natural Aphrodisiacs
Chemistry of Love

Buvat J, Bou JG (2006) Significance of hypogonadism in erectile dysfunction. *World J. Urol* 24(6): 657-67.

Giuliano F, Allard J (2001) Dopamine and male sexual function. *Eur Urol* 40(6): 601-8.

Hollander E, Bartz J, Chaplin W, *et al.* (2007) Oxytocin increases retention of social cognition in autism. *Biol Psychiatry* 61 (4): 498-503.

Kosfeld M *et al.* (2005) Oxytocin increases trust in humans. *Nature* 435:673-676.

Mikhail N (2006) Does testosterone have a role in erectile function? *Am J. Med* 119(5): 373-82.

Oh WK (2002) "The evolving role of estrogen therapy in prostate cancer." *Clin Prostate Cancer* **1** (2): 81-9.

Schmid RE (2009) Kisses unleash chemicals that ease stress levels, Academic Press, Feb 13, 6.22pm.

Chapter Eight
Aroma as Natural Aphrodisiacs

Berwick A (1994) Holistic Aromatherapy: Balance of the Body and Soul with Essential Oils, Liewellyn Publications, St.Paul, MN.

Buchbauer G (1998) "The detailed analysis of essential oils leads to the understanding of their properties"-paper presented at 20th IFSCC Congress Sept 14-18th 1998 Cannes, France and reproduced in *Perf & Flav* 25: 64-67 (2000).

Damian P, Damian K (1995) Aromatherapy: Scent and Psyche, Healing Art, Rochester, VT.

Eyres G, Marriott PJ, Dufour JP (2007) The combination of gas chromatography-olfactometry and multidimensional gas chromatography for the characterisation of essential oils. *J Chromatogr A* 1150(1-2): 70-7.

Hirsch AR (2004) Sexually Exciting Odors. *Chicago Image*, August/September, pp. 14-15.

Huenberger E, Hongratanaworakit T, Bohm C, Weber R, Buchbauer G (2001) Effects of chiral fragrances on human automomic nervous system parameters and self-evaluation. *Chem Senses* 26(3): 281-292.

Marriot PJ, Shellie R, Cornwell C (2001) Gas chromatographic technologies for the analysis of essential oils. *J Chromatogr A* 936(1-2):1-22.

Maury M, Higley C, Higley A (2004) Eds., Reference Guide for Essential Oils, Abundant Health Publishers, 8th Edition. UT, USA.

McCormack EA (2000) Aromatherapy and the Treatment of the Reproductive System, *AGORA* Web Page, December.

Ryman D (1989) The aromatherapy handbook: The Secret healing of Power of Essential oils, The C.W. Daniel Company Ltd, UK.

Sugawara Y, Hara C, Aoki T, Sugimoto N, Masujima (2000) "Odour distinctiveness between enantiomers of linalol: difference in perception and responces elicited by sensory test and forehead surface potential wave measurement." *Chem Senses* 25: 77-84.

Traynor S (2001) The Musk Dilemma *Perf & Flav* 26: Sept/Oct 2001 p31.

Willard P (2001) *Secrets of Saffron*, Beacon Press, MA, USA.

Chapter Nine
Human Pheromones as Natural Aphrodisiacs

Bensafi M, Tsutsui T, Khan R, Levenson RW, Sobel N (2004) Sniffing a human sex-steroid derived compound affects mood and autonomic arousal in a dose dependent manner. *Psychoneuroendocronology* 29: 1290.

Berglund H, Linderstrom P, Slavic I (2006) Brain Response to Putative Pheromones in Lesbian Women. *Proc Natln Acad Sci USA* 103: 11098

Berliner D, *et al* (1996) The Functionality of the Human Vomeronasal Organ (VNO): Evidence for Steroid Receptors. *Journal Steroid Biochem Molecular Bio* 58: 259-65.

Brennan PA, Keverne EB (2004) Something in the air? New insight into mammalian pheromones. *Curr Biol* 14: R81-9.

Cohn BA (1994) In search of human skin pheromones. *Arch Dermatolo* 130: 1048-51.

Copeland P and Link A (2001) Sexual Magnetism: Pheromones-The scent of Sex, Urban Male Magazine, Winter.

Cowley JJ, Brooksbank BWL (1991) Human Exposure to putative pheromones and changes in aspect: social behavior. *J. Steroid Biochem Mol Biol* 39: 647-59.

Cutler WB (1999) Human Sex Attractants Pheromones: Discovery, Research, Development and Application in Sex Therapy. *Psychiatric Annals* 29: 54-59.

Cutler WB, Friedmann E, McCoy NL (1998) Pheromonal influences on sociosexual behavior in men. *Arch Sex Behav* 27: 627

Cutler WB, Genovese-Stone E (1998) Wellness in women after 40 years of age: the role of sex hormones and pheromones. *Dis Mon* 44: 421-546.

Cutler WB, Genovese-Stone E (2002) Pheromones, sexual attractiveness and quality of life in menopausal women. *Climacteric* 5: 105-6.

Cutler WB, Preti G, Krieger A, Huggins GR, Garcia CR, Lawley RJ (1986) Human Auxillary Secretions Influence Women's Menstrual Cycles: The Role of Donor Extract from Men. *Hormones and Behavior* 20: 474-482.

Etienne B (2002) Pheromones, in context, *Monitor on Psychology* 33: 9.

Friebely J, Rako S (2004) Pheromonal influences on sociosexual behavior in postmenopausal women. *J. Sex Res* 41: 372-80.

Gorner P (2000) Sixth Sense Detects Pheromones: University of Chicago Research Show, *Chicago Tribune,* March 17.

Grammer K (1993) 5-a-androst-16en-3a-on: A male pheromone? A brief report. *Ethology and Sociobiology* 14: 201-207.

Grammer K, Fink B, Neave N (2005) Human pheromones and sexual attraction. *Eu J. Obstet Gyenacol Reprod Biol* 8: 135-42.

Grammer K, Jutte A (1997) Bottle of odors: significance of pheromones for human. *Gynakol Geburtshilfliche Rundsch* 37: 150.

Grosser BI, Monti-Bloch L, Jennings-White C, Berliner DL (2000) *Psycho-neuroendocrinolgy* 25: 289-99.

Hitti M (2006) How Women Respond to Pheromones, www.webmd.com/news/20060508

Huggins GR, Preti G (1976) Volatile constituents of human vaginal secretions. *Am J Obstet Gynecol* 126(1): 129-36.

Karlson P, Luscher M (1959) Pheromones: a new term for a class of biologically active substances. *Nature* 183: 55-6.

Knowlton L (1994) Elixirs of Life. *The Los Angeles Times*, July 15.

Kohl JV (2006) The Mind's Eyes: Human Pheromones, Neuroscience and Male Sexual Preference. *J. Psycholology & Human Sexuality* 18: 313-369.

Kohl JV, Francocur RT (2002) The Scent of Eros-Mysteries of Odor in Human Sexuality, Authors Choice Press, NY.

Maclintock MK (1971) Menstrual Synchrony and Suppression. *Nature* 229: 244-5.

Martinez-Macros A (2001) Controversies on the human vomeronasal system. *Eur. J. Anat* 5: 47-58.

Martins Y, Preti G, Crabtree CR, Tamar R, Vainius AA, Wysocki CJ (2005) Preference for Human Body Odors is Influenced by Gender and Sexual Orientation. *Psychological Sciences* 16: 694.

McCollough PA, Owen JW, Pollak EL (1981) Does Androstenol affect emotion? *Ethol Sociobilo* 2: 85-88.

McCoy NL, Pitino L (2002) Pheromonal Influences on Sociosexual Behavior in Young Women. *Physiology and Behavior* 75: 367-375.

Michael RP, Bonsall RW, Kutner M (1975) Volatile Fatty Acids, "Couplins" in human vaginal secretions. *Psychoneuroendocrino* 1: 153-162.

Michael RP, Bonsall RW, Warner P (1974) Human vaginal secretions: Volatile Fatty Acid content. *Science* 186: 1217-1219.

Monti-Bloch L, Grosser BI (1991) Effect of putative pheromones on the electrical activity of the human vomeronasal organ and olfactory epithelium. *J Steroid Biochem Mol Biol* 39: 573-582

Monti-Bloch L, Jennings-White C, Dolberg DS, Berliner DL (1994) The human vomeronasal system. *Psychoneuroendocrinology* 19: 673-683.

Morofushi M, Shinohara K, Funabashi T, Kimura F (2000) Positive Relationship between Menstrual Synchronomy and Ability to Smell 5a-Androst-16-en-3a-ol. *Chem Senses* 25: 407-411.

Preti G, Crabtree CR, Tamar Runyan, Vainius AA, Wysocki CJ, Martins, Y (2005) Preference for Human Body Odors is influenced by Gender and Sexual Orientation, *Psychological Science* 16: 694.

Preti G, Wysocki CJ, Kurt T, Barnhart KT, Sondheimer, SJ, Leyden, JJ (2003) Male Axillary Extracts Contain Pheromones that Affect Pulsatile Secretion of Luteinizing Hormone and Mood in Women Recipients, *Biology Reproduction* 68: 2107-2113.

Psychic Daily. com, May 09, 2005; posted 11:04 pm EDT (03:04 GMT).

Ragelosn W (2002) Pheromones-Understanding the Mtstery of Sexual Attraction, Smart Publications, Petaluma, CA

Rowland R (1998) Study finds proof that human react to pheromones.CNN.Com (CNN News). March 11.

Shinora K, Morofushi M, Funabashi T, Mitsushima D, Kimura F (2000) Effects of 5α-andrst-16 en-3α-ol on the pulsatile secretion of luteinizing hormone in human females *Chem Senses* 25: 465-67.

Sobel N, Brown WM (2001) The Scented Brain: Pheromonal Responses in Humans. *Neuron* 31: 512-514.

Sokolov JJ, Harris RT, Hecker, MR (1976) Isolation of substances from human vaginal secretions previously shown to be sex attractant pheromones in higher primates. *Arch Sex Behav* 5(4): 269-74.

Spencer NA, McClintock MK, Sellergren SA, Bullivant S, Jacob S, Mennella JA (2004) Social chemosignals from breastfeeding women increase sexual motivation. *Horm Behav* 46: 362-70.

Stern K, Mclintock MK (1998) Regulation of ovulation by human pheromones. *Nature* 401: 232.

Stoddart DM (1990) The scented ape: The Biology and Culture of Human Odor, Cambridge University Press, Cambridge.

Stowers L, Morton TF (2005) What is a pheromones? Mammalian pheromones reconsidered. *Neuron* 46: 699-702.

Thorne F, Neave N, Scholey A, Moss M and Fink B (2002) Effects of Putative Male Hormones on Female ratings of Male attractiveness: Influence of Oral Contraceptives and the Menstrual Cycles. *New Endocrinology Lett* 23: 291-7.

Trevathan WR, Burleson MH, Gregory WL (1993) No Evidence for Menstrual Synchronomy in Lesbian Couple. *Psychoneurone Endocrinology* 18: 425-35.

Utton T (2002) Secret perfume ingredients that may boot your love life. *Daily Mail* (UK) March 20.

Villemure C, Bushnell MC (2007) The effects of the steroid androstadienone and pleasant odorants on the mood and pain perception of men and women. *Eur J. Pain* 11: 181-191.

Waltman R, Tricomi V, Wilson GE.Jr, Lewin AH, Goldberg NL, Chang MM (1973) Volatile Fatty acids in Vaginal secretions: Human Pheromones. *Lancet* 2: 496.

Willis MT (2002) Pheromones increases sexual attractions. ABC News, March 21.

Winman A (2004) Do perfume additives termed human pheromoes warrant being termed pheromones? *Physiol Behav* 82: 697-701.

Wyart C, Webster WW, Chen JH , Wilson SR, McClary A, Khan RM, Sobel N (2007) Smelling a Single Component of Male Sweat Alters Levels of Cortisol in Women. *The Journal of Neuroscience* 27(6): 1261-1265.

Wysocki CJ, Preti G (2000) Human odors and their perception. *Jpn J Taste Smell Res.* 7: 19-42

Wysocki CJ, Preti G (2004) Facts, fallacies, fears and frustrations with human pheromones. *Anat Rec A Discov Mol Cell Evol Biol* 281: 1201-11.

Zeng X-N, Leyden JJ, Brand JG, Spielman AI, McGinley K, Preti G (1992) An investigation of human apocrine gland secretion for axillary odor precursor. *J chem Ecol* 18: 1039-1055.

Chapter Ten
Natural Aphrodisiacs: Interaction with Drugs, Herbs and Food

Almeida JC, Grimsley, EW (1996) Coma from the health food store: Interaction between kava kava and alprazolam. *Ann Inernt Med* 125: 940.

Anderson LA, Philipson JD (1985) Herbal medicine, education and the pharmacist. *Pharm J* 236: 303.

Ang HH, Lee El, Matsumota K (2003) Analysis of lead content in herbal preparations in Malaysia. *Hum Exp Toxicol* 22 (8): 445-51.

Baily DG, Arnold JM, Spence, JD (1994) Grapefruit juice and drugs. How significant is the interaction? *Clin Pharmacokinet* 26: 91(1994).

Baily DG, Dresser GR (2005) Natural products and adverse drug interactios. *CMAJ* 170: 1531-1532.

Blumenthal M, Gruenwald J, Hall T, Riggins C, Rister R (1998) The Complete E Monographs Therapeutic Guide to Herbal Medicine, American Botanical Society, Austin, TX.

But P P-H (1993) Need the correct identification of herbs in herbal poisoning. *Lancet* 341: 637.

Chan K (2003) Some aspects of toxic contaminants in herbal medicines. *Chemosphere* 52(9): 1361-71.

Chaudhri RD (1999) Herbal Drug Industry-A Practical Approach to Industrial Pharmacognosy, Eastern Publishers, New Delhi, India,

Clement YN, Williams AF, Khan K, Bernard T, Bhola S, Fortune M, Medupe O, Nagee K, Seaforth CE (2005) A gap between acceptance and knowledge of herbal medicines by physicians: the need for educational intervention. *BMC Compliment Altern Med* 18: 20.

Conn JW, Rovner DR, Cohen, EL (1968) Licorice-induced pseudoaldosteronism, hypertension, hypokalemia, aldosteronopenia and suppressed plasma renin activity. *JAMA* 205: 492.

Cowley G (1996) Herbal Warning, Newsweek 6 May, 60.

Cupp MJ (1999) Herbal remedies: adverse effects and drug interactions. *Am Fam Physician* 59: 1239.

D'Arcy PF (1991) Adverse reactions and interactions with herbal medicine Part 1, Adverse reactions. *Adverse Drug React Toxicol Rev* 10: 189.

D'Arcy PF (1993) Adverse reactions and interactions with herbal medicine, Part 2. Drug interactions. *Adverse Drug React. Toxicol Rev* 12: 147.

de Smet Pagm (1997) An introduction to herbal pharmacovigilance. In *Adverse Effects of Herbal Drugs,* Vol.3, Eds. de Smet Pagm, Keller K, Hansel R, Chandler R.F, Berlin, Heidelburg, Springer-Verlag, p. 1-13.

de Smet Pagm, Dukes MN, Eds (1992) Drugs used in non-orthodox medicine, in Dukes MN, Ed, Meyler's Side Effects of Drugs, 12th ed., Amsterdam, Elsevier, p.1209.

Fetrow CW, Avila JR (1999) Professional Handbook of Complementary and Alternative Medicine, Springhouse Corporation, Spring, Pennsylvania.

Friedman RA, Natural does not mean safe (editorial), *The New York Times* 19th April, 1996, A29.

Fugh-Berman A (2000) Herb-drug interaction. *Lancet* 355: 134.

Hardy ML (2000) Herbs of special interest to women. *J. Am Pharm Assoc* (Wash) 40: 234.

Howard CR, Lawrence CA (1999) Drugs and breastfeeding. *Clin Perinatol* 26: 447.

Kempin SJ (1983) Warfarin resistance caused by broccoli. *N.Engl J Med* 308: 1229 .

Koletzo S, Sherman P, Corey M (1989) Role of infant feeding practices in development of Crohn's disease in childhood. *Brit J. Med* 298**:** 1617.

Kopec K (1999) Herbal medications and breastfeeding**.** *J. Hum Lact* 15: 157.

Krauss C (1996) Pataki outlaws herbal stimulant linked to deaths. The New York Times, 24 May, 1996, B1

Lambreecht JE, Hamilton W, Rabinovich A (2000) A Review of Herb-Drug Interactions: Documented & Theoretical, *US Pharmacist* 25: 42.

Miller *LG* (1998) Herbal Medicinal, selected clinical considerations focusing on known or potential drug herb interaction. *Arch Intern Med* 156: 2200.

Philp RB (2004) Herbal-Drug Interactions And Adverse Effects, McGraw-Hill, NY.

Ridker PM, McDermott, WV (1989) Comfrey herb tea and hepatic veno-occlusive disease. *Lancet* 334: 657.

Ryan A (1997) The resurgence of breastfeeding in the United States, *Pediatrics* 99, E12

Saper RB, Kales SN, Paquin J, Burns MJ, Eisenberg DM, Davis RB, Phillips RS (2004) Heavy metal content of ayurvedic herbal medicine products. *JAMA* 292: 2868-73.

Schmidt DN (1997) Vitamin E: What should the pharmacist know? *Pharm Times* 63: 243.

Shoup J, Carson DS (1999) Anticoagulant use during lactation. *J. Hum Lact* 15: 255.

Wheaton AG, Blanck HM, Gizlice Z, Reyes M (2005) Medicinal herb use in a population-based survey of adults: prevalence and frequency of use, reason for use, and use among their children. *Ann Epidemol* 15: 678-85.

Wright A, Bauser, M, Naylor A (1998) Increasing breastfeeding rates to reduce infant illness at the community level. *Pediatrics* 101: 837

Chapter Eleven
Natural Aphrodisiacs: Myth or Reality

Albertazzi P, *et al.* (1999) The effect of dietary soy supplementation on hot flushes. *Obstet Gynecol* 91(1): 6-11.

Ang HH, Cheang HS (1999) Studies on the anxiolytic activity of Eurycoma longifolia Jack roots in mice. *Jpn J Pharmacol* 79(4): 497-500.

Ang HH, Cheang HS, Yusof AP (2000) Effects of Eurycoma longifolia Jack (Tongkat Ali) on the initiation of sexual performance of inexperienced castrated male rats. *Exp Anim* 49(1): 35-8.

Ang HH, Ikeda S, Gan, EK (2001) Evaluation of the potency activity of aphrodisiac in Eurycoma logifolia Jack. *Phytother Res* 15: 435-6.

Ang HH, Lee KL (2002) Effect of Eurycoma longifolia Jack on orientation activities in middle-aged male rats. *Fundam Clin Pharmacol* 16(6): 479-83.

Ang HH, Lee KL, Kiyoshi M (2004) Sexual arousal in sexually sluggish old male rats after oral administration of Eurycoma longifolia Jack. *J Basic Clin Physiol Pharmacol* 15 (3-4): 303-9.

Ang HH, Ngai TH (2001) Aphrodisiac evaluation in non-copulator male rats after chronic administration of Eurycoma longifolia Jack. *Fundam Clin Pharmacol* 15(4): 265-8.

Ang HH, Sim MK (1997) Eurycoma longifolia Jack enhances libido in sexually experienced male rats. *Exp Anim* 46(4): 287-90.

Ang HH, Sim MK (1998) Eurycoma longifolia increases sexual motivation in sexually naive male rats. *Arch Pharm Res* 21(6): 779-81.

Ang HH, Sim MK (1998) Eurycoma longifolia JACK and orientation activities in sexually experienced male rats. *Biol Pharm Bull* 21(2): 153-5.

Arletti R, *et al.* (1999) "Stimulating property of *Turnera diffusa* and *Pfaffia paniculata* extracts on the sexual-behavior of male rats *Psychopharmacology* 143(1): 15-19.

Armanini D, Bonanni G, Palermo M, *et al.* (1999) Reduction of serum testosterone in men by licorice. *New England Journal of Medicine* 341: 1158.

Baird DD, *et al.* (1995) Dietary intervention study to assess estrogenicity of dietary soy among postmenopausal women. *J Clin Endocrinol Metab* 80(5): 1685-90.

Bogani P, Simonini F, Iriti M, Rossoni, M, Faora F, Poletti A, *Visioli F* (2006) Lepidium meyenii (Maca) does not exert direct androgenic activities. *J Ethnopharmacol* 104(3): 415-7.

Carraro J C, *et al.* (1996) Comparison of Phytotherapy (Permixon) with Finasteride in the Treatment of Benign Prostate Hyperplasia: A Randomized International Study of 1,098 Patients. *Prostate* 29: 231-40, discussion 241-2.

Chen KK, Chiu JH (2006) Effect of Epimedium brevicornum Maxim extract on elicitation of penile erection in the rat. *Urology* 67(3): 631-5.

Chen X, Lee TJ-F (1995) Ginsenosides-induced nitric oxide-mediated relaxation of the rabbit corpus cavernosum. *Br J Pharmacol* 115: 15-18.

Cicero AF, Piacente S, Plaza A, *et al.* (2002) Hexanic maca extract improves rat sexual performance more effectively than methanolic and chloroformic maca extracts. *Andrologia* 34: 177-179.

Fisher G, Shmerling RH (2005) "Oyster May Be an Aphrodisiac Afterall". American Chemical Society Annual Meeting, San Diego, March 16.

Gauthaman K, Adaikan PG, Prasad RN (2002) Aphrodisiac properties of Tribulus Terrestris extract (Protodioscin) in normal and castrated rats. *Life Sci* 71: 1385.

Gauthaman K, Ganesan AP, Prasad RN (2003) Sexual effects of puncturevine (*Tribulus terrestris*) extract (protodioscin): an evaluation using a rat model. *Journal of Alternative and Complementary Medicine* 9 (2): 257-265.

Gillis CN (1997) Panax ginseng pharmacology: a NO link? *Biochem Pharmacol* 54(1): 1-8.

Hadidi KA, Aburjai T, Battah AK (2003) A comparative study of Ferula root extracts and sildenafil on copulatory behaviour of male rats. *Fitoterapia* 74: 242-246.

Hirata JD, Swiersz LM, Zell B, *et al.* (1997) Does Dong quai have estrogenic effects in postmenopausal women? A double-blind, placebo-controlled trial. *Fertil Steril* 68(6): 981-986.

Ilayperuma I, Ratnasooriya WD, Weerasooriya TR (2002) Effect of Withania somnifera root extract on the sexual behavior of male rats. *Asian J Androl* 4(4): 295-8.

Iuvone T, Esposito G (2003) Induction of nitric oxide synthase expression by *Withania somnifera* in macrophages. *Life Sci* 72(14): 1617-25.

Kang BJ, Lee SJ, Kim MD, Cho MJ (2002) A placebo-controlled, double-blind trial of Ginkgo biloba for antidepressant-induced sexual dysfunction. *Hum Psychopharmacol* 17: 279-284.

Kenjale R, Shah R, Sathaye S (2008) Effects of Chlorophytum borivilianum on sexual behaviour and sperm count in male rats. *Phytother Res* 2008 Apr 15 [Epub ahead of print].

Komesaroff PA, Black CV, Cable V, Sudhir K (2000) Effects of Wild Yam Extract on Menopausal Symptoms, Lipids and Sex Hormones in Healthy Menopausal Women. *Climacteric* 4(2): 144-150.

Lee J (1996) What Your doctor May Not Tell You About Menopause. Warner Books, New York, 1996.

Meyer C (1993) Herbal Aphrodisiacs from World Sources, Meyerbooks, Publishers, Glenwood, IL

Murray MT (1995) *The Healing Power of Herbs*, Prima Publishing. Rocklin, California.

Neychev VK, Mitev VI (2005) "The aphrodisiac herb Tribulus terrestris does not influence the androgen production in young men." *Journal of Ethnopharmacology* 101 (1-3): 319-323.

Puri RK, Bhatnagar JK (1974) *Solanum platanifolium*: A New Source of Solasodine. *Lloydia* 37: 318.

Wang S, Zheng Z, Weng Y, *et al.* (2004) Angiogenesis and anti-angiogenesis activity of Chinese medicinal herbal extracts. *Life Sci* 74: 2467-78.

Waynberg J (1995) Male Sexual Asthenia-Interest in a Traditional Plant-Derived Medication. *Ethnopharmacology,* March.

Waynberg, J (1990) "Contributions to the Clinical Validation of the Traditional Use of Ptychopetalum guyanna." Presented at the First International Congress on Ethnopharmacology, Strasbourg, France, June 5-9.

Werbach MR, Murray MD, Michael ND (1994) *Botanical Influences on Illness, A Sourcebook of clinical research,* Third Line Press, Tarzana, California p. 200.

Wheatley D (1999) Ginkgo biloba in the treatment of sexual function due to antidepressant drugs. *Human Psychopharmacol* 14: 512-513.

Willard P (2001) Secrets of Saffron, Beacom Press, Boston, Massachusetts.

Zafar R, Lalwani M (1989) Tribulus terrestris Linn-a review of the current knowledge. *Indian Drugs* 27(3): 148-153.

Chapter Twelve
Conclusion: What is a good Aphrodisiac?

Marty AT (1999) Potentially Fatal Natural Remedies, Letters to the Editor, *American Academy of Family Physicians*, March 1.

Mondaini N, Gontero P (2005) Idiopathic short penis: myth or reality? *BJU International* 95 (1): 8-9.

Phillips DP, Christenfield N, Glynn LM (1998) Increase in US medication-error deaths between 1983 and 1993. *Lancet* 351: 643-4.

Slatkin J (2008) New York City Issues Firm Warning Over Fatal Toad-Venom Aphrodisiac. *Consumerist*, May 7.

Chapter Thirteen
Future of Natural Aphrodisiacs

Ang *et al*. references from 1997-2004. See Chapters six-*Eurycoma* references.

Ang HH, Lee KL (2006) Contamination of mercury in Tongkat Ali hitam herbal preparations. *Food Chem Toxico* 44(8): 1245-50.

Awang DVC (1993) Feverfew fever: a headache for the consumer. *HerbalGram* 29: 34.

Betz JM, White KD, Der Marderosian A (1995) Gas chromatographic determination of yohimbine in commercial yohimbe products. *J AOAC Int* 78(5): 1189-1194.

Bisset NG, (1994) editor, Herbal Drugs and Phytopharmaceuticals, Medpharm Scientific Publishers, Stuttgart

Blok-Tip L, Zomer B, Bakker F, Hartog KD, Hamzink M, Ten Hove J, Vredenbregt M, De Kaste D (2004**).** Structure elucidation of sildenafil analogues in herbal products. *Food Addit Contam* 21(8): 737-48.

Bogani P, Simonini F, Iriti M, Rossoni M, Faora F, Poletti A, *Visioli F* (2006) Lepidium meyenii (Maca) does not exert direct androgenic activities. *J Ethnopharmacol* 104(3): 415-7.

Bogusz MJ, Hassan H, Al-Enazi E, Ibrahim Z, Al-Tufail M (2006) Application of LC-ESI-MS-MS for detection of synthetic adulterants in herbal remedies. *J Pharm Biomed Anal* 41(2): 554-64.

Chan TY, Chan JC, Tomlinson B, Critchley, JA (1993) Chinese herbal medicine revisited: a Hong Kong perspective. *Lancet* 342: 1532.

Cicero AF, Piacente S, Plaza A, *et al*. (2002) Hexanic maca extract improves rat sexual performance more effectively than methanolic and chloroformic maca extracts. *Andrologia* 34: 177-179.

Cupp MJ (2000) Ed**.,** Toxicology and Clinical Pharmacology of Herbal Products, Humana Press, Totowa, NJ.

Dalen JE (1999) Is integrative medicine the medicine of the future? A debate between Arnold S. Relman and Andrew Weil. *Arch Intern Med* 159: 2122-6.

Durgant JM, Heuser J, Andrey D, Perrin C (2005) Quality and Safety Assessment of Ginseng Extracts by determination of the contents of pesticides and metals. *Food Addit Contam* 22: 1224-30.

Ernst E (1998) Harmless herbs? A review of the recent literature. *Am J Med* 104: 170-8.

Ernst E (2000a) Interactions between synthetic and herbal medicinal products. Part 2: a systematic review of the indirect evidence. *Perfusion* 13: 60-70.

Ernst E (2000b) Possible interactions between synthetic and herbal medicinal products. Part 1: a systematic review of the indirect evidence. *Perfusion* 13: 4-6.

Ernst E (2000c) Risks associated with complementary therapies. In: Dukes MN, Aronson JK, Eds. Meyler's Side Effects of Drugs,14th ed., Amsterdam.

Ernst E, De Smet PA, Shaw D, Murray V (1998) Traditional remedies and the "test of time." *Eur J Clin Pharmacol* 54: 99-100

Fetrow CW, Avila JR (1999) Professional's Handbook of Complementary and Alternative Medicines, Springhouse, PA, Springhouse.

Fisher G, Shmerling RH (2005) "Oyster May Be an Aphrodisiac Afterall," American Chemical Society Annual Meeting, San Diego, March 16.

Gonzales GF, Cordova A, Vega K, Chung A, Villena A, Gonez CJ (2003) Effect of Lepidium meyenii (Maca), a root with aphrodisiac and fertility-enhancing properties, on serum reproductive hormone levels in adult healthy men J. *Endocrinol* 176(1): 163-8.

Good Housekeeping Institute: New Good Housekeeping institute study finds drastic discrepancy in potencies of popular herbal supplements. News release, Consumer Safety Symposium on Dietary Supplement and Herbs, New York City, March 3, 1998

Gurley BJ (2000) Contents versus label claims in ephedra-containing dietary supplements. *American J. Health-System Pharmacists* 57: 96.

Harkey MR, Henderson GL, Gersswin MF, Stern JS, Heckman RM (2001) Variability in commercial ginseng products: an analysis of 25 preparations *Am J. Clin Nutr* 73: 1101-6.

Hirsch AR (2004) Sexually Exciting Odors. *Chicago Image*, August/September, pp.14-15.

http://www.consumerlab.com. Accessed 22 October 2001 Consumer Reports, Herbal Roulette, Nov 1995.

Klotz T, Mathers MJ, Braun M, Bloch W (1999) Effectiveness of oral L-arginine in first-line treatment of erectile dysfunction in a controlled crossover study. *Urol Int* 63:220-223.

Ko RJ (1998) Adulterant in Asian Patent Medicine. *The New England JMed* 339(12): (letter to the editor).

Kong JM, Goh NK, Chia LS, ChiaTF (2003) Recent advances in traditional plant drugs and orchids. *Acta Pharmacol Sin* 24 (1): 7-21

Lebret T, Herveda JM, Gornyb P, Botto H (2002) Efficacy and Safety of a Novel Combination of L-Arginine Glutamate and Yohimbine Hydrochloride: a New Oral therapy for Erectile Dysfunction. *European Urology* 41 (6) 608-613.

McGuffin M, Hobbs C, Upton R, Goldberg A (1997) Botanical Safety Handbook, CRC Press, Boca Raton, FL.

Oh SS, Zou P, Low MY, Koh HL (2006) Detection of sildenafil analogues in herbal products for erectile dysfunction. *J Toxicol Environ Health A* 69(21): 1951-8.

Roffman GE (2000) Herbal remedy rip-offs, *D Magazine*, April.

Segasothy M, Samad S (1991) Illicit herbal preparation containing phenylbutazone causing analgesic nephropathy. *Nephron* 59: 166.

Stanislavov R, Nikolova L (2003) Treatment of erectile dysfunction with Pycnogenol and L-Arginine *J.Sex and Marit Ther* 29(3): 207-213.

Wilks J, Ferrett M (1999) Who needs regulation or research? *International Journal of Alternative and Complementary Medicine,* Sep: 17-8.

Yakubu MT, Akanji MA, Oladiji AT (2007) Male Sexual Dysfunction and Methods in Assessing Medicinal Plants with Aphrodisiac Potentials. *Pharmacognosy Reviews 1: 49-57.*

Yang SY, Kim HK, Lefeber AW, Erkelens C, Angelova N, Choi YH, Verpoo R (2006) Application of two-dimensional nuclear magnetic resonance spectroscopy to quality control of ginseng commercial products. *Planta Med* 72(4): 364-9.

Zheng, BL, *et al.* (2000) "Effect of a lipidic extract from lepidium meyenii on sexual behavior in mice and rats." *Urology* 55(4): 598-602.

Ziglar W (1979) The Ziglar report: an analysis of 54 ginseng products. *Whole Foods* 2:48.

Appendix A: Natural Aphrodisiacs for Male on the Market

Product	Ingredients	Recommended Dosage	Additional Information**
Lust	*Sida cordifolia, Saw palmetto,* L-arginine.	1-3 capsules a day.	Legal-highs@legal-highs. co.uk
Bois Bande	Bark-*Roupala montana,* constituents unknown.	Alcoholic extract 5-10 ml.	www.sexualtonics.com boisbandebark.com
IntimX	*Muira puama, Tribulus terrestris,* Macca, Yombine, Damiana, Guarana, Ginkgo, Cordyceps, DHEA, Vitamin E, Zinc.	1-2 capsule a day.	www.intimax.com
Cobra	Horny Goat Weed, Ginseng, Yohimbine, *Muira puama, Saw palmetto.*	1-2 capsule a day.	www.naturalbalance.com
Virility patch	Ginseng, Gotu Kola, Fo Ti, *Saw palmetto,* Damiana, Menthol.	One patch a day.	www.shopping.msn.com
Chinese Virility pills	Catuaba, *Saw palmetto, Muira puama, Dioscorea opposita, Fructus crataegus,* Cuscuta, Horny Goat Weed.	1-2 capsules.	www.greatwallherbal.com
Sexstacy	*Saw palmetto,* Siberian Ginseng, GABA, L-arginine, 4-OH-2-Furanone.	3capsule, 30minutes before sex.	www.herbalhighs.co.uk
Libidoex	Yohimbine, Damiana, *Saw palmetto,* Passion flower, Chinese Hibiscus and Tea tree oil.	1-2 capsules.	www.malesexualenhancement. nt.com
VP-RX	Yohimbine, *Ginkgo biloba, Panax ginseng, Muira puama,* Velvet Deer Antler, Cayenne.	1-2 tablets.	www.herbalpillsonline.biz
***Magna Rx**	Yohimbe Bark Extract (4%), L-arginine HCL, Oyster meat extract, *Muira puama* Extract Peruvian, Maca Root, Korean & Siberian Ginseng, Oat Straw, Nettles Leaf, Orchic, Zinc oxide, Cayenne, Boron Citrate, Licorice Root, Pumpkin Seeds, Sarsaparilla Root.	2 capsules per day.	www.magnarx.com
Vigorax	*Avena sativa*	300mg 30days.	www.doctorg.com
Inferno	Deer Antler Velvet, Ginseng, Dioscorea, *Cordyceps sinensis,* White Peony Root,	1-2 capsules.	www.herbalshop.com

Product	Ingredients	Recommended Dosage	Additional Information**
Libido Fort	*Radix ginseng*, Folium Ginkgo, Dioscorea.	1 capsules.	www.shaman.nl.com
Potensan	*Capsella bursa-pastoris, Similax officinalis*, Damiana, Nutmeg, Cassia Bark.	3 pills before sex.	www.potesan.com
*VigRx oil	Epimedium leaves ext, Korean Red Ginseng, *Ginkgo biloba, Muira puama*, Catuaba Bark, Cuscuta Seeds.	Few drops before sex.	www.vigrxoil.com
Nirvana	*Tribulus terrestris, Mucuna puriens, Withania somnifera, Matthiola incana, Salvia haematodes, Centaurea behen, Cheiranthus cheri.*		www.herbal-nirvana.com
Boom	*Panax ginseng*, Yohimbe bark, Deer antler, *Cordyceps sinensis*, Astragalus, Dong quai, *Ganoderma lucidum*, L arginine.	1-2 capules per day.	www.takeboom.com
SexitivaA	Siberian Ginseng, *Muira puama, Avena sativa, Saw palmetto*, Zallouh root.	1-2 capsule.	Mcjosee@worldwideherbs.com
Zallouh Root	Ferulic acid	2 capsules a day.	www.worlwideherbs.com
*Stamanex tm or Stamazide	Niacin, Horny Goat Weed (Whole Herb) Icarin, Maca 4:1 Extract (*Lepidium meyenii root*), German *Tribulus terrestris* Extract (fruit),70%, Furostanolic Saponins Fenusterols 70%, Furostanolic Saponins (7.5% Protodioscin), *Avena sativa* 10:1 extract (Oat Straw), *Eurycoma longifolia, Ginkgo biloba.*	2 tablets per day with meal.	www.esourcenutrition.com www.liposlimsystems.com
Damiana Male Potential	Ginseng, *Saw palmatto*, Dibasic cal phosphate, Acacia Gum, Stearic acid.	2 tablets per day.	IHealth Tree.com
Testrol	Horny Goat Weed, L-arginine, *Ginkgo biloba*, Tongkat Ali, Damiana, Selenium.	1-2 capsules per day.	USAsupplement.com
Kama Raja	*Withania somnifera, Asparagus racemosus, Mucuna puriens, Tribulus terrestris, Pueraria tuberosum, Emblica officinalis, Asparagus adcendens, Ginger officinale, Piper longum, Anacyclus pyrethrum, Vitex negundo.*	One capsules a day.	www. kamaraja.com

283

Product	Ingredients	Recommended Dosage	Additional Information**
*Male Boost	Yohimbe Bark Extract (4%), L-arginine HCL, Oyster Meat Extract, *Muira puama* Extract, Peruvian Maca Root, Korean & Siberian Ginseng, Oat Straw, Nettles Leaf, Orchic, Zinc oxide, Cayenne, Boron citrate, Licorice Root, Pumpkin Seeds, Sarsaparilla Root.	2 capsules per day.	www.maleboost.com
*Avela	Niacin, Zinc glutonate, L-arginine HCL, *Butea superba* 4:1 extract, Maca 4:1 extract, *Tribulus terrestris* (45% Active Saponins), *Ginkgo biloba,* GABA, *Eurycoma longifolia* 100:1 extract, *Avena sativa* 10:1 extract, *Xanthoparmelia scabrosa* 15:1 extract, *Cnidium monnieri* (35% Active Saponins), *Mucuna pruriens* (10% Active Saponins), *Epimedium sagittatum* (20 % Active Saponins), Bioperine (piperine).	2 capsules per day.	Avela.us
*Viacyn	*Tribulus terresteris,* Yohimbe extract, Niacin, *Epimedium sagittatum , Avena sativa,* Zinc oxide, Maca, Potency Wood, *Ginkgo biloba,* L-arginine, *Saw palmet*to.	2 capsules per day.	www.cleavageonline.com/
Sizepro	Vitamin E, Soya protein conc., Cuscuta seeds, *Epimedium sagittatum, Ginkgo biloba leaves, Ginseng panax roots, Tribulus terrestris aerial part, Saw palmetto,* Hawthorne Berry, *Fructose crataegi,* Inosine, Oat Straw, Cayenne.	2 tablets a day.	Amazon.com
*Enzyte	*Tribulus terrestris* extract (10% saponins, stem and fruit), *Panax ginseng* root extract (4% ginsenosides), *Ginkgo biloba* leaf 6:1 extract, Vegetable Cellulose, Swedish flower pollen extract, *Avena sativa* (aerial parts),	One tablet a day.	www.enzyte.com

	Zinc gluconate, L-Arginine HCl, Maca Root, *Saw palmetto* Berry, Niacin, Zinc citrate, Zinc oxide, Horny Goat Weed extract (10%), Flavonoids as icarian (whole plant), Copper oxide, *Muira puama* stem 4:1 extract, Copper gluconate, Octacosanol.		
*Ziozpher	*Lepidium meyenii, Tribulus terrestris* extract, Maca, *Mucuna pruriens,* Chinese Wolfberry, Horse Chestnut.	2 capsule a day	www.slimstore.com/ziozpher. htm
*Veromax	L-arginine, Zizyphi fructus, Rg 1 Ginsenoside (Ginseng)	2 capsules per day	www.altnetmedwork. net
*Maxoderm	Methyl salicylate, L-arginine, White Nettle extract, Catuaba extract, Maca extract, *Saw palmetto,* Zinc oxide, *Muira puama, Panax ginseng,* Methyl nicotinate	Cream to apply	nutrazone.net
*Vipra	*Tribulus terrestris, Muira puama,* L-arginine, octacosanol, *Ginkgo biloba, Avena sativa,* Horney Goat Weed, *Saw palmetto,* Zinc & Copper gluconate	2 tablets a day	www.enhancementexperts. com
Vitastat	Horny Goat Weed, *Avena sativa,* D-arginine malate, Maca, Damiana, *Eurycoma longifolia, Tribulus terrestris, Ginkgo biloba, Cnidium monnieri* ext., Bioperine.	2 capules a day.	www.bodestore.com
*Procyclon	L-arginine, *Zizyphus jujuba,* Cinnamon bark, *Cnidium monnieri,* Korean ginseng, *Ginkgo biloba,* Maca root, Green tea, Goat Rue Powder, NADH, *Angelica pubescens.*	2 tablets per day.	Nutrazone.net
Erect Pills	*Ginkgo biloba, Muira puama, Tribulus terrestris,* Dodder seeds,Vit E, Yombine, Zinc.	1-2 capsules.	www.upforlife.com

* These products have been reviewed (2006) by consumerHealth Digest and ranked among top25 products available on the market.

** Some of the websites may not exsit since these companies come and go and stay only for a short time in business.

Appendix B: Natural Aphrodisiacs for Female on the Market

Product	Ingredients	Recommended Dosage	Additional Information
*Fematril	Clove vine powder, Damiana Leaf, D-arginine malate, *Avena sativa*, Catawba Bark, *Cinidium monnier*, Bioperine (piperine ext)	3 capsules per day.	www.fematril.com
*Femtrex	Sage Leaf, Raspberry Leaf, Black Cohosh, Kudzu Roots, *Avena sativa*, β-sitosterol, *Chrysanthemum morifolium, Ligusticum wallichii, Acanthopana gracilistylus*.	2 capsules per day.	www.vitamindeal.com
*Stamina-Rx	*Muira puama* Root ext, Epimedium Extract, leaves and stems, Velvet Beans Extract, Yohimbe ext, Cnidium Ext, *Xanthoparmelia scabrosa* Ext, Herba Cistanche, Ageratum Roots, Ginkgo Leaves ext, L-aginine	One capsules a day.	www.herbalshopcompany.com
Arginine Max	L-arginine, Ginseng, Ginkgo, **Damiana**, Vitamins, Calcium and Minerals.	3 tablets a day	www.shopica.com
*Zestra	PA-Free Borage Seed Oil, Evening Primrose Oil, Angelica extract, Coleus extract, Vitamin C, Vitamin E.	Apply the oil	www.tntnutritioncenter.com
Kama Rani	*Tinospora cordifolia, Allium cepa, Glycine max, Withania somnifera, Asparagus racemose, Centella asiatica, Pueraria tuberosa, Chlorophytum arundinaceum, Mucuna pruriens, Celastrus depedens, Myristica fragrans*.	One capsule a day	*www.Kamarani.com*

Product	Ingredients	Recommended Dosage	Additional Information
*Veromax for Women	*Zizyphi fructus,* Amino acid blend, Glutamic acid, L-Lysine HCl, L-arginine HCl, *Ginkgo bilob*a, Siberian ginseng, Soya isoflavones.	2 tablets a day	*www.veromax.com*
Lioness™	Catuaba bark, *Muira puama,* Wild Yam, Maca, *Avena sativa*, Chasteberry, Ipriflavone, Vitamine E, Tribulus, Ginseng, Androstenedione.	3 capules a day.	www.drugstore.com
Erostat	*Turnera diffusa, Ginkgo biloba, Zingiber officinalis, Panax ginseng, Alium sativa, Hypericum perforatum, Valeriana officinalis, Echinacea purpurea, Ptychopetalum olacoids, Ananas comosus, Filipendula ulm*aria.	3 capsules per day.	Nutrazone.net
*Sentia	Red Raspberry leaf, *Ginkgo biloba*, Epimedium, Dodder Seeds, Capsicum, Licorice, *Morella conifera,* Damiana, Valeriana, Ginger, Black Cohosh Root.	One tablet a day.	Sentiaforwomen. com
*Provestra	Red raspberry Leaf, Licorice Root, Damiana leaf, Valerian Root, Ginger Root, Black cohosh.	One tablet a day.	www.provestra.com
*Avenavin	Maca, L-Taurine *Muira puama*, L-arginine, *Tribulus terrestris, Avena Sativa, Epimedium, Cnidium monnieri, Mucuna pruriens.*	Two pills per day.	www.avenavin.com
*Therafem	Tongkat Ali root, *Cnidium monnieri*, Epimedium, L-arginine, Bayberry, Licorice Root, Ginger, Caumiana, Valeriana, Cayenne.	One capsule a day.	www.therafem.com

Product	Ingredients	Recommended Dosage	Additional Information
*Xzite	*Acanthopanax gracilistylus, Ligusticum wallichii, Chrysanthemum morifolium.*	One capsule a day.	xzite.org
Triggerin	*Pontedaria vaginallis, Solanum indicum, Saraca indica, Withania somnifera.*	One capsules a day.	www.ayurlife.org
ExoticaT-6	Horny Goat Weed, Maca roots, *Eurycoma longifolia, Avena sativa, Catuaba bark, Muira puama,* Damiana.	2 capsules per day.	www.china-satibo.com

* These products have been reviewed (2008) by consumerHealth Digest and ranked among top25 Female sexual enhancers out of large numbers of product available in the market.

Appendix C: Chemistry of Natural Aphrodisiacs

Plants	Nature of Constituents	Active Constituents
Alstonia scholaris (Saptparan)	Alkaloids and N-Oxide Alkaloids	Alstonine, alschomine, alstonamine, astonidine, ditamine, detaine or echitamine, echitenine, nareline, picrinine, reserpine, rhazimanine, strictamine, scholarine, vincamajine, vilastonine, yohimbine.
Angelica sinensis (Dong Quai)	Coumarins, Flavonoids, Ligustilides.	Safrole, scopoletin, ß-sitosterol, umbeliferone, carvacrol, terpenes, falcarindiol,falcarinol, ferulic acid.
Avena Sativa (Oat)	Alkaloids, Saponins, Sterols and flavonoids,	Gramine, trigonelline, avenine and saponins such as avenacosides A and B. sitosterol, stigmastadienol, cholesterol, brassicasterol, campesterol, and stigmasterol
Capsicum annuum (Cayenne)	Alkaloids, Flavonoids, Volatile oil, Sesquiterpenes.	Capsaicin, capsanthin, capsorubin and capsaicinoids.
Chlorophytum borivillianum (Safed Musli)	Steroidal alkaloids, Saponins, Spirostanol glycosides	Asparanin A & B, furastanol glycosides.
Crocus sativus (Saffron)	Flavonoids, Terpenes and Carotenoids.	Picrocrocin, crocin, safranal carotene & lycopene.
Epimedium sagittatum (Horny Goat Weed)	Alkaloids, Flavonoids, Terpenes.	Magnoflorine, sesquiterpenes, lignins, phenylated flavonoid, icarin
Erythroxylon catuaba (Catuaba)	Alkaloids, Sesquiterpenes, Flavolignans, Volatile oil, Flavonoids.	Catuabin A, B &C, phytosterols, cyclolignins.
Eurycoma longifolia (Tongkat Ali)	ß-carboline alkaloids, Quassinoid Glycosides, Indole alkaloids.	Quassin, neoquassin, sedrin, eurycomanol
Ferulis harmonis (Zallouh Root)	Sesquiterpenes, Oleo gum resins.	Ferulic acid, ferulosides, ferutinine, feroline, daucane-sesquiterpenes.
Ginkgo biloba (Maidenhair Tree)	Flavoglycosides, Dimeric flavones, Diterpenes.	Quercetin, kaempferol, diterpenes, ginkoglides A,B,C,J,M and dimeric flavones-bilobetin, ginkgetin, sciadopitysin
Glycyrrhiza glabra (Licorice)	Flavonoids, Biflavonoids, Flavone-c-glycosides, prenylated Flavonoids, Saponoids, Sterols, Essential oils.	Glycyrrhizin, glycyramarin, licoagrodin, bisprenyl flavones. licoagrochalcones, A, B,C and D.
Hypericum perforatum (St. John's Wort)	Anthraquinones, Flavonoids, Biflavonoids, Xanthones, Terpenes.	Hypericin, pseudo hypericin, protohypericin, hyperoside, quercetin, rutin, hyperforin, xanthones, terpenes.

Plants	Nature of Constituents	Active Constituents
Lepidium meyenii (**Maca Root**)	Imidazole alkaloids, Alkamides and Volatile oil.	Lepidiline A & B, alkamides, Phenyl acetonitrile.
Mandragora officinarum	Tropane alkaloids.	Tropane, scopolamine, hyoscyamine, mandragorine
Muira puama (**Potency Wood**)	Alkaloids, Terpenes, Coumarins, Resinic acid.	Muirapuamanine, phlobaphenes, lupeol, sterols
Panax quinquefolium	Ginsenosides, Sterols, Organic Acids, Peptides.	Rg1, Re, Rd, Rc, Rb1, Rb2, Rd, Rf, sterols, α, ß-amyrins
Panax ginseng (**Korean Ginseng**)	Ginsenosides.	Rg1, Re, Rd, Rc, Rb1, Rb2, Rd, Rf, sterols.
E. senticosus (**Siberian Ginseng**)	Glycosides, Steroidal Saponins, Flavonoids.	Eleuthrosides B,C,D,E,F , sterols.
Pausinystalia yohimbe (**Yohimbe Bark**)	Indole alkaloids.	Yohimbine, alloyohimbine, ajmaline, corynanthine.
Piper methysticum (**Kava-Kava**)	Kava-lactones, Alkaloids.	Methysticin, kavahin, yangonin, piperidine.
Polygonum multiflorum	Glycosides-anthraquinones.	Polygonimitins, chrysophanol, emodin
Roupala montana (**Bois Bonde**)	Essential oil.	No information available
Serenoa repens (**Saw palmetto**)	Phytosterols, flavonoids, Tannins, Volatile oil	ß-sitosterol, cycloartenol, stigmasterol, lupeol, steroidal saponins, flavonoids, ferulic acid, terpenes.
Tribulus terrestris (**Gokshura**)	Saponins, Glycosides, Alkaloids, Sterols	Protodioscin, furostanol, Terrestris A,B,C,E & D. diosgenin, hecogenin, neotigogenin.
Turnera diffusa (**Damiana**)	Tannins, flavonoids, Glycosides, Volatile. oil	Luteolin, sterols, damianin, the glycosides gonzalitosin, arbutin, and cyanoglycoside tetraphyllin B. Some aliphatic hydrocarbons such as hexacosanol, triacontane1, 8-cineole, cymae, p-cymene, alpha- and beta-pinene, thymol, α-copaene.
Withania somnifera (**Ashwagandha**)	Steroidal alkaloids, Steroidal lactones known as withanolides.	Withanine, somniferine, cuscohygrine, anahygrine, pseudowithanine and their glycosides sitoindosides. At present 12 alkaloids, 35 withanolides and some sitoindosides have been reported.
Deer Antler	Amino acids, Free Fatty acids, Minerals and Steroids	15 Free amino acids, free fatty acids, gangliosides, lecithin, phospholipids, steroids, prostaglandins, glycosamino-glycans, chondroitin, monoamine oxidase inhibitors, some minerals like magnesium, potassium, selenium, calcium and phosphorus.
Spanish Fly	Benzofuran	Cantharidin

Appendix D: Anticoagulant Herbs[1]

Common Name	Botanical Name (Family)
Alfa-alfa*	*Medicago sativa* (Fabaceae/Leguminose)
American Ginseng	*Panax quinquefolia* (Araliaceae)
Angelica*	*Angelica sinensis* (Apiaceae)
Anise*	*Anisum vulgaris* (Apiaceae)
Arnica*	*Arnica montana* (Asteraceae/ Compositae)
Asafoetida*	*Ferula asafoetida* (Apiaceae)
Aspen Bark	*Populus tremula* (Salicaceae)
Bilberry	*Vaccinum myrtillus* (Vacciniacea)
Birch	*Betula pendula* (Betulaceae)
Black Cohosh	*Cimicifuga racemosa* (Ranunculaceae)
Borage seed oil	*Borago officinalis* (Boraginaceae)
Capsicum	*Capsicum annuum* (Solanaceae)
Cat's claw	*Uncaria tomentosa* (Rubiaceae)
Celery	*Apium graveolens* (Apiaceae)
Chamomile	*Chamaemelum nobile* (Asteraceae)
Clove	*Eugenia caryophyllata* (Myrtaceae)
Cordyceps	*Cordyceps sinensis* (Clavicipitaceae)
Danshen	*Salvia miltiorrhiza* (Lamiaceae)
Devil's Claw	*Harpagophytum procumbens* (Pedaliaceae)
Eleuthero	*Eleutherococcus senticosus* (Araliaceae)
Fenugreek,	*Trigonella foenum-graecum* (Fabaceae)
Flaxseed/flax powder	*Linum usitatissimum(Linaceae)*
Feverfew	*Tanacetum parthenium* (Compositae)
Ginkgo	*Ginkgo biloba* (Ginkgoaceae)
Garlic	*Allium sativum* (Alliaceae)
Grapefruit juice	*Citrus paradisi* (Rutaceae)
Green Tea	*Camellia sinensis* (Theaceae)
Guggul	*Commiphora mukul* (Burseraceae)
Chestnut*	*Aesculus hippocastanum* (Hippocastanaceae)
Horse radish*	*Radicula armoracia* (Brassicaceae)
Licorice root*	*Glycyrrhiza glabra* (Fabaceae)
Male Fern	*Dryopteris filixmas* (Polypodiaceae)
Onion	*Allium cepa (Alliaceae)*
Parsley	*Petroselinum crispum*(Apiaceae)
Prickly Ash	*Zanthoxylum americanum* (Rutaceae)
Quassia*	*Quassia amara (*Simaroubaceae)
Red Clover*	*Trifolium pratense* (Fabaceae)
Sweet Clover	Melilotus officinalis (Fabaceae)
Wild Carrot	*Daucus carota* (Apiaceae)
Wintergreen	*Gaultheria procumbens (Ericaceae)*
Yucca	*Yucca filamentosa (Agavaceae)*

* These herbs contain coumarin that acts as anticoagulant. Herb[1] - The term is used in this book for both herbaceous and woody plants.

Asteraceae - previously called compositae, Apiaceae as Umbelliferae and Fabaceae as Leguminose.

Appendix E: Hypoglycemic Herbs

Common Name	Botanical Name (Family)
Agrimony leaves	*Agrimonia eupatoria* (Rosaceae)
Adiantum plant	*Adiantum capillus* (Pteridaceae)
Aloe	*Aloe vera* (Asphodelaceae/ liliaceae)
Bilberry	*Vaccinium myrtillus* (Ericaceae)
Bitter Melon	*Momordica charantia* (Cucurbitaceae)
Bitter Gaurd	*Momordica charantia* (Cucurbitaceae)
Burdock	*Arctium lappa* (Asteraceae)
Damiana	*Turnera diffusa* (Turneraceae)
Fenugreek	*Trigonella foenum graecum* (Fabaceae/Leguminosae)
Fo-Ti roots	*Polygonum multiflorum* (Polygonaceae)
Garlic cloves	*Allium sativum* (Alliaceae)
Gurmar leaves	*Gymnema sylvestris* (Asclepiadaceae)
Guargum seeds	*Cyamopsis tetragonolobus* (Fabaceae)
Gulancho	*Tinospora cordifolia* (Manispermaceae)
Ginseng	*Panax ginseng* (Araliaceae)
Horse Chestnut	*Aesculus hippocastanum* (Spindaceae)
Jambul seeds	*Syzygium cumin* (Myrtaceae)
Lotus roots	*Nymphaea lotus* (Nymphaeaceae)
Lupin seeds	*Lupinus albus* (Fabaceae)
Psyllium seeds	*Plantago ovata* (Plantaginaceae)
Sacred Basil plant	*Ocimum sanctum* (Lamiaceae)
Spinach leaves	*Spinacea oleracea* (Amaranthaceae)
Mushroom	*Ganoderma lucidum* (Gonadermaceae)
Milk Thistle	*Silybum marianum* (Asteraceae)
Rosemary	*Rosmarinus officinalis* (Lamiaceae)
Stinging nettle	*Urtica species* (Urticaceae)
White Horehound	*Marrubium vulgare* (lamiaceae)
White Mulberry leaves	*Morus alba* (Moraceae)

Appendix F: Hyperglycemic Herbs

Common Name	Botanical Name (Family)
Amato seeds	*Bixa orellana* (Bixacea)
Cocoa seeds	**Theobroma cacao* (Malvaceae/Sterculiaceae)
Coffee seeds	*Coffea arabica* (Rubiaceae)
Cola seeds	*Cola acuminata* (Malvaceae/Sterculiaceae)
Ma Huang	*Ephedra sinica* (Ephedraceae)
Rosemary leaves	*Rosmarinus officinalis* (Lamiaceae)
Tea leaves	*Camellia sinensis* (Theaceae)

*Previously Sterculiaceae

Appendix G: Hypotensive Herbs

Common Name	Botanical Name (Family)
Aconite	*Aconitum napellus* (Ranunculaceae)
Baneberry	*Actea rubra* (Ranunculaceae)
Betal Nut	*Areca catechu* (Arecaceae)
Bilberry	*Vaccinium myrtillus* (Ericaceae)
Black Cohosh	*Cimicifuga racemosa* (Ranunculaceae)
Bryony	*Bryonia alba* (Cucurbitaceae)
Calendula	*Calendula officinalis* (Asteraceae)
California Poppy	*Eschscholtzia californica* (Papaveraceae)
Coleus	*Coleus forskohlii* (Lamiaceae)
Curcuma	*Curcuma longa* (Zingiberaceae)
Evening Primrose	*Oenothera biennis* (Onagraceae)
Eucalyptus	*Eucalyptus globulus* (Myrtaceae)
Flaxseed	*Linum usitatissimum* (Linaceae)
Garlic	*Allium sativum* (Alliaceae))
Ginger	*Zingiber officinale* (Zingiberaceae)
Ginkgo	*Ginkgo biloba (Ginkgoaceae)*
Goldenseal	*Hydrastis canadensis* (Ranunculaceae)
Green Hellebore	*Helleborus viridis* (Rutaceae)
Hawthorn	*Crataegus oxycantha* (Rosaceae)
Indian Tobacco	*Lobelia inflata* (Campanulaceae)
Jaborandi	*Pilocarpus jaborandi* (Rutaceae)
Mistletoe	*Viscum album* (Loranthaceae)
Oleander	*Nerium oleander* (Apocynaceae)
Periwinkle	*Vinca minor* (Apocynaceae)
Pleurisy root	*Asclepias tuberosum* (Asclepiadaceae)
Shepherd's purse	*Capsella bursapastoris* (Cruciferae)
Texas Milkweed	*Asclepias texana* (Asclepiadaceae)
Wild Cherry	*Prunus serotina* (Rosaceae)

Appendix H: Hypertensive Herbs

Common Name	Botanical Name (Family)
Arnica	*Arnica montana* (Asteraceae)
Bayberry	*Myrica cerifera* (Myricaceae)
Betal Nut	*Areca catechu* (Arecaceae)
Blue Cohosh	*Caulophyllum thalictrum* (Berberidaceae)
Broom	*Strophanthus scoparius* (Apocynaceae)
Cayenne	*Capsicum annuum* (Solanaceae)
Cola	*Cola nitida* (Malvaceae)
Coltsfoot	*Tussilago farfara* (Asteraceae)
Ephedra	*Ephedra vulgaris* (Ephedraceae)
Ginger	*Zingiber officinalis* (Zingiberaceae)
Licorice	*Glycyrrhiza glabra* (Fabaceae)
Polypodium	*Polypodium vulgare* (Polypodiaceae)
Yerba mate	*Anemopsis californica* (Saururaceae)

Appendix I: Herbs that Enhance Sedative Effects

Common Name	Botanical Name (Family)
Ashwagandha	*Withania somnifera* (Solanaceae)
Black Cohosh	*Cimicifuga racemosa* (Ranunculaceae*)*
Calendula flowers	*Calendula officinalis* (Asteraceae)
Catnip	*Nepeta cataria* (Lamiaceae*)*
Damiana	*Turnera aphrodisiaca* (Turneraceae)
Hops	*Humulus lupulus* (Cannabacee)
Kava kava	*Piper methysticum* (Piperaceae)
Lavender leaves	*Lavandula officinalis* (Lamiaceae)
Muira puama	*Ptychopetalum olacoides* (Olacaceae)
Siberian Ginseng	*Eleutherococcus senticosus* (Araliaceae)
St John's Wort	*Hypericum perforatum* (Clusiaceae)
Valeriana	*Valeriana officinalis* (Valerianaceae)
Passion flower	*Passiflora incarnata* (Passifloraceae)
Yerba Mansa	*Anemopsis californica* (Saururaceae)

Appendix J: Herbs that Stimulate Stomach Acid Secretion

Common Name	Botanical Name (Family)
Angelica	*Angelica angularis* (Apiaceae)
Barberry bark	*Berberis vulgaris* (Berberidaceae)
Black Pepper	*Piper nigrum* (Piperaceae)
Cayenne fruits	*Capsicum annuum*(solanacea
Cinnamomum bark	*Cinnamomum zeylanicum* (Lauraceae)
Coffee seeds	*Coffea arabica* (Rubiaceae)
Cola seeds	*Cola nitida* (Malvaceae)
Gentian root	*Gentiana lutea (Gentianaceae)*
Ginger	*Zingiber officinale* (Zingiberaceae)
Goldenseal root	*Hydrastis canadensis (Ranunculaceae)*
Goldthread root	*Copius trifolia* (Ranunculaceae)
Horseradish root	*Armoracia rusticana* (Cruciferae)
Mustard seeds	*Brassica nigra* (Cruciferae)
Tobacco leaves	*Nicotiana tabacum* (Solanacea)
Tea leaves	*Camellia sinensis* (Theaceae)
Onion bulbs	*Allium cepa* (Alliaceae)*
Garlic Cloves	*Allium sativum (Alliaceae)*

*Alliaceae formerly Liliaceae

Appendix K: Drugs-Herbal Aphrodisiacs Interaction

Drugs	Aphrodisiacs	Aphrodisiac-Drug Interaction	Symptoms due to Interaction
NSAIDs Appendix O	*Epimedium sagittatum* (Horny Goat Weed), Zallouh root, *Muira puama*, Ginseng	Synergistic effect (coumarins)	Increase blood clotting time
Antidepressants Appendix S	Kava-Kava, Hypericum (St John's wort), Damiana, Valeriana, Withania, Catuaba, *Eurycoma longifolia* (Tongkat Ali), Mandrake root, Yohimbine	Synergistic effect	Increased sedation leading to coma
Antiplatelet & anticoagulant Warfarin, Aspirin, Heparin	Epimedium (Horny Goat), *Ginkgo biloba*, Garlic, Capsicum, *Angelica sinensis*, Zallouh root, *Muira puama*	Contain coumarin, ginkgolide B May inhibit platelet activity increased plasma concentration of warfarin	Increased potential for bleeding
Steroids and Diuretics, Cardiac Glycosides-Digitalis	Licorice, Yohimbine, Ginseng	Increased Potassium loss in GI tract	Increased Potassium loss can be fatal
Antihypertensive Drugs-ACE inhibitors, Beta-Blockers, Alpha Blockers, Ca++ Chanel blockers Appendix P,Q,R,Y,V	Horny Goat Weed, Licorice, Capsicum Yohimbine	May increase sodium chloride and water retention	Increase or decrease blood pressure
Hypoglycemic agents Appendix U	Ginseng species, Safed Musli, Damiana	Decreased glucose level	Coma & can be fatal
Immuno-suppressants	Vitamine E, Licorice, Ginseng	Stimulate the immune system	Decreased immuno-suppressants effect
MAOIs Appendix S	*Hypericum, Valeriana, Withania, Capsicum, Zallouh root, Licorice, Ginseng*	Synergistic effect	Change BP, heart rate, cause Insomnia
Sedatives-Barbiturates, Benzodiazepines, alcohol	*Valeriana, Withania, Hypericum*, Ginger, Garlic, Kava-Kava	Synergistic effect	Increased sedation leads to coma (can be fatal).
Oral contraceptive	*Licorice, Hypericum, Ginseng, Saw palmetto*	Potassium loss	Coma (can be fatal)
Anticancer Drugs such as Bleomycin.	*Angelica sinensis Capsicum, Muira puama*	Increase time of blood clot	Cause bleeding

Appendix L: Solvents and Perfumery Materials as Adulterants in Aromatic oils

Name	Solvent
Abitol	It is used to increase resin contents
Benzyl benzoate	It is used to increase resin contents
Benzyl alcohol	Solvent
Carbitol	Solvent
Diacetone	Solvent
Dipropylene	Solvent
Diethyl phthalate	Solvent
Polyethylene glycol	Solvent
Isooctyl phthalate	Solvent

Appendix M: Addition of Cheap Volatile oil as Adulterant in Costly V. Oil

Aromatic oils	Adulterant
Basil exotic oil	Linalol
Bergamot oil	Lemon oil
Bitter orange oil	Sweet orange
Cedarwood oil Virginia	Cedarwood oil Chinese variety
Cinnamon bark oil	Cinnamon leaf oil
Cinnamon bark oil	Inferior Fraction of clove oil
Clove bud oil	Clove stem oil + eugenol
Geranium oil Chinese	Geranium oil inferior variety
Grape fruit oil	Orange oil
Lavandula	Cheap Lavandula sp oil
Lemon oil	Lemon and orange terpenes
Nutmeg oil	Pine terpenes
Patchouli oil	Gurjun Balsam or Chinese Pat species
Peppermint oil	Cornmint oil
Sandalwood oil	Sandalwood terpenes
Rosemary oil	Eucalyptus oil, Camphor oil
Ylang -Ylang	Cananga oil (Cheap)

Appendix M-1: Cheap Synthetics as Adulterants in Aromatic oils

Oils	Synthetic adulterants
Anise oil	Anethole
Basil oil Exotic	Methyl chavicol, linalol
Bergamot oil	Linalol and linalyl acetate
Bitter Almond oil	Benzaldehyde
Cassia oil	Synthetic Cinnamic aldehyde
Chamomile oil	Isobutyl angelate, bisabolols
Cinnamon bark oil	Benzaldehyde, cinnamic aldehyde, eugenol
Citrus oils	Monoterpenes alcohols
Caraway seed oil	Limonene and racemic carvone
Cardamom oil	Linalyl acetate, 1. 8-cineole, a terpinyl
Cognac oil	Ethyl o enanthalate
Coriander seed oil	Lonalol
Garlic oil	2 propenyl disulphide
Jasmine oil	Reconstructed oil
Lavender oil	Eucalyptus oil, Spanish sage oil
Lemongrass	Citral
Mentha oil	Linalol +linalyl acetate
Neroli oil	Lemon oil from leaves
Rose oil	β-ionon + citronellol, cheaper rose oil
Rosemary oil	Camphor, isobornyl acetate
Rosewood oil	Linalol, methyl heptenone, heptenol
Spearmint oil	(-) Carvone
Thyme oil	Para cymene, thymol
Wintergreen oil	Methyl salicylate
Ylang-Ylang oil	Benzyl acetate, geranyl acetate, Benzyl cinnamate, cedarwood oil

Appendix M-2: Photosensitive Aromatic Oils

Aromatic Oils	Source
Ammi visnaga oil	*Ammi visnaga*
Angelica root oil	*Angelica archangelica*
Bergamot oil	*Citrus aurantium*
Cumin seed oil	*Cuminum cyminum*
Grapefruit oil	*Citrus paradisi*
Lemon oil cold pressed	*Citrus limonum*
Mandarin oil cold-pressed	*Citrus reticulata*
Myrrh oil/resin	*Commiphora species*
Orange oil	*Bitter Citrus aurantium*
Parsley leaf oil	*Petroselinum crispum*
Tagetes oil	*Tagetes minuta*
Tangerine oil cold-pressed	*Citrum reticulata*

Appendix N: Essential Oils Used as Aphrodisiacs

Essential Oil	Compatibility	Main Constituents
Angelica oil (*A. archangelica*) Family: Apiaceae Source: roots and leaves	Blends well with patchouli, clary sage, citrus oils, frankincense	α,β-Phellandrene, bornyl acetate, coumarins, bergaptene, cryptone, myrcene, p-cymene, penta and heptadecanolide, pinene, limonene, linalool, borneol, humulene oxide, sabinene and terpenes.
Basil oil (*Ocimum basilicum*) Family: Lamiaceae Source: leaves	Lavender, bergamot, clary sage and geranium	Linalool, β-elemene, α, β-pinene, β-caryophyllene, phenol, methyl chavicol, eugenol, limonene, geraniol, citronellal and citronellol.
Bergamot oil (*Citrus bergamia*) Family: Rutaceae Source: Fruit rind	Chamomile, coriander, cypress, geranium, juniper, lavender, lemon, neroli, ylang-ylang	α-pinene, limonene, α-bergaptene, bergapten, bergptol, β-bisabolene, linalool, linalyl acetate, nerol, neryl acetate, geraniol, geraniol acetate and α-terpincol and furocoumarins
Chamomilla oil (*Matricaria chamomile*) Family: Asteraceae Source: Flowers Roman Chamomile German Chamomile	Lavender, neroli oil	Roman chamomile consists of bergamotine, bergaptol, α-pinene, camphene, β-pinene, sabinene, myrcene, 1, 8-cineole, y-terpinene, caryophyllene, and propyl angelate and butyl angelate. German chamomile oil contains chamazulene, α-bisabolol, bisabolol oxide A, bisabolol oxide B and bisabolone oxide A. Chamazulene, produced during steam distillation, not present in the fresh plant.
Clary sage oil (*Salvia sclarea*) Family: Lamiaceae Source: leaves	Lavender, basil, bergamot, geranium	Linalool, linalyl acetate, caryophyllene, α-terpineol, geraniol, neryl acetate, sclareol and Germacrene D
Cinnamon oil (*C. zeylanicum*) Source: Bark Family: Lauraceae	Blends with all Citrus and spicy Volatile oils. caraway, clove, myrtle, nutmeg, olibanum, frankincense, lavender oils.	Bark oil contains cinnamaldehyde, eugenol, β-caryophyllene, borneol, benzaldehyde, pinene, cineole, phellandrene, furole, cymene, linalool. Leaf oil comprises eugenol, eugenol acetate, cinnamaldehyde, benzyl benzoate, linalool, borneol,

Essential Oil	Compatibility	Main Constituents
Frankincense oil *(Boswellia carteri)* Family: Burseraceae Source: Oleogum resin from trunk	Basil, bergamot, cardamom, cedar wood, chamomile, cinnamon, clary sage, coriander, geranium, ginger.	α-pinene, phellandrene, cymene, α-gurjuene, α-guaiene, actanol, linalool, octyl acetate, bornyl acetate, farnesol, incensole, incensole-oxide and incensyl acetate.
Jasmine oil *(Jasminum officinale)* Family: Jasminaceae Source: Flower	Blends with bergamot, rose, sandalwood and all citrus.	Benzyl acetate, linalool, benzyl alcohol, indole, benzyl benzoate, cis-jasmone, geraniol, methyl anthranilate and trace amounts of p.-cresol, farnesol, cis-3-hexenyl benzoate, eugenol, nerol, creosol, benzoic acid, benzaldehyde, γ-terpineol, nerolidol, isohytol, phytol etc. cis-jasmone, farnesol.
Lavender oil *(Lavendula angustifolia)* Family: Lamiaceae Source: Flower and leaves	Blends with bergamot, clove, rosemary, eucalyptus, patchouli, clary sage, rose, jasmine	Lavender flower contains 1.5-3% volatile oil, of which 25-55% is linalyl acetate, linalool, cis-b-ocimene, octanone, cineole, a-terpineol, camphor, limonene; caryphyllene, lavandulol, lavundyl acetate tannins (5-10%); coumarins; flavonoids (luteolin); phytosterols; and triterpenes
Musk oil *(Moschus species)* Family: Moschidae Source: Deer	Personifiers, enhancers and modifiers	Muscone, isolated in 1926. Since then, more than 300 musk substitutes have been synthesized in the laboratory
Patchouli oil (*Pogostemon cablin* also known as *Pogostemon patchouli*) Family : Lamiaceae Source: Flower	Blends with sandalwood, bergamot, cedar wood, rose, sweet orange, cassia, myrrh, and clary sage.	α-pinene, α-guaiene, β-patchoulene, caryophyllene, α-patchoulene, seychellene, α-bulnesene, norpatchoulenol, patchouli alcohol, pogostol and sesquiterpene alkaloids

Essential Oil	Compatibility	Main Constituents
Rose Oil *(Rosa centifolia)* Family: Rosaceae Source: Flowers	Blend with personifiers, enhancers and modifiers	About 300 componets have ben identified. Main chemical components are-citronellol, phenylethanol, geraniol, nerol, farnesol and stearpoten with traces of nonanol, linalool, nonanal, phenyl acetaldehyde, citral, carvone, citronellyl acetate, 2-phenylmenthylacetate, methyl eugenol, eugenol and rose oxide.
Rosemary Oil *(Rosmarinus coronarium)* Family: Lamiaceae Source: Fresh flowering top	It blends with lavender, origanum vulgare, thyme, pine, basil, peppermint, cedar wood, cinnamon and other spice oils	Pinenes, camphene, limonene, cineole, borneol, camphor, linalool, terpineol, octanone, and bornyl acetate.
Sandalwood Oil *(Santalum album)* Family: Santalaceae Source: The heart- wood and roots of a mature tree	Ylang-Ylang, patchouli, spruce, lavender, frankincense, cyprus,	The main constituents (90%) are α, β-santalols, 6 per cent sesquiterpene hydro carbons, santene, teresantol, borneol, santalone, tri-cyclo-ekasantalal and other minor constituents.
Vanilla oil *(Vanilla planifolia or Vanilla fragrans)* Family: Orchidaceae Source: vanilla pods	Almost all the Volatile. oils	Vanillin along with about 130 constituents such as phenols, phenol ether, alcohols, carbonyl compounds, acids, ester, lactones, aliphatic and aromatic carbohydrates and heterocyclic compounds have been identified in vanilla extract.
Ylang-Ylang Oil The name means 'flower of flowers' *(Cananga odorata var. Unona odorantissimum)* Family : Anonaceae Source: Flowers	Blend with bergamot, lavender, neroli, rosewood and sandalwood	The main chemical components are linalool, geranyl acetate, caryophyllene, p-cresyl methyl ether, methyl benzoate, benzyl acetate, benzyl benzoate and other sesquiterpenes. Germacrene-D, farnesol, α-farnesene, cinnamyl acetate.

The terms essential oil and volatile oil in this text have been used for Aromatic oil.

Appendix O: Commonly Used Nonsteroidal Anti-inflammatory Drugs

Generic Name	Brand Name
Salsalate	Disalcid, Salflex
Diflunisal	Dolobid
Ketoprofen	Orudis
Nabumetone	Relafen
Naproxen	Naprosyn
Diclofenac	Voltaren
Indomethacin	Indocin
Sulindac	Clinoril
Tolmetin	Tolectin
Etodolac	Lodine
Ketorolac	Toradol
Oxaprozin	Daypro
Celecoxib	Celebrex
Ibuprofen	Motrin
Aspirin	Ecotrin
Piroxicam	Feldene
Meloxicam	Mobie

Appendix P: Commonly Used ACE Inhibitors

Generic Name	Brand Name
Captopril	Capoten
Benazepril	Lotensin
Enalapril	Vasotec
Lisinopril	Prinivil, Zestril
Fosinopril	Monopril
Ramipril	Altace
Quinapril	Accupril
Perindopril	Aceon
Trandolapril	Mavik
Moexipril	Univasc
Fosinopril	Monopril

Appendix Q: Commonly Used Beta (β)-Blockers

Generic Name	Brand Name
Sotalol	Betapace
Timolol	Blocadren
Esmolol	Brevibloc
Carteolol	Cartrol
Carvedilol	Coreg
Nadolol	Corgard
Propranolol	Inderal
Betaxolol	Kerlone
Penbutolol	Levatol
Metoprolol	Lopressor
Labetalol	Normodyne
Acebutolol	Sectral
Atenolol	Tenormin
Metoprolol	Toprol
Pindolol	Visken
Bisoprolol	Zebeta

Appendix R: Commonly Used Alpha (α) Blockers

Generic Name	Brand Name
Doxazosin	Cardura
Prazosin	Minipress
Terazosin	Hytrin
Tamsulosin	Flomax
Alfuzosin	Uroxatrol
Phentolamine	Regitine
Phenoxybenzamine	

Appendix S: Antidepressant Drugs (Monoamine Oxidase Inhibitors)

Generic Name	Brand Name
Isocarboxazid	Marplan
Moclobemide	Moclodura, Manerex, Aurorix
Phenelzine	Nardil
Tranylcypromine	Parnate, Jatrosom
Selegiline	Eldepryl, Emsam
Rasagiline	Azilect
Nilamide	
Iproniazid	Marsilid, ipronid, Rivivol,
Toloxatone	

Appendix T: Commonly Used SSRIs

Generic Name	Brand Name
Citalopram	Celexa, Cipramil, Emocal, Sepram
Escitalopram oxalate	Lexapro, Cipralex, Esertia
Fluoxetine	Prozac, Fontex, Seromax, Saronil, Sarafem, Fluctin
Fluvoxamine maleate	Luvox, Faverin
Paroxetine	Paxil, Seroxat, Aropax, Deroxat, Rexetin Xeetanor, Paroxat
Sertraline	Zoloft, Lustral, Serlain
Dapoxetine	No known trade name

Appendix U: Commonly Used Antidiabetic Drugs

Generic Name	Brand Name
Insulin	Humulin, Novolin
Sulfonylureas group	
Chlorpropamide	Diabinese
Tolazamide	Tolinase
Glipizide	Glucotrol
Glimepiride	Amaryl
Alpha-glucosidase Group	
Acarbose	Precose
Miglitol	Glyser
Biguanides Group	
Metformin	Glucophage
Meglitinides Group	
Repaglinide	Prandin
Nateglitinide	Starlix
Thiazolidinediones Group	
Rosiglitazone	Avandia
Pioglitazone	Actos

Appendix V: Commonly Used H2/PP Blockers

Generic Name	Brand Name
Cimetidine	Tagamet
Famotidine	Pepcid
Nizatidine	Axid
Ranitidine	Zantac
Esomeprazole	Nexium
Omeprazole	Prilosec

Appendix W: Commonly Used Triptans Group of Drugs

Generic Name	Brand Name
Frovatriptan	Frova
Naratriptan	Amerge
Rizatriptan	Axert
Sumatriptan	Imitrex
Zolmitriptan	Zomig

Appendix X: Commonly Used Calcium Channel Blockers

Generic Name	Brand Name
Nisoldipine	Sular
Nifedipine	Adalat, Procardia
Nicardipine	Cardene
Bepril	Vascor
Isradipine	Dynacire
Nimodipine	Nimotop
Felodipine	Plendil
Amlodipine	Norvasc
Diltiazem	Cardizem
Verapamil	Calan, Isopyin

Appendix Y: Commonly Used Anticonvulsants

Generic Name	Brand Name
Carbamazepine	Tegretol
Phenytoin	Dilantin
Valproic acid	Depakote
Barbiturates	
Phenobarbital	
Benzodiazepines	
Alprazolam	Xanax
Lorazepam	Ativan
Diazepam	Valium

Appendix Z: Plants under Investigation for Aphrodisiac Activity

Plant Species	Investigators/Institution
Butea superba Enhances Penile Eerection in Rats.	Chainarong T, Sitsari Y, Jeenapongsa R, *Phytotherapy Res* 20(6): 484-489 (2006). Faculty of Pharmaceutical Sciences, Naresuan University, Phitsanulok, Thailand.
Butea frondosa Aphrodisiac activity of *Butea frondosa* Koen. ex Roxb. extract in male rats was studied. The results were encouraging.	Ramachandran S, Sridhar Y, Sam SK, Saravanan M, Leonard JT, Anbalagan N, Sridhar SK, *Phytomedicine*. 11(2-3): 165-8 (2004). Department of Pharmacology and Toxicology, C. L.Baid Mehta College of Pharmacy, Thorapakkam, Old Mahabalipuram Road, Chennai, India.
Boesenbergia rotunda Effects of *Boesenbergia rotunda* (L.) on sexual behavior of male rats were studied.	Sudwan P, Saenphet K, Aritajat S, Sitasuwan N, *Asian J Androl* 9(6): 849-55 (2007). Department of Biology, Faculty of Science, Faculty of Medicine, Chiang Mai University, Chiang Mai 50200, Thailand.
Casimiroa edulis *Casimiroa edulis* seed extract verses sildenafil citrate (Viagra(tm)) on mating behavior in normal male rats was studied. The research provide preliminary evidence that the aqueous seed extract of *casimiroa edulis* possesses aphrodisiac activity and may be used as an alternative drug therapy to restore sexual functions.	Ali ST, Rakkah NI., *Pak J Pharm Sci* 21(1): 1-6 (2008). Department of Physiology, Faculty of Medicine, Umm-Al-Qura University, P.O. Box 7607, Makkah, Saudi Arabia.
Curculigo orchioides Effect of *Curculigo orchioides* rhizomes on sexual behavior of male rats was studied. They found an increase in mount frequency and enhanced attractability towards female. Penile erection index was also high in treated group. The results were encouraging**.**	Chauhan NS, Rao ChV, Dixit VK, *Fitoterapia* 78(7-8): 530-4(2007). Department of Pharmaceutical Sciences, Dr. H.S. Gour University Sagar (MP), India.
Fadogia agrestis Aphrodisiac potentials of the aqueous extract of *Fadogia agrestis* (Schweinf. Ex Hiern) stem in male albino rats were studied. The aqueous extract of *Fadogia agrestis* stem increased the blood testosterone concentrations. It may be used to modify impaired sexual functions in those arising from hypotestosteronemia.	Yakubu MT, Akanji MA, Oladiji AT, *Asian J Androl* 7(4): 399-404(2005). Medicinal Plants Research Laboratory, Department of Biochemistry, University of Ilorin, PMB 1515.

Plant Species	Investigators/Institution
Kaempferia parviflora **Effect of *Kaempferia parviflora* Wall. ex. Baker on sexual activity of male rats and its toxicity.**	Sudwan P, Saenphet K, Saenphet S, Suwansirikul S, *Southeast Asian J Trop Med Public Health* 37(3): 210-5 (2006). Department of Biology, Faculty of Science, Faculty of Medicine, Chiang Mai University, Chiang Mai, Thailand.
Effects of Kaempferia parviflora extracts on reproductive parameters and spermatic blood flow in male rats.	Chaturapanich G, Chaiyakul S, Verawatnapakul V, Pholpramool C. *Reproduction* 136(4):515-22 (2008). Department of Physiology, Faculty of Science, Mahidol University, Rama VI Road, Bangkok 10400, Thailand
Litsea chinensis The ethanolic extract of the bark showed encouraging results in male rats.	Ageel AM, Islam MW, Ginawi OT, Al-Yahya, *Phytotherapy Res* 8 (2): 103-105 (1994). Department of Pharmacology and Pharmacognosy College of Pharmacy, King Saud Univesrsity, Riyadh, Saudhi Arabia.
Microdesmis Keayana Aqueous extract and pure alkaloid of the root stimulate sexual parameters in rats.	Zamble A, Martin-nizard F, Sahpaz S, Reynaert M, Staels B, Bordet R, Duriez P, Gressier B, Bailleul F, *Phytotherapy Res* 23(6): 892-5 (2009). Faculte de Pharmacie, laboratoire de Pharmacognosie, Little cedex, France.
Montanoa tomentosa **Aphrodisiac properties of *Montanoa tomentosa* aqueous crude extract in male rats.** Results showed that acute oral administration of crude extracts of *M. tomentosa* facilitated expression of sexual behavior in sexually active male rats and significantly increasesd mounting behavior.	Carro-Juárez M, Cervantes E, Cervantes-Méndez M, Rodríguez-Manzo G, *Pharmacol Biochem Behav* 78(1): 129-34(2004). Laboratorio de Comportamiento Reproductivo, Escuela de Medicina Veterinaria y Zootecnia, Universidad Autónoma de Tlaxcala, CP 90000, AP. 484, Tlaxcala, Mexico.
Myrstica fragrans **Experimental study of sexual function improving effect of *Myristica fragrans* Houtt (Nutmeg).** The result showed an increase in the sexual activity of normal male rats without any conspicuous adverse effects. 50% ethanolic extract of nutmeg possessed aphrodisiac activity, increasing both libido and potency.	Ahmad TS, Latif A, Qasmi IA, Amin KMY, *BMC Complement Altern Med* 5:16 (2005). Department of Ilmul Advia (Unani Pharmacology), Faculty of Unani Medicine, Aligarh Muslim University, Aligarh-202002, India.
Orchis maculata (Orchidaceae) The ethanolic extract of tubers showed enhanced sexual arousal in male rats.	Ageel AM, Islam MW, Ginawi OT, Al-Yahya, *Phytotherapy Res* 8 (2): 103-105(1994). Department of Pharmacology and Pharmacognosy College of Pharmacy, King Saud Univesrsity, Riyadh, Saudhi Arabia.

Plant Species	Investigators/Institution

Passiflora incarnata
Aphrodisiac activity of methanol extract of leaves of *Passiflora incarnata* Linn in mice.
The methanol extract of *P. incarnata* exhibited significant aphrodisiac activity in male mice.

Dhawan K, Kumar S, Sharma A, *Phytother Res* 17(4): 401-3 (2003). Pharmacognosy Division, University Institute of Pharmaceutical Sciences, Panjab University, Chandigarh-160014, India.

Piper guineense: **Effects of *Aframomum melegueta* and *Piper guineense* on sexual behavior of male rats.**
The effects of aqueous extracts of *Aframomum melegueta* and *Piper guineense* on the sexual behaviour of male rats were studied Both plant extracts stimulated male sexual behaviour.

Kamtchouing P, Mbongue GY, Dimo T, Watcho P, Jatsa HB, Sokeng SD, *Behav Pharmacol* 13(3): 243-7(2002). Laboratoire de Physiologie Animale, Faculté des Sciences, Universitéde Yaoundé I, Yaoundé, Cameroun.

Terminalia catappa
Effects of *Terminalia catappa* seeds on sexual behavior and fertility of male rats were studied.

The kernel of *T. catappa* seeds has aphrodisiac activity and may be useful in the treatment of certain forms of sexual inadequacies, such as premature ejaculation.

Ratnasooriya WD, Dharmasiri MG, *Asian J Androl* 2(3): 213-9 (2000), Department of Zoology, University of Colombo, Sri Lanka.

Allium tuberosum
Aphrodisiac properties of *Allium tuberosum* seeds extract.
Present findings provide experimental evidence that the n-BuOH extract preparation of *Allium tuberosum* seeds possesses aphrodisiac property.

Guohua H, Yanhua L, Rengang M, Dongzhi W, Zhengzhi M, Hua Z J. *Ethnopharmacol*.122(3):579-82 (2009). College of Life and Environment Science, Shanghai Normal University, Shanghai, PR China.

Plants used for *MSD in Nigeria
Carpolobia alba
Euphorbia hirta
Garcinia kola
Musa paradisiaca
Massularia acuminata
Maytenus senegalensis
Pausinystalia johimbe
Terminalia catappa
Tribulus terrestris
Rauwolfia vomitoria
**Mondia whitei

*These plants are used in the management of male sexual dysfunction in Nigeria as described by Yakubu *et al. Pharmacognosy Reviews 1: 49-57* (2007).

**Fanuel L, Krom D, du Plessis SS, The *in vitro* effects of *Mondia whitei* on human sperm motility parameters. *Phytotherapy Res* 22: 1272-1273 (2008). Division of Medical Physiology, Department of Biomedical Sciences, University of Stellenbosch Tygerberg, South Africa.

*MSD -Male Sexual Dysfunction

GLOSSARY OF TERMS

Adaptogen: Natural product proposed to increase the body's resistence to stress, trauma, and fatigue.

Alkaloids: Nitrogenous substances basic in nature derived from plants and have a definite pharmacological action.

Alopecia: Hair loss due to skin disorder.

Anaphylaxis: Acute systemic and severe type allergic reaction.

Androgens: Male hormone.

Aphrodite: Greek goddess of love.

Atherosclerosis: Hardening of the arteries.

Ayurveda: Ancient Indian science of life healing and medicine.

Biotin: Vitamin H, or B7

Corpus spongiosum: A column of erectile tissues in the center of the penis and surrounding the urethra, together with the corpora cavernosa produces erection when filled with blood.

Crohn's disease: An inflammatory bowel disease causes ulcers in the digestive system.

Dioecious: Male and female flowers on different trees.

Dyspareunia: Pain during intercourse.

Estrogenic: Like estrogen-female sexual hormone.

Extract: To derive the active constituents of a drug by the use of suitable solvents

Gynecomastia: Development of mammary glands in male.

Hallucination: Profound distortion in a person's perception of reality delusions and false notions.

Hallucinogen: Drugs that produce hallucination.

Hemopoietic: Blood forming.

Hermaphrodite: Bisexual.

Hirsutism: Excessive hair growth.

Hypercysteinemia: High concentration of cysteine in the blood.

Hyperglycemic: Higher concentration of sugar in the blood.

Hyperlipidemia: Higher concentration of lipids in the blood.

Hypertonia: Upper motor neuron dysfunction marked by increase in tightness of muscle tone.

Hypoglycemic: Low blood sugar.

Hypokalemia: Low concentration of potassium in the blood.

Hypomania: Mood disorder bipolar II.

Hypoproaccelerinemia: Abnormally low concentration of blood-clotting factor V, proaccelerin, in the circulating blood.

Inflorescence: A characteristic arrangement of flowers on a stem

Libido: Sexual desire

Luteinizing hormone: A hormones produced by the anterior lobe of the pituitary gland that stimulates ovulation and development of the corpus luteum in the female and the production of testosterone by the interstitial cells of the testis in the male.

Macular degeneration: Blurring vision of the eyes due degeneration of the macula.

Multiple sclerosis: It is a disease that affects the CNS-the brain and spinal cord. Demyelination occurs.

Neurotransmitters: A chemical substance such as acetylcholine or dopamine that transmits nerve impulses across a synapse.

Noncoital sexual pain: Genital pain following stimulation during foreplay.

Obesity: Excess body fat

Oligospermia: Low sperm count

Oophorectomy: Surgical removal of the ovary

Phytoestrogen activity: Naturally occurring nonsteroidal plant products their structure similar to estradiol and have the ability to cause estrogenic effects.

Postpartum: The term refers to the period shortly after child birth.

Postprandial: After meal time

Priapism: Persistent erection of the penis.

Prolactin: A peptide hormone associated with lactation.

Prostaglandin E1: One of the lipid compounds derived from fatty acids contains 20 carbon atoms including a 5-carbon ring.

Racemic: The mixtures of D and L isomers.

Rasayana: An elixir of life - rejuvenation potion from herbs and minerals.

Renaissance Humanists: European intellectual movement that was a crucial component of the *Renaissance*, beginning in *Florence* in the latter half of the 14th century.

Scrofulous tumors: Tumors affected with scrofula

Semicomatose state: A partial or mild coma; a coma from which a person may be roused by various stimuli.

Serotonin: It is a monoamine neurotransmitter synthesized in serotonergic neurons in the CNS.

Spermatogenesis: The process of producing sperms

Spermatogenic: Produces sperms

Sympathomimetic: Producing physiological effects resembling those caused by the activity or stimulation of the sympathetic nervous system.

Tachyarrhythmia: An excessive rapid heartbeat accompanied by arrhythmia.

Teratogenic: An agent that causes malformation of an embryo or fetus.

TGF-β1 (Transforming Growth Factor): Disulfide-linked homodimeric protein

Vaginismus: Involuntary vaginal spasm that interfere with penetration.

Index

B

C

G

H

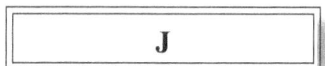

oleic, 114
oligospermia, 54
Opium, 8
Orchis maculata, 194
Oriental Palm Tree, 69
Oshun, 6
osthole, 46
oxypeucedanin, 46
oxytocin, 138, 145

P

Padmani, 5
Palmaceae, 114
palmitic, 114
Panax ginseng, 96, 97, 98, 195, 211,
    230, 253, 254, 277, 283, 284,
    285, 287, 290, 292
  schinseng;, 96
panaxtriol, 96
Pantocrin, 128
Papyrus Ebers,, 57
para gigantocellular nucleus (PGN),
    29
Passiflora incacarnata, 310
patchouli, 149, 161, 300, 301, 302
Pausinystalia yohimbe, 195, 256, 290
PDE-5, 60
Pepcid, 187, 306
Peptide Hormones, 138
Perfume-Garden, 7
periwinkle, 184
Peru Ginseng, 79
Peruvian Ginseng, 79
pesticides, 153, 196, 216, 217, 218
Petroselinum crispum, 291
Pfaffia paniculata, 121, 212
Pfizer, 2
Pharmacology, 3
Phellandrene, 300

phenelzine, 185, 186
phenobarbital, 188
Phenyl acetonitrile, 79
phenytoin, 188
pheromones, 170, 171, 172, 173, 174,
    175, 176, 177, 269, 270, 271,
    272, 273
phlobaphene, 88
Phoenicians, 6
photoanethole, 151
phytoestrogen activity, 195
phytol, 301
picrinine, 289
picrocrocin, 56, 238
pinene, 46, 56, 120, 290, 300, 301
Piperaceae, 106
Piper guineense, 310
Piperidine, 258
Piper methysticum, 58
  cubeba, 106, 215, 258, 290, 295
p-methoxy benzyl isothiocyanate, 79
Pogostemon cablin, 301
pogostol, 301
Polygonum multiflorum, 259, 260, 290,
    292
porphyrin, 289
posterior pituitary, 137
Potency Wood, 88
prednisolone, 185
prenylated flavonoids, 73, 289
prenylflavones, 59
Prilosec, 187, 306
primrose oil, 184
proanthocyaninidins, 69
progesterone, 68, 129, 143, 144, 145,
    173, 196, 197
proline, 80
Propulsid, 187
prostaglandins, 128, 290
Prostaglandins, 18, 313
protodioscin, 262, 277
prurenidine, 86
prurenine, 86

pseudowithanine, 290
ptychonal, 88
ptychonolide, 88
Ptychopetalum olacoides, 88, 230, 251, 295
Pueraria tuberosa(Indian Kudzu), 10
Purple Coneflower, 190

Quackery, 1
quassin,, 64
quassinoid-type glycosides, 64
quercetin,, 69, 289

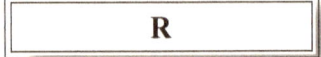

racemic, 153
Rasayana, 53
Rauwolfia serpentina, 10, 190
Red Pepper, 51
reserpine,, 289
retrochalcones, 73
Rhamnus, 75
rhazimanine, 289
riboflavin, 80
rifampin, 188
RigiScan, 98
Rosa centifolia, 302
rosemary, 149, 161, 184
Rosemary oil, 161
Rosmarinus coronarium, 161, 302
Roupala montana
    dentata, 260, 290
Rubiaceae, 291
rutin, 289

S

Sabal, 114
Sabal serrulata, 114
sabinene, 300
Safed Musli, 10, 53, 54, 212, 237, 297
Saffron, 56, 57, 58, 161, 233, 237, 239, 269
safranal, 56
safrole, 46
Sagittatins A and B, 59
salicylic acid, 96, 189
Salisburia adiantifolia, 69
Salvia miltiorrhiza, 189, 291
Salvia sclarea, 300
sandalwood, 149, 162, 301, 302
santalone, 302
Santalum album, 149, 302
sarcosine, 80
Sarpgandha, 10
Saw palmetto, 72, 114, 196, 204, 212, 260, 261, 282, 283, 284, 285, 290
scholarine, 289
sciadopitysin, 69, 289
sclareol, 300
scopolamine, 290
scopoletin, 46, 64, 289
Sea Beans, 86
sedrin, 64
selective serotonin reuptake inhibitors (SSRIs), 70
Senna, 75
Serenoa repens, 114, 215, 260, 261, 290
serine, 80
serotinin reuptake inhibitors, 29
serotonin, 29, 30, 36, 60, 70, 71, 86
sesquiterpenes, 46, 59
seychellene, 301
shinflavone, 74
shingles, 51

T

U

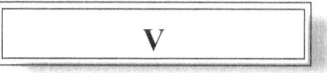

V

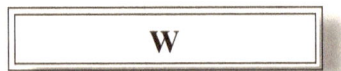

W

Y

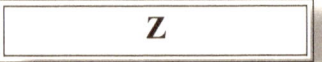

Z

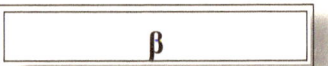

β

www.ingramcontent.com/pod-product-compliance
Lightning Source LLC
Chambersburg PA
CBHW031820170526
45157CB00001B/122